D1233300

Rural Sustainable Development *in* America

Rural Sustainable Development *in* America

edited by

Ivonne Audirac
Florida State University
Tallahassee, Florida

John Wiley and Sons, Inc.
New York • Chichester • Weinheim • Brisbane • Singapore • Toronto

Library of Congress Cataloging in Publication Data:

Rural sustainable development in America / edited by Ivonne Audirac
p. cm.
Includes bibliographical reference and index.
ISBN 0-471-15233-1 (cloth : alk. paper)
1. Sustainable development—United States—Case studies.
2. Rural development—United States—Case studies.
I. Audirac, Ivonne.
HC110.E5R868 1997
333.7—dc20 96-34519

Printed in the United States of America

10 9 8 7 6 5 4 3 2 1

To my parents Henri and Hella

Contents

Preface

In recent years, the concept of rural sustainable development has engaged many academics; town planners; rural, small-town, and even urban development people; all levels of government; politicians; architects; landscape architects; and community activists. This engagement has stimulated a great deal of rhetoric, research, and experimentation in a field of endeavor that varies widely in its content and practice. This book is both a thoughtful and promising celebration as it joins the dialogue.

It is interesting that this attention to rural development is so widespread, when urban populations dominate the landscape. It could be caused by the realization of the interdependency of rural and urban populations, that survival for both is inextricably related to environmental, economic, and social interchange. Such relationships do not permit a discrete treatment of sustainability. It may appear that curiosity and wistfulness direct attention to the rural countryside. The authors of this book are not wistful nor do they completely separate the concept of sustainable development into rural and urban dichotomies, only so far as it is necessary to focus on rural ecological and man-built economic and sociopolitical systems.

The importance of this work is revealed by the authors as they explore the evolution of environmental, economic, and social systems in America. I have witnessed Florida evolve and change from a rural and small-town state in the 1930s to a state that today counts itself among the most urbanized. In 1930 its rural and urban populations were 48.3% and 51.7%, respectively. In 1990 the census reported that urban populations had increased to 84.8%, while the total population had grown tenfold (Shermyen 1992, 4). This growth has brought with it the high price of environmental destruction, economic stress among minority populations, and the perennial incapability of local and state governments to cope with increasing demands for education, transportation, and urban services. Agriculture and rural life in Florida reflect national trends in the growth of capital-intensive agribusiness, the loss of family farms, the spread of suburbs and proliferation of hobby farms, and increasing sociopolitical discord between urban and rural agendas.

Recent in-state population movements may reflect the motivations of people seeking space for recreation, retirement, and perceived relief from urban life. These movers—coming from very urbanized parts of south and southwest Florida to rural, less populated central and northern counties—

have imposed new, often unwelcome demands and responsibilities on local rural governments and enterprises unprepared to cope with such growth and change. The emerging and often urgent demands and environmental stresses have had consequences that resulted in the real and perceived deterioration of the ineffable quality of life in small towns and rural counties.

Herman Daly and John Cobb (1989), in their book *For the Common Good*, address the redirection of the economy toward community, the environment, and a sustainable future. They cite "the recent surge of interest in 'sustainable growth' or 'sustainable development' within development agencies and Third World countries" (1981, 71). Daly and Cobb clearly seek to distinguish the terms *growth* and *development*. They conclude that "'growth' should refer to quantitative expansion in the scale of the physical dimensions of the economic system, while 'development' should refer to the qualitative change of the physically nongrowing economic system in dynamic equilibrium with the environment." It is their definition of *growth* that describes the Florida phenomenon of the last 65 years. This differentiation of the terms *growth* and *development* is a thoughtful departure and provides a basis for understanding the purposes and contents of this book.

The history and theory of the concepts of sustainability and sustainable development form the basis of Part One of this book. Throughout this first part, the concepts of rurality, or *rural*, and sustainability, or *sustainable*, are explored, discussed, and developed in explaining the paradigmatic shift that is evolving in this field of inquiry. The reader will discover the underlying principles of a sustainable future for rural America in which rural development policies must embrace holistic and interactive views of ethics, ecology, and economic and sociopolitical systems.

In Part Two several authors develop the transition from concepts and principles of rural sustainability to implementation of development policies. The critical and genuine involvement of citizens in choosing and making basic changes in community life is explored. Information is gathered and reported, suggesting various approaches to the regional or local scope in policy and planning strategies. The vitality and diversity of community life are fostered in the context of citizen-based community analysis and citizen initiatives that often drive redevelopment activities.

Part Three takes several views of the symbiotic relationship between citizens of rural and urban communities by examining such alternatives as greenways and trails, farming at the urban edge, waste management, and mutually developed enterprises. A strong theme emerges in these chapters as the authors recognize and report outstanding community-based efforts in several case studies. In addition, a diversity of approaches and strategies become evident in new rural initiatives. The reports of these studies emphasize that external help, in the form of technical and monetary assistance, may come from nongovernmental and/or governmental sources. These resources are often merged and leveraged innovativly. However, in each case, assistance is tailored to community traditions, citizens' interpretations of community needs, and systems of choice that tend to build a community's social and economic strength and ecological health.

This book is written to appeal to all who are interested in the future of rural and urban America and, within that context, explores exciting opportunities and accounts of successes and failures, all of which should serve the reader well. We, the authors and editor, commend it as an important and very useful work that focuses on rural sustainability, the vital and hopeful future of rural America, and its inextricable relationship to urban America.

EARL M. STARNES

REFERENCES

Daly, H. E., and J. B. Cobb. 1989. *For the Common Good.* Boston Beacon Press.

Shermyen, A., ed. 1992. *Florida Statistical Abstracts.* Gainesville: University Presses of Florida.

Acknowledgments

In preparing this volume several people were of special help. Cavell Kyser, assistant editor of the *Journal of Planning Education and Research,* not only provided invaluable copyediting and technical assistance, but also pleasantness and optimism. Ilen Zazueta-Audirac, general editor of New College's *The Catalyst,* provided patient copyediting assistance and review for wide-audience readability. Earl Starnes was always willing to offer a generous second opinion in editing decisions when needed. My gratitude goes also to John Crotts, Ed Malecki, Fred Buttel, Owen Furuseth, and Jay Stein, who reviewed early versions of the prospectus of this volume, and to the various academics and practitioners who offered opinions and reviews on some of the chapter contributions. My appreciation is extended to Gary Greenberg, who, while a graduate student in the Department of Urban and Regional Planning at the University of Florida, unselfishly contributed assistance in the early phases of this project. Finally, I want to thank the students in my Sustainable Development Planning in the Americas seminar at Florida State University for their insightful discussions of some of the ideas presented in this volume. Deep gratitude is due each chapter contributor and our editors at John Wiley & Sons, Philip Manor and Milagros Torres.

PART ONE

Rural Sustainable Development: Frameworks and Underpinnings

One

Rural Sustainable Development: A Postmodern Alternative

Ivonne Audirac

INTRODUCTION

Sustainable development eludes precise definition, connoting a medley of meanings that vary with each interest group that adopts or reinterprets it. From CEOs of multinational corporations avowing "green" convictions about the latest biotechnology innovation, to the rural ecotourism entrepreneur reviving Schumacher's "small is beautiful"; from deep-ecologists' pleas for cultural and moral self-renewal, to the promotion of ecocommunities; *sustainable development* has seeped into our language and permeated our sensibilities. To make sense of the apparent contradictions in meaning connoted by *sustainable development*, we must view it in light of the emerging ecological worldview. Sustainable development is a postmodern alternative; its roots are planted in the environmental critique of modern industrial society. This critique has ripened into the emerging ecological worldview, from which several shades of "greenness"[1] color the sustainable development debate, including the contributions to this volume. This chapter explores the policy ramifications of the sustainable development debate in a similarly contended terrain: that of rurality and rural development. It offers a conceptual backdrop and an overview of some of the major issues at the junction of rurality and sustainability. This overview frames the volume's plan and purpose.

SUSTAINABLE DEVELOPMENT AND THE ECOLOGICAL WORLDVIEW

According to Ralph Metzner (1993), we are in the midst of a transition to an ecological age grounded in the postmodern transformation of the natural sciences, the social sciences, philosophy, and religion.[2] The shift was spearheaded by ecological thinkers such as Rachel Carson, Lynn White Jr., and Wendell Berry; scientists such as Aldo Leopold, James Lovelock, and Herman Daly; philosophers such as Arnae Naess, Holmes Rolston III,

George Sessions, and Paul Taylor; theologians such as Thomas Berry; and activists such as Chico Mendez (Sessions 1993; Metzner 1993). Collectively, their indictment of the practices, values, attitudes, sciences, and institutions of our modern industrial-technological paradigm have gradually undermined many of the paradigm's intellectual foundations, furthering its imminent transformation (see Table 1.1).

The emerging ecological worldview (EEW) is premised on the fact that we are in the midst of a global ecological crisis that threatens both human and nonhuman life (see Rees 1995 and Rees, Chapter 3). The culprit of the crisis is none other than our own modern industrial worldview, the core values of which include a mechanistic and deterministic model of science, and linear and atomistic economic thinking; an exploitative utilitarian view of nature; an economic mentality which seeks GDP growth as a cure for poverty and environmental damage; a consumerist outlook on life; top-down, centralized institutions; open-cycle, capital-intensive technologies with high loads of pollution and toxic and nonrecyclable wastes in rivers, seas, and the atmosphere. The agroindustiral model of development epitomizes all of these values. Excessive reliance on monoculture farming and agroindustrial inputs, such as capital-intensive technology, pesticides, and chemical fertilizers, has negatively impacted the environment and rural society (see Clugston, Chapter 4). Topsoil has been lost, genetic diversity has eroded, and toxic and nonrecyclable agricultural wastes have damaged rivers and lakes. Furthermore, the economic and political domination of the rural development agenda by agribusiness has thrived at the expense of the interests of consumers, farm workers, small family farms, wildlife, the environment, and rural communities (Paarlberg 1980).

In contrast, in the emerging ecological worldview, organismic and holistic models of science are superseding previous mechanistic and determininistic scientific paradigms. Hence, economic linear models are increasingly penetrated by ecological holistic thinking, exemplified in the rise of environmental and ecological economics. The arrogance of man over nature is beginning to be tamed by biocentric philosophies that stress the "intrinsic" over the "utility" value of nature. Exploitative uses of natural resources are gradually being sensitized to notions of ecological stewardship and coevolutionary symbiosis—notions that include ecosystem and biodiversity protection. Prevailing economic values—the least likely to be relinquished—are nonetheless being influenced by social and political concerns about pollution, waste disposal, and species extinction. Natural resource accounting and economic models that recognize biophysical limits to all human activity (see Davies and Penfold, Chapter 5) are slowly taking over previous models that assumed limitless economic growth. On the technology front, the greening of industrial ecosystems—an engineering perspective of working with rather than against nature—aims at redesigning currently open-loop industrial infrastructures into closed-loop systems via massive reuse and re-

Table 1.1 The Modern Industrial Versus the Emerging (Postmodern) Ecological Worldview

	Modern Industrial	*Emerging Ecological*
Scientific Paradigms	• Mechanistic, determinism, linear causality	• Organismic, indeterminacy, chaos: nonlinear dynamics
	• Atomism	• Holism
dominant discipline	• Economics	• Ecology
Human/Nature Relation	• Conquest of nature	• Living in nature
	• Domination over nature	• Co-evolution symbiosis
	• Resource management	• Ecological stewardship
	• Nature has utility value	• Nature has intrinsic value
	• Anthropocentric	•Biocentric
Social System	• Cultural Homogeneity	•Cultural diversity
	• High-consumption life-styles	• Minimize superfluous consumption
	• Metropolitan oriented	• Community, bioregion oriented
decision-making	• Top-down	• Bottom-up
Economic System	• Competition	• Cooperation
	• Limitless growth	• Biophysical limits to grow
	• Export and trade	• Self-reliance
Technology	• Open system: waste disposal	• Closed system: reuse, recycle, recover
	• Large scale, capital intensive	• Small, appropriate
Agriculture	• Monoculture farming	• Poly- and permaculture
	• Agribusiness, industrial farms	• Community and family farms
	• High input: chemicals, pesticides	• Low input: organic
	• High-yield hybrids	• Preserve genetic diversity

Adapted from Metzner (1993)

cycling of wastes and materials (Richards et al. 1994; Ayers 1994). Socially and politically, a public yearning for more cooperative, community-based social relations and control of impersonal bureaucratic institutions are forcing a change from top-down, centralized decision making to bottom-up cooperative partnerships. The move to protect ecological diversity is inclusive of all species, including humans. Hence, the cultural heritage, land use, knowledge, and traditional ways of relating to the environment of many indigenous communities are increasingly being vindicated, in spite of powerful opposing economic and political interests (see Seymoar, Chapter 20). Grass-roots-driven initiatives for community revitalization, by those still conscious of their economic dependence on local ecosystems, are sprouting across the landscape (Clugston, Chapter 4; Flaccavento, Chapter 19; Luther, Chapter 8; and Segedy and Lyons, Chapter 16). Systems-oriented community development models designed to guide such efforts are being adapted to the needs of specific communities (see Chiras and Herman, Chapter 6). In agriculture, organic farming and landscape-regenerative practices—subscribing to the holistic land-ethic of Wendel Berry and others (see Lapping, Chapter 2)—are steadily gaining adherents. The movement seeks a return to traditional—Amish-inspired—small-scale farming. For the first time, policy-making in agriculture is feeling the impact of environmental, farm labor, animal rights, and consumer lobbies. Issues such as the effects of agricultural chemicals and highly processed foods on human health; agriculture's impact on water quality, wildlife, and ecosystems; rural poverty; the decline of rural communities; farm laborers' rights, and humane treatment of animals are gaining political recognition (Beus and Dunlap 1993).

Given this delineation of the emerging ecological worldview, we can appreciate that *sustainability* is shorthand for "commitment to ecology".[3] Thus, *sustainable agriculture, sustainable economics, sustainable forestry, sustainable communities,* and *sustainable development* refer to the specific discipline or intellectual domain under the transformational influence of the emerging ecological worldview. However, not all formulations of sustainable development agree with each other. These incompatibilities result from the varying degrees to which individual ideologies buy into the critique of the modern industrial worldview (MIW) and differences in disciplinary, professional, and philosophical commitments to various environmental ideologies. T. O'Riordan (1981, 1995) identified four major environmental ideologies, or shades of greenness (Turner et al. 1993). The first two, deep-ecologists and communalists are part of the *ecocentrist* camp, which is politically egalitarian and community oriented, welcomes soft technologies, and is holistic in scope (see Table 1.2). As a whole, ecocentrists are guided by bioethics, are active detractors of the MIW, and are the most committed to the EEW. Deep-ecologists are the greenest of all. They lead the strongest challenge to the MIW and are devoted preservationists. By extending to nonhuman species the right to remain unmolested, their view of economy and technology stresses

Table 1.2 Shades of Greenness

	Ecocentrists		*Technocentrists*	
	Deep-Ecologists	*Communalists*	*Accommodators*	*Cornucopians*
Critique of MIW	Highest	High	Low	Lowest
Approach to Nature	Deep-preservationist	Preservationist/ Stewardship	Stewardship/ conservationist	Conservationist/ exploitative
Economy	Minimal resource take	Community/small scale, self-reliant	Economic growth through trade but internalize environmental externalities (polluters-pay principle)	Maximize economic growth and global competitiveness
Technology (substitutability for natural capital)	Minimal, nonintrusive, no substitutability	Soft/appropriate/small scale Some substitutability	Accommodating to conditions Substitutability	High-tech Perfect substitutability
Politics	Egalitarian-decentralized	Community-based/decentralized	Concessions/conflict resolution	Top-down Based on expertise
Ethics/Rights	Bio-egalitarian	Collective over individual	Compensation to private rights	Private rights
Greenness	Deep green	Green	Light green	Very light green
Sustainability	Very strong	Strong	Weak	Very weak

(Adapted from Metzner 1993, O'Riordan 1995, and Turner et al. 1993) minimizing all human impact on the earth by extracting only the minimum resources necessary to satisfy basic human needs. Communalists (or soft-technologists) are committed to small-scale production units, community-oriented organizations, and self-reliant steady-state economies in harmony with the ecosystem. They stress participation and emphasize the rights of the collectivity (including ecosystems) over private property rights.

In the second camp, the *technocentrist* outlook on nature is managerialist and values economic efficiency over equity (O'Riordan 1995). It holds an optimistic attitude toward technological solutions and is generally blind to the social critique of the MIW. In this camp, accommodators differ from cornucopians in that the first are willing to accommodate, to the extent possible within existing resource management structures (e.g., EPA), a conservationist approach to natural resource use. For instance, this group supports green accounting and EIA (environmental impact assessments) and believes in market mechanisms or taxes to address resource use and extraction. Although it may be sympathetic to some of the critiques of the MIW (e.g., consumerism), its role is limited by existing institutions and professional ambits. The accommodator believes in the legal right to a minimum level of environmental quality and compensation for those who experience adverse environmental effects (O'Riordan 1995). Cornucopians, on the other hand, celebrate scientific and technological expertise. They believe that scarcity constraints or biophysical limits can always be overcome through the technological substitutability of natural capital for manmade capital (see Rees, Chapter 3). Thus, technological progress, the hallmark of our civilization, will be fully realized only when economic and market forces are unleashed from state regulation (O'Riordan 1995; Turner et al. 1993). Cornucopians bear the lightest shade of green and the weakest commitment to sustainability as embodied in the EEW. Not surprisingly, their views are the least sympathetic to the critique of the MIW.

In reality, no policy, plan, or project fits neatly into any of these categories. The environmental leanings of even one author often have considerable overlap. Nevertheless, the typology is helpful in identifying various degrees of endorsement of the critique of the MIW, in discerning the levels of acceptance of the EEW, and in identifying the potential conflicts and controversies among the many definitions and variants of sustainable development. However, beyond this heuristic exercise of classification lies the inherently postmodern realization[4] that there is no hegemonic definition of sustainable development, let alone a universally agreed-upon agenda—this book provides no exception.

THE SUSTAINABLE DEVELOPMENT DEBATE

Policy statements on sustainable development are multifarious, reflecting the political compromises negotiated within the spectrum of green and brown[5] agendas rather than a true consensus. Not surprisingly, most policy documents on sustainable development reveal a persistent accommodator bent, agreeable to technocentrists but troublesome to ecocentrists, who decry the policy's debilitated power to drastically revamp the development agenda. Even the most popular definition of *sustainable development*: "to meet the needs of the present without compromising the ability of future generations to meet their own needs" (Brundtland Commission Report, WCED 1987), is a nonbinding protocol. It is ritualistically invoked to remind us that sustainable development was the first international attempt to frame domestic and international development issues as environmental imperatives and as a global compact for future generations. By itself, the Brundtland Commission Report's (BCR) optimistic endorsement of world trade and increased economic growth captures little of the critique of the MIW and has spurred two leading controversies—one between economists, the other between nations.

The conflict between ecocentrists and cornucopians concerns the earth's capacity to support the impact of the present world population if current levels of per-capita consumption in developed countries were extended to the rest of the developing world, as the BCR proposes. The disagreement has come from the no-growth advocates. They deplore the BCR's notion of sustainable development as a bad oxymoron (Daly 1991; Pearce et al. 1989).

> The term sustainable development makes sense for the economy, but only if it is understood as development without growth—that is, qualitative improvement on a physical economic base that is maintained in a steady state by a throughput of matter-energy which is within the regenerative and assimilative capacities of the ecosystem (Daly 1991; 39).

According to Daly, the limits to economic growth have been reached and the present scale of growth is unsustainable. Hence, multiplying the present scale of industrial output by five or ten as the BCR suggests, would move us from unsustainability to imminent collapse. For Herman Daly, the crucial question raised by the BCR (but not faced) is to what extent can we alleviate poverty with development without growth. If the five- to tenfold expansion of the world economy is meant to alleviate world poverty, then industrial output would have to consist of things needed by the poor—food, clothing, shelter—rather than information services and superfluous goods. In his view, the policies implied by sustainable development in the developed world would require heavy taxation of resource extraction—especially energy—and resource severance taxes while reducing taxes on the lower end of the income distribution. Sustainable development involves generational, as well as intergenerational, equity and is tantamount to cultural adaptation produced by society as it realizes the need for nongrowth (Daly and Towsend 1993).

The second controversy, between North and South nations, concerns the South's contention that the issue of the earth's sustainability is not only one of population growth versus resources, as the BCR surmises, but also of the Western standard of living. A child born in a rich industrialized nation puts a much greater burden on the planet than one born in a poor country (MacNeill 1991; de la Court 1990; Schmidheiny 1992)[6]. Developing nations charge that for sustainable development to occur, the greatest reforms should come from developed countries, which consume the most resources, produce the most pollution, and have the greatest capacity to make the necessary changes (Schmidheiny 1992, 6). In the North-South debate, developing countries have protested that wealthy countries risk reversing the relationship between production and needs satisfaction, inasmuch as production in wealthy nations has shifted from primarily satisfying needs to the creation of superfluous consumption.

Extreme viewpoints within these two controversies polarized the sustainable development debate at the 1992 Earth Summit in Rio. On one hand, GATT supporters—criticizing the more radical policy proposals as "ecoimperialism"—echoed the BCR's approach, asserting their faith in free trade and multilateralism's ability to resolve global environmental problems. On the other hand, steady-state ecocentrists advocated a slowdown in economic growth to save the earth (Friedmann and Rangan 1993). Agenda 21—Rio's failed attempt to put teeth in the BCR's global environmental concerns (Hecht and Cockburn 1992)—gives little hope for international consensus on sustainable development. However, despite these controversies, sustainable development remains a powerful ecological metaphor and vision, which harbor multiple political economic perspectives on how the integration of environment, economy, and equity (social justice) should proceed.

Rural Sustainable Development

Rurality, the traditional locus of food production, enters the sustainable development debate in three policy contexts: (1) the international farm crisis and its global environmental and trade repercussions; (2) domestic policies attempting to redefine rurality beyond the confines of traditional agricultural policy; and (3) policies for sustaining decent rural livelihoods, which focus on promoting the viability of rural communities' infrastructure, employment, education, health, social services, and intergovernmental coordination (Wimberly 1993).

A common theme within international and domestic policy debates on rurality is a conspicuous criticism of the Fordist,[7] or agroindustrial, model of development, which is viewed as the primary cause of global food imbalances, environmental problems, and rural decay in the United States (Goodman and Redclift 1989; Lyson and Welsh 1993; Beus and Dunlap 1993). Proponents of this view see the persistent poverty and decline of many rural areas in North America as an indication of the unsustainability of the agroindustrial model, which for more than 70 years concentrated on producing cheap food and fiber staples for urban consumption at the expense of rural America's farm diversity and community viability.

> "Fordism" in US agriculture is typified by the corn-soybean-livestock equation as the source of cheap food for the industrial working class and the burgeoning suburban white-collar labor force. This model sustained the expansion of the off-farm sectors of the food and fiber system as technological competition provided markets for agroindustrial inputs and low cost supplies for downstream food processing industries. . . . The cheap food policy, centered on basic food and feed grains and livestock products, is the rural complement to the mass production of industrial "Fordism" (Goodman and Redclift 1989, 8).

The Fordist Diet and the Hamburger Connection

The international dissemination of the agroindustrial model, a product of scientific and technological innovation, depended on diffusion of the Fordist diet, which consisted of increased consumption of animal protein and intensive production of grain-fed livestock. For instance, in the 1980s, Western nations fed 400 million tons of grain to livestock—an amount equal to total Third World consumption (de la Court 1990). The apparently inexhaustible demand for feed grain in the developed world and the urge to earn foreign currencies encouraged debt-ridden developing countries to divert agricultural production from staple food crops to animal feed (e.g., soybean, sorghum, and cassava) for export markets (Goodman and Redclift 1989). The dedication of land to livestock production and forage crops has increased pressure on natural resources, intensifying the global rate of deforestation, soil erosion, and chemical pollution of air, soil, and water resources.

The worldwide spread of fast-food chains epitomizes the global dimension of the Fordist diet and the high stakes involved in the industrialization of their products. A multimillion dollar business, hamburger franchising has attracted multinational corporations other than McDonald's and Burger King[8] into beef and potato production, processing, agricultural research, and transportation. According to de la Court (1990), a hamburger is an international fast-food product[9] that requires large volumes of beef cattle and cattle feed, increasingly grown in Third World countries at the expense of rain forests, global biodiversity, and other natural resources.

The International Farm Crisis

Technological and scientific modernization of agriculture became the cornerstone of the postwar model of agroindustrial development. It involved a package of inducements to increase food production, supported by fiscal and credit incentives, public research and extension services, agroindustrial promotion, and commodity price supports. Highly subsidized food surpluses, the result of agroindustrial policies adopted in the countries of North America and Europe, precipitated the international farm crisis of the 1980s. Mounting competitive pressures in these countries to dispose of farm surpluses in overseas markets prompted the familiar United States-(European Union) (EU) trade wars involving "bankrupt Mid-Western farmers in the US; milk quotas and the plight of British hill farmers; the burning of imported lamb carcasses by irate French farmers; tractorcade protests in Canberra; and GATT negotiations for a new system of international agricultural trade" (Goodman and Redclift 1989, 4). As food surpluses depressed world market prices and the spread of the Fordist diet among the world's urban middle classes increased the demand for beef products, perverse market incentives in many Third World nations made it cheaper for these countries to import their food from the United States and Europe rather than produce it themselves. Hence, the international farm crisis resulting from highly leveraged agricultural surpluses is not only a symptom of global imbalances in food consumption, but also a cause of Third World dependence on agriculture imports.

With the aim of eliminating these perverse market incentives and their corresponding ecologically destructive effects, both neoliberal and BCR-based agricultural policy proposals agreed on the need to phase out export and farm subsidies. However, there is little ground for agreement on how this should be accomplished. The Uruguay Round to reform the GATT broke down largely over this issue. Canada, Japan, and European countries feared the domestic political fallout of the American proposal for rapid elimination of all subsidies unrelated to production. Nonetheless, as MacNeill (1991, 35) points out, if subsidies cannot be eliminated, "why not pay farmers to build up rather than deplete their basic farm capital? The greening of agricultural subsidies rather than their complete elimination might be more acceptable to all parties." The structural adjustments in agriculture that these reforms imply, pose pressing societal questions that lie at the core of the rural sustainable development debate and are reflected in recent farm bills. Which farmers should be allowed to remain in agriculture? Those who make the most profit (e.g., large-scale, high-tech farmers) or those who are most valued for their social and community characteristics (e.g., small-scale, organic, family farmers) or environmental practices (e.g., farmers engaged in soil and wetlands conservation) (Beus and Dunlap 1993, Goodman and Redclift 1989)?

SUSTAINABLE AGRICULTURE AND RURAL DEVELOPMENT

In North America, the emphasis in mainstream policy is slowly shifting from maximizing food production to mitigating farmers' environmental impacts. In Canada, strategies to support the transition from conventional to sustainable agriculture include the redesign of food production, processing, and distribution. Recent policy frameworks have focused on research, extension, market development, tax provisions, and farmers' safety net programs (Agriculture Canada 1989). However, a few Canadian rural sustainability proposals (MacRae et al. 1990) have gone beyond the greening of farm subsidies to proposals for a complete overhaul of the national food production and distribution system on the basis of the *optimal diet*, which would entail a thorough institutional redesign to promote the production of quality food without additives, pesticides or antibiotic residues; decrease consumption of polysaturated fats; and discourage dependence on livestock diets and imports of forage and other crops. Presumably, an optimal diet system of food production should stimulate a more self-reliant agricultural and rural sector, because a more diverse diet, from the perspective of sustainability, requires a more diversified production system. Moreover, in order to meet rural sustainability goals (i.e., ensuring the well-being of rural communities and the preservation of ecological integrity), these proposals call for diversification of food production within farm units and bioregions so that the agricultural self-sufficiency that Canada once enjoyed, but lost to international imports, may be reclaimed[10] (MacRae et al. 1990).

In the United States the sustainable agriculture movement, encompassing a variety of alternatives to the conventional agroindustrial model (such as organic, regenerative, and low-input agriculture, permaculture, agroecology, etc.) pursues not only a change toward ecologically sensitive farming systems, but also a change in values toward permanence and care for the land, resource stewardship, small-scale family farms, and rural community life (see Lapping, Chapter 2; Clugston, Chapter 4). Like its Canadian counterpart, the American sustainable agriculture movement is deeply influenced by communalist environmental beliefs. It seeks the regeneration of small-scale family farming communities, increased regional ecological self-reliance, and the reshaping of the entire food system in ways that are economically viable for farmers and consumers.

While status quo agriculture celebrates the fact that the productivity of one farmer feeds 200 people and that persistent decline in the number of small farms is the result of technological and economic efficiencies, detractors of the agroindustrial model of development decry such boasting, pointing at numerous rural sociological studies that confirm the Goldshmidt hypothesis. This hypothesis asserts that changes in the structure of agriculture have been accompanied by negative social consequences to agriculture-dependent communities (Lasley et al. 1993; Goldshmidt 1978; Lobao 1990; Hefferman 1982). Concentrating production into fewer corporate farms reduces local agricultural diversity and the number of farms, increases reliance on low-paid wage labor, and erodes the business diversity and economic vitality of rural towns. The decline in economic power of the local population, resulting from small farm attrition, negatively affects local businesses and retail enterprises as well as democratic governance. Lack of remunerative farm employment promotes community outmigration. Moreover, soil, water, and air quality suffer from monoculture production, soil loss from high tillage practices, and soil and water contamination from chemical fertilizers and pesticides. The sustainable agriculture movement aspires to reverse such trends by the promotion of low-input, ecologically sensitive farm practices that are more labor intensive and applicable to small-scale farms. In short, sustainable agriculture is believed to have the power to revitalize rural America, not only because a proliferation of small family farms means more people on the land and in the community, but also because sustainable agricultural practices require a new community ethos and a new package of inputs, which, if locally provided, can result in new community businesses and jobs (Flora 1990).

The sustainable agriculture movement endeavors to be a rural development alternative. In order to affect food and farm policy it has spearheaded a new rural coalition (Fisher 1993) made of such diverse and dissimilar public interests as regional and urban land use planners who support agricultural land use preservation; suburban residents in increasingly close proximity to agriculture who lobby for environmentally safe farming practices; farm labor activists who demand socially just and safe employment conditions; community development analysts who deem farm structure changes—from industrial corporate farms to a rebirth of family farms—desirable for enhancing the rural quality of life; and consumer organizations that seek to improve the food system by linking sustainable producers with processors, retailers, and consumers (Lasley et al. 1993; Bird and Ikerd 1993; Feenstra 1995). In the last two decades, the movement has spread to sectors of the USDA and the land grant university system and has gained increased acceptance within the policy mainstream, as reflected in the 1985 and 1990 Farm Bills. However, whereas the last two farm bills addressed groundwater pollution, water quality, and sustainable agriculture and revoked commodity program benefits to farmers who failed to manage adequately wetlands and highly erodible lands, the 1996 Farm Bill reauthorizes most conservation provisions. In response to the congressional leadership's view that conservation measures are burdensome to farmers' private property rights, however, this bill allows for compensation for undue economic hardship. The new bill's Environmental Quality Incentive Program provides assistance to crop and livestock producers in dealing with environmental and conservation improvements. Yet it seems to be deliberately silent on sustainable agriculture issues raised in past farm bills, such as the 1990 Farm Bill that mandated the USDA to conduct research and education in support of ecologically and economically sound alternative systems of agricultural production (Bird and Ikerd 1993).

The 1996 Farm Bill provides funding for restoration of the Everglades but does not assess the proposed one-cent-per-pound tax on Florida sugar producers. Despite past political gains attained by the sustainable agriculture movement and the President's Council on Sustainable Development endorsement of sustainable agriculture (PCSD 1996), current agriculture policy is saying yes to conventional crop subsidies benefiting the most profitable farmers,[11] a qualified yes to farmers voluntarily adopting soil and wetland conservation practices, and no to socially and environmentally desirable small-scale farming systems. This outcome confirms the notion that present political conservatism tends to abrade ecocentrist-communalist policy gains in favor of the technocentrist-cornucopian sustainable development agenda[12] (Buttel 1992).

POSTINDUSTRIAL RESTRUCTURING AND SUSTAINING OF RURAL LIVELIHOODS

Changes in the structure of agriculture engendered by agroindustrial development are not the only factors to have affected the decline of rural communities. Because agriculture, logging, and mineral and natural resource extraction and processing are no longer the economic mainstays of rural America, postindustrial global economic factors, such as intense international economic competition, technological change, flexible manufacturing, and the international division of labor (Henderson and Castells 1987; Hoffman and Kaplinsky 1988)[13] have also contributed to rural economic decline. Branch plants—which in the past sought rural locations for their comparative advantages (e.g., cheaper labor, lower property taxes, and hefty local government inducements)—are relocating overseas for similar comparative advantages offered by developing countries. Even though the ensuing loss of manufacturing and service jobs also affects urban areas, rural areas that have characteristically lower levels of human capital resources are faced with a triple-edged predicament. How can rural communities generate new jobs in a postindustrial society that places a premium on highly skilled human capital? How can local entrepreneurship thrive when the rural-to-urban brain drain continues to work against the development of human and community capital that will stay in the rural community? How can rural economic development enhance the quality of life in rural communities in an ecologically sustainable way and without depending on branch plants or urban expansion?

Rural Development: More Than Sustainable Agriculture and Forestry

Even with sustainable forest management, the viability of rural logging communities in the Northwest and elsewhere remains threatened by the aforementioned changes. In addition to the depletion of old-growth forests (more than 87% of old growth has been cut), technological change and global competition affecting the timber industry have reduced its capacity to absorb jobs previously supported by old saw mills and unsustainable harvest practices (Levi and Kocher 1995). Misguided blame for the woes of Northwest logging communities has been placed on the spotted owl controversy, which prohibits timber harvesting of old-growth forest on federal lands. However, the effects of this controversial ban include a total 10% drop in jobs, while global restructuring factors account for a steady 1% to 2% per year drop in jobs since the mid-1980s—a much larger and more decisive factor in the fate of these communities (Levi and Kocher 1995).[14] Although sustainable agriculture and forestry are necessary to stem the tide of global economic trends, they are not by themselves sufficient to reverse that tide (Lasey et al. 1993, Dobbs and Cole 1992).

The critical development issue for many rural North American communities is how to sustain the evolving livelihoods and well -being of rural locales and regions increasingly vulnerable to global economic restructuring. Warren's (1972) "Great Change," affecting the rural American community, has increased in scale and complexity. Its supranational and global dimensions underscore the celebrated dictum: "Think globally, act locally." However, the typical progrowth, "smoke-stack chasing," industrial development prescribed as antidote to rural malaise has been proven more afflictive than curative (Summers and Clemente 1976; Meyer and Burayidi 1991; Flora et al. 1991) and as controversial as Popper and Popper's (1987) community abandonment and Daniels and Lapping's (1987) community triage (see Luther, Chapter 8).[15]

When rural towns become dependent on a single business or branch plant—ready to leave or to extract the next round of concessions on wages, environmental regulations, and local property taxes at every economic or technological shake-up—the community's social, human, and natural capital steadily dwindles. This development strategy—antithetical to sustainability—typifies urban and corporate free-riding on rural America, which, as the "shadow ecology" (MacNeill 1991) of urban America, continues to provide food, fiber, minerals, and natural and human resources to the nation's urban and suburban centers, while it receives their accumulated industrial wastes at great social, environmental, and economic cost. This unequal rural-urban exchange is further exacerbated by the conundrum that post-Fordist society poses to many North American towns. As Lester Thurow (1996, 75–77) puts it, in a global (postindustrial) economy, "Those with third world skills will earn third world wages, even if they live in the first world. Unskilled labor will simply be bought wherever in the world it is cheapest . . . [and] American firms don't have to hire an American high school graduate if that graduate is not world-class. . . . Investing to give the necessary market skills to a well-educated Chinese high school graduate may well look like a much more attractive (less costly) investment than having to retrain an American high school dropout." If, according to Thurow, skills are the only source of long-run, sustainable global competitive advantage, some rural areas are clearly at a loss, primarily in the South[16] where 55% of the rural poor reside and nearly 97% of the rural black poor live (Dudenhefer 1993, 8). Investments in community social and human capital are fundamental to promoting sustainable livelihoods and sustainable rural communities (see Beaulieu and Israel, Chapter 10). However, the Persistent Rural Poverty Task Force (RSS 1993) rejected the explanation of rural poverty based solely on the paucity of human capital. Instead, it estimated that rural America suffers primarily from a shortage of good jobs, not only of good workers. Although improving the education levels of rural residents—as well as the rest of the nation—is necessary, just raising the educational attainment of ruralites would not by itself lower persistent poverty levels. The Task Force recommended a place-specific development strategy centered on small-scale enterprises that would help stem the tide of outmigration among the rural well educated (Dudenhefer 1993).

Furthermore, according to Thurow (1996, 315), the "builder" mentality embodied in the values and savings habits of the self-employed must be promoted in order to rescue American capitalism from its own destructive myopia, inasmuch as the builder mentality stresses longer time investment horizons than those typically exhibited by absentee capitalists or consumers. Perhaps rural North America, as the bastion of such values as self-help, mutual aid, family solidarity, entrepreneurial skills, and resourcefulness (Meyer and Burayidi 1991, 11), needs to look no further. The builder mentality is palpably present in communities pursuing grass-roots self-development (Flora et al. 1991) (see also Green, Chapter 9). Whether pursuing Main Street programs, ecotourism, historical preservation, or small business incubators, self-reliant development initiatives seem to emanate from private-public partnerships where the local government unit plays an important leadership role in the procurement of federal funding.[17] Given that commercial banks seldom find rural projects attractive, federal and state funds remain the most important source of capital for community development. In addition, the availability and cost of credit, lack of skilled labor, and need for technical assistance are the three major constraints in starting and sustaining rural self-help initiatives (Flora et al. 1991 and Green, Chapter 9). It is not surprising that these constraints are also the greatest challenges repeatedly overcome by the rural builder mentality.

Rural Sustainable Communities: Two Alternatives

In postindustrial society, two alternative strategies for rural sustainable communities vie for policy attention. One is biocentric-communalist, which seeks to diversify the local economy by asserting control over local resources. It promotes the adoption of green, ecologically sensitive technologies and strategic small-scale rural-urban enterprises beneficial to the local community (see Flaccavento, Chapter 19). The other, dominated by technocentrist-accommodator values, upholds the conviction that a rural renaissance will unfold for remote and isolated communities as the new telecommunications and information technologies permit them to overcome the friction of space. Telemedicine, distance learning, and improved communications among local units of government (see Seroka, Chapter 11) facilitate multicommunity networks and alliances and permit communities and individuals to access the global village's information, knowledge, and extra-local labor markets by bypassing regional and local constraints.

The first alternative asserts the value of sense of place and localism by preserving the uniqueness of local culture and landscape and ecosystem diversity. It is a bottom-up community-based initiative that makes strategic choices about the types of imports and exports to pursue. For instance, it may opt for community composting instead of urban garbage imports and for the niche marketing, on the right scale, of its sustainably produced exports. As in the rest of the world, where not-for-profit sustainable development initiatives are being undertaken, to ensure durability these projects depend on a support network of domestic and international NGOs (nongovernmental organizations) that aid in procurement of capital, training, and technical assistance as well as in securing the moral and political support necessary to withstand the skepticism and intolerance of mainstream professional, educational, and political institutions. Echoing the sustainable development concerns of biocentric-communalist perspectives in integrating ecology, economy, and social equity, this strategy emphasizes social equity and ecology over economic efficiency.

The second vision is also community based. However, it relies on existing regional and national networks of chambers of commerce and local governments to gain access to capital, training, and technical assistance. It defines *rural* as nonmetropolitan or unincorporated, and *sustainable* as enduring economic growth via the diversification of local enterprises (beyond retirement, residence, and recreation). This vision's notion of ecological obligation is tantamount to enforceable compliance with environmental regulation. It supports a rural-urban regionalism whereby the urban dictates the choices available for rural development. Solid waste streams, back-office data processing, and telemarketing jobs are the urban exports frequently available. Nevertheless, the power of such a vision to crystallize into a rural renaissance is limited (see Shapek, Chapter 17) and vulnerable to competition from lower-wage countries (Howland 1991). Bearing the imprint of technocentrist-accommodator perspectives, this alternative puts efficiency before ecology and social equity.

The 1996 Farm Bill, through its Fund for Rural America, continues support to the Rural Development Administration created by the 1990 Farm Bill. This fund provides $300 million for agricultural research and rural development, emphasizing value-added enterprises and international exports as well as community sewer and water infrastructure projects. It is premised on the optimistic notion that "these investments will ensure that all Americans, regardless of how remote an area they live in, will have the opportunity to better their lives and share in the economic growth spurred by the revolution in information technology" (Clinton 1996, 1). Furthermore, according to advocates for the small and medium-size family farm, this meager fund and other titles contained in the bill that remove all safety nets for farmers have the potential to divert assistance from low-income rural populations to aid large-scale businesses (Knutson et al. 1996; UMC 1996; Siedenburg 1996). The national vision on sustainable development as articulated in *Sustainable America: A New Consensus* by the President's Council on Sustainable Development (PCSD 1996, iv) adopts the policy and public discourses associated with the current global ecological crisis, the declining quality of life of many regions and places resulting from global economic restructuring, and a concern for future generations as the ethical justification for action.

> Our vision is of a life-sustaining Earth. We are committed to the achievement of a dignified, peaceful, and equitable existence. A sustainable United States will have a growing economy that provides equitable opportunities for satisfying livelihoods and a safe, healthy, high quality of life for current and future generations.

However, the vision itself, as well as the policies and actions designed to implement it, including the farm bill—the major piece of legislation affecting rural America—resonate as a series of pragmatic compromises between biocentrist and technocentrist positions concerning the integration of ecology, economy, and social equity. In the aftermath of the policy debate the outcome concedes front stage to economic efficiency through markets, trade and technological innovation; middle stage to ecology through environmental protection and conservation; and back stage to social equity through minimal funding for rural development. At a glance, the policy outcome also reflects which groups have the power to set the agenda and define whose present and future generations should count.

A MULTIDISCIPLIANRY PERSPECTIVE ON RURAL SUSTAINABLE DEVELOPMENT

For the most part, this book is a reflection of the sustainable development debate as seen through the eyes of rural and small town planners, rural sociologists, rural public administrators, agricultural economists, community researchers, and sustainable development advocates and practitioners. It garners a variety of positions and experiences on the topic, but in no measure does it claim to present a consensus or a set of ecumenical prescriptions. The hallmark of Part One is the conspicuous espousal of the emerging ecological worldview by all the authors, albeit with different emphases. These range from a secular overview of Amish principles of agriculture and community and alternative visions of rural development, to ecologically principled positions on planning for the region and community.

In Chapter 2, Mark Lapping invites us to revisit the various rural myths and agrarian and arcadian leitmotifs embodied in discussions of rurality and sustainability. He argues that such myths must be confronted, because they mask the reality of much of the American countryside. Through the eyes and writings of Gene Logsdon, David Kline, and Wendell Berry, Lapping explores Amish principles of rural sustainability and argues that as we craft a definition of the concept, the Amish way of life stands as test and metaphor for rural renewal. Because the Amish way provides us not only with principles for rural sustainability but also a living paragon, other definitions and secular implementations of the concept must meet the Amish test: that they be family based, small-scale, environmentally benign and convivial, and supportive of communitarian values and structures.

William Rees (Chapter 3) argues persuasively against the myth that humankind is the master of nature—an illusion constructed on technological arrogance and conventional economics and blind to biophysical reality. Based on ecological economics criteria and a modern interpretation of the second law of thermodynamics, Rees develops the concept of human ecological footprints and contends that achieving rural sustainability is both an ecological and moral imperative for the survival of our citified, high-income way of life. The more urban we become, the greater our dependence on protecting the ecological integrity of the countryside.

Richard Clugston (Chapter 4) presents two alternative visions for rural revitalization. The first requires the further depopulation of rural America to create increasingly automated agribusiness megafarms. The second emphasizes a shift toward diversified sustainable agriculture. Clugston argues that current policies support almost exclusively the first scenario and that this trend must be balanced by policies and consumer choices in support of sustainability, lest rural America risks becoming a depleted, denuded "third world." He presents examples of initiatives inspired by the second alternative as evidence that the movement is expanding. He also outlines the agenda for sustainable rural revitalization.

In Chapter 5, Erik Davies and George Penfold propose an exploratory model for evaluating the sustainability of rural and agricultural change. Inspired by the work of William Rees and others, Davies and Penfold present a definition of sustainable development and use it as a basis for exploring the implications of applying the concepts of biophysical, natural resource, built, social, and management environments. Using an example of agricultural change in Ontario, Canada, the authors suggest how the model could be applied and provide implications for research and rural planning.

In Chapter 6, the last chapter in Part One, Daniel Chiras and Julie Herman outline the ecological principles of sustainable development that provide the ideological foundation for integrating ecology, economics, and social equity. From these principles they derive several key operational directives to guide the planning for sustainable community development. The authors present a systems approach designed to guide communities to an understanding of the root causes of their environmental unsustainability and the planning alternatives available to overcome them.

Part Two focuses on the *community* and is introduced by O. J. Furuseth and D. S. K. Thomas, who alerts us to the practical intricacies of planning for sustainable development. In this section, the community is both the focus and locus of rural sociological research. In Chapter 9, G. Green concentrates on the rural experience of community self-development, while L Beaulieu and G. Israel's in Chapter 10 explore the contribution of education and community social capital to community sustainability. Chapter 8, by J. Luther—written from the perspective of community development planning—and Chapter 9, by J. Seroka—representing a view from rural public administration—offer somewhat dissimilar philosophies on rural community revitalization. While Luther argues for a "New Localism" based on self-determination and cooperative behavior among adjacent rural communities, Seroka advocates a similar approach, but argues that geographical proximity cannot longer be the basis of community cooperative linkages; that at the end of this century, telecommunication technologies have opened the possibility for creating rural community networks beyond the state's confines. Also on the theme of rural community revitalization, J. Segedy offers in Chapter 12 a hands-on approach to the community-based visioning process and details how community-based charrette workshops can be an effective vehicle for public involvement and community redevelopment.

The contributions to Part Three, prefaced by M. Boswell and I. Audirac, converge on the *region* as rural-urban symbiosis and as the rural-urban interface of agriculture, recreation, recycling, and partnerships for rural sustainable development. Chapter 14 by R. Heimlich and C. Barnard identify three types of farms coexisting at the urban fringe. They discuss how has farming adapted to increased urbanization, and how farmland preservation programs may affect the survival of the three types of metro farms. E. M. Starnes, M. Benedict, and M. S. Sexton (Chapter 15) identify the benefits of green ways and trails for rural sustainability and present three case studies: the St. Marks Trail in north Florida, the Heritage Trail in eastern Iowa, and the Scuppernong River Greenway in northeastern North Carolina. Waste management and recycling—a major opportunity for rural-urban symbiosis—is addressed by R. Shapek in Chapter 17 and D. Derr and P. Dhillon in Chapter 18. Shapek examines solid waste management initiatives in rural and sparsely populated areas, concentrating on Florida's efforts, while Derr and Dhillon delve into the economic benefits of on-farm composting of leaves which otherwise would go into municipal landfills.

Chapters 16, 19, 20 each present a rural community development experience whereby environmental preservation is intrinsically a part of community self-determination and strong and determined leadership is an indispensable factor for long-term project durability. In assessing the experience of the SURE (Sustainable Urban/Rural Enterprise) partnership between the city of Richmond, Indiana, and its rural hinterland in Wayne County, J. Segedy and T. Lyons (Chapter 16) argue that the durability of these initiatives depends on continued high levels of commitment from citizens and the elected and appointed leadership. A. Flaccavento (Chapter 19) writes about development from-within and describes the innovative responses to the problem of jobs versus the environment promoted by the Clinch Powell Sustainable Development Initiative in the Central Appalachian regions of Northeast Tennessee and Southwest Virginia. In Chapter 20, the final chapter of this section, N. K. Seymoar uses the perspective of the "self-empowerment cycle" to describe three Canadian indigenous communities whose struggle for self-determination and political recognition led them to overcome internal strife and to forge strong external partnerships that allowed them to preserve their local ecosystem in spite of strong industrial and federal government interests.

NOTES

1. The word greenness is used in its most general sense, representing both theory and practice that is environmentally conscious. As argued in this chapter there are differing shades of greenness each corresponding to a different environmental ideology.
2. Ralph Metzner (1993, 164–168) argues that postmodern transformations in the natural sciences correspond to a shift from "mechanistic" views of the universe to "organismic" views. Hence, the earth, previously conceived as inert matter, is currently seen in Lovelock's Gaianism as "living earth," a superorganism in mutual metabolic interaction between living organisms and its biochemical en-

vironment. In the physical sciences, linear causality has been superseded by chaos theory. A holistic view of reality as a nested hierarchy of systems (holarchy) interconnected at all levels is replacing the atomistic view of the universe. For a similar view in planning, see Rees 1995.

In the social sciences "critical realism" and "constructivism" have emerged as postmodern alternatives to logical positivism and operationalism. In philosophy and religion, the human/nature relationship has been amply scrutinized and Judeo-Christian theology criticized for its anthropocentric attitude toward nature. Deep ecology and ecophilosophies call for a change in values from "human on nature" to "human in nature." All living beings are valued intrinsically rather than for their human utility. The modern materialistic, nihilistic worldview is giving way to the resurgence of various religious spiritualities and animistic ethics.

3. Both as ultimate reality (ontology) and essence of knowledge (epistemology).
4. In postmodern relativism, all theories are accorded equal validity and no single definition is dominant.
5. In the development policy arena, the green agenda has traditionally dealt with global issues such as climate change, ozone depletion and loss of biodiversity. The brown agenda, on the other hand, focuses on localized problems of cities and communities related to pollution, poverty and environmental hazards. In general the green agenda is associated with rural resource protection, while brown agenda with cities and urban pollution.
6. At current trends of energy consumption, by 2025 the expected global population of more than 8 billion would be using the equivalent of 14 billion metric tons of coal per year. But if, instead, all the world used energy at industrialized levels, by 2025 the world would consume 55 billion metric tons (Schmidheiny 1992, 5).
7. Fordism as a general term derives from the new industrial organization of mass automobile production devised by Henry Ford at the beginning of this century. As a sociological concept, Fordism is a socioeconomic and technopolitical regime that emerged in the United States and Western Europe after 1920. It evolved into a social contract between organized labor, industrial corporations (e.g., automobile manufacturers), and national governments. At the heart of this contract is the role of the welfare state as regulator of market forces and provider of a minimum social safety net. The economy is oriented toward mass production and consumption of homogeneous products, while the dominant technological clusters are in electronics and petrochemicals. The regime's internal order disintegrated in the 1970s, coinciding with the oil embargo, the decline of American hegemonic power, and the United States' abandonment of the Bretton Woods gold parity.

In the rural sector, the agricultural establishment, which monopolized rural policy, obtained important commodity subsidies and mandated the land grant institutions and the USDA to concentrate in the scientific modernization of agriculture. New seeds and high-yield varieties were developed, and the green revolution was launched and diffused throughout the world. This constituted the industrialization of agriculture, accompanied by impressive productivity and efficiency gains. However, agroindustrialization was also accompanied by a strong restructuring of rural economies. With fewer farms, considerable portions of farm labor became redundant and migrated to the cities, while those remaining increased the ranks of the rural poor (Dudenhefer 1993).

The new post-Fordist regime—also known as Toyotist (associated with the Toyota Company's industrial organization of automobile production and its industrial complex headquartered in Toyota City, Japan)—coincides with the emergence of Japan as an economic power and a lean and flexible system of production. The new system, based on microelectronics, robotics, and a decentralized network of subcontractors and outsources, permits domestic and multinational corporations to move their plants and production to lower-wage countries. With the ensuing weakening of labor unions and the rise of political neoconservatism (Reaganism and Thacherism), the welfare state is gradually dismantled and its role transformed into market deregulator (Lipietz 1987; Hoffman and Kaplinsky 1988; Lash and Urry 1987; Buttel 1992; Swyngendouw 1992).

In postindustrial and post-Fordist analysis, the symptoms of economic distress are similar. The difference is in the prognosis each one offers. Post-industrial analysts—mostly American authors—share an optimistic perspective of technology both as the cause and remedy of current economic woes. The observed growth in the information and services industries, accompanied by increased technological unemployment in the industrial and agrarian sectors, are seen as transitory growing pains necessary in the path to a postindustrial society. They believe that such pains will be outgrown with advances in science and information technology and the steady unleashing of human creativity. In Bell's (1976) formulation, information professionals and technical experts attain the highest status in society, while the most important factors of production become data and knowledge.

Post-Fordist formulations—primarily by European authors—in contradiction to postindustrial analyses, are less sanguine about the future and raise social equity issues related to these changes. Their focus is on explaining the social transformation from Fordism to post-Fordism.

8. For instance, ITT is involved in large-scale bakeries and in the sale of seeds; Greyhound has moved from buslines to meat processing; Proctor & Gamble does soya production R&D, and Boeing, the airplane maker, invests in potatoes (de la Court 1990, 42).
9. In 1990, the typical meal of a cheeseburger and fries in West Germany required each day 500 cows to be processed for hamburger meat. The white bread, made from American wheat, was brought from Pfungstadt, West Germany. The gherkins, mustard, and mayonnaise were supplied by a West German company, cheese was imported from the Netherlands and lettuce flown in from California in winter or from Spain in summer. Onions came from the United States, potatoes from the Netherlands, the ketchup from an American company, the raw materials for the cups and packaging from Canada and Scandinavia. The packaging was printed in West Germany and cut to size in France (de la Court 1990, 42).
10. In the 1950s Canada was a self-sufficient producer of plums, peaches, apricots, strawberries, and pears, but by 1980, 28% to 57% of these fruits were im-

ported. Canada has also lost much of its processing capacity for tomatoes and is now a net importer of apple juice concentrate, albeit a major producer of apples (MacRae et al. 1990,86).

11. According to Mark Epstein in *Public Voice* (Van Nostrand Reinhold, 1996,5; http://www.public voice), the 1996 Farm Bill is pro status quo rather than for alternative agriculture. Wealthy producers, rather than consumers, will be the beneficiaries. The largest 11% of corn farmers will receive 60% of all corn subsidies; the top 7% of cotton producers will receive an average subsidy of $24,000 in 1996. Moreover, total crop subsidies are not really phased out. They will just decline 30% from a total of $5.6 billion in 1996 to $4 billion in 2002, and costly sugar and peanut prices support programs will persist despite a strong challenge from consumer, environmental, and taxpayers groups.
12. Buttel (1992, 23–24) warns that the greening of the rural agenda in a new neoconservative political era could result mainly from alliances between environmentalists and corporate elites to the neglect of rural development. In this regard, Du Pont has joined the sustainable agricultural movement, but with a cornucopian outlook on science and technology. It developed the Environmental Respect Award, a national program that recognizes agricultural dealers for adopting improved standards for storing and handling pesticides. Co-opting the grass-roots, care-for-the-land movement spurred by Midwestern family farmers, Du Pont began in 1993 the No-Till Neighbors Program, a forum for farmers to teach other farmers about residue management and conservation tillage. In cooperation with the EPA and USDA, it has launched an on-farm demonstration project in sustainable agriculture at its Remington Farms (Bird and Ikerd 1993, 96).
13. See note 5.
14. Regardless of the spotted owl controversy's effect on employment, the region will experience a minimum of 1% to 2% decline per year in employment, resulting primarily from economic and technological restructuring factors (Levi and Kosher 1995, 633). Technological restructuring of the timber industry in the Northwest has been the result of increased competition from newer, more technologically advanced sawmills in the Southeast. In addition, foreign markets for raw lumber have reduced the need to process timber in the Northeast; the depletion of old-growth forests has forced the adoption of renewable forest practices in second-growth forests; and, in order to pay off debts from ill-fated mergers and acquisitions of the 1980s, the industry has resorted to expansion of logging on public lands (Levi and Kocher 1995).
15. The Poppers' article made headlines by recommending the Buffalo Commons Plan, whereby the federal government would assist the depopulation of the Great Plains, reintroduce the bison and natural grasses, and turn a portion of the plains into a national park. Daniels and Lapping (1987) argued that shrinking federal and state assistance should be redirected from "dying" rural towns with little hope for economic survival to those whose survival seemed more promising.
16. The South comprises Alabama, Arkansas, Delaware, the District of Columbia, Florida, Georgia, Kentucky, Louisiana, Maryland, Mississippi, North Carolina, Oklahoma, South Carolina, Tennessee, Texas, Virginia and West Virginia.

17. In a national survey of rural self-development projects, 55% of 160 cases identified the municipal or town government as an active promoter, and in about 37% of the cases a county government was involved. Private business and city governments were identified as key initiators of these projects (Flora et al. 1991, 21).

REFERENCES

Agriculture Canada. 1989. *Growing Together: A Vision for Canada's Agri-food Industry.* Ottawa: Agriculture Canada Supply and Services.

Ayres, R. U. 1994. "Industrial Metabolism: Theory and Policy." In *The Greening of Industrial Ecosystems,* edited by D. J. Richards, B. R. Allenby, and R. A. Frosh. Washington, D.C.: National Academy Press, 23–37.

Bell, D. 1976. *The Coming of Post-Industrial Society.* New York: Basic Books.

Beus, C. E., and R. E. Dunlap. 1993. "Agricultural Policy Debates: Examining the Alternative and Conventional Perspectives." *American Journal of Agriculture* 8(3):98–106.

Bird, J. W., and J. Ikerd. 1993. "Sustainable Agriculture: A Twenty-First Century System." In *Rural America: Blueprint for Tomorrow,* edited by W. E. Gahr. Philadelphia: Sage Publications, 92–102.

Buttel, F. H. 1992. "Environmentalization: Origins, Process, and Implications for Rural Social Change." *Rural Sociology* 57(1):1–27.

Clinton, W. 1996. *Statement by the President on the Farm Bill Signing.* Washington, D.C.: The White House Office of the Press Secretary (http//www.usda.gov/farmbill/state.htm).

Daly, H. E. 1991. "Sustainable Growth: A Bad Oxymoron." *Grassroots Development* 15(3):39.

Daly, H. E., and K. N. Townsend, eds. 1993. *Valuing the Earth.* Cambridge, Mass.: MIT Press.

Daniels T. L., and M. B. Lapping. 1987. "Small Town Triage: A Rural Settlement Policy for the American Midwest." *Journal of Rural Studies* 3(3):273–280.

de la Court, T. 1990. *Beyond Brundtland: Green Development in the 1990s.* London: Zed Books Ltd.

Dobbs L. T., and J. D. Cole. 1992. "Potential Effects on Rural Economies of Conversion to Sustainable Farming Systems." *American Journal of Alternative Agriculture* 7(1 and 2):70–80.

Dudenhefer, P. 1993. "Poverty in the Rural United States." *Focus* 15(1):4–13.

Feenstra, G. 1995. "What Is Sustainable Agriculture?" Davis: University of California, Sustainable Agriculture Research and Education Program (http://www.sarep.ucdavis.edu/SAREP/Concept.html).

Fisher, D. U. 1993. "Agriculture's Role in a New Rural Coalition." In *Rural America: Blueprint for Tomorrow,* edited by W. E. Gahr. Philadelphia: Sage Publications, 103–112.

Flora, C. B. 1990. "Sustainability of Agriculture and Rural Communities." In *Sustainable Agriculture in Temperate Zones,* edited by C. A. Francis, C. B. Flora, and L. D. King. New York: John Wiley & Sons, 343–359.

Flora, C., J. L. Flora, G. P. Green, and F. E. Scmidt. 1991. "Rural Economic Development Through Local Self-Development Strategies." *Agriculture and Human Values* 7(3):19–24.

Friedmann, J., and H. Rangan, eds. 1993. *In Defense of Livelihood: Comparative Studies on Environmental Action*. West Hartford, Conn.: Kumarian Press.

Goldschmidt, W. 1978. *As You Sow: Three Studies in the Social Consequences of Agribusiness*. Montclair, N. J.: Allanheld, Osmun and Co.

Goodman, D., and M. Redclift, eds. 1989. *The International Farm Crisis*. New York: St. Martin's Press.

Hecht, S. B., and A. Cockburn. 1992. "Realpolitik, Reality and Rhetoric in Rio." *Environment and Planning D* 10:367–375.

Hefferman, W. D. 1982. "Structure of Agriculture and Quality of Life in Rural Communities." In *Rural Society in the US: Issues for the 1980s*, edited by D. A. Dillman and D. J. Hobbs. Boulder, Col.: Westview Press, 337–346.

Henderson, J. W., and M. Castells. 1987. *Global Restructuring and Territorial Development*. London: Sage.

Hoffman, K., and R. Kaplinsky. 1988. *Driving Force: The Global Restructuring of Technology, Labor, and Investment in the Automobile Components Industries*. Boulder, Col. and London: Westview Press.

Howland, M. 1991. "Are Export-Oriented Services a Source of Growth for Rural Economies?" *Proceedings of the Rural Planning and Development: Visions of the 21st Century Conference*, Orlando, Florida, February, 1991. Gainesville: University of Florida.

Knutson R. D., E. G. Smith, J. L. Outlaw, and W. Fred Woods. 1996. "New Farm Bill: Watershed Chance in Policy." 4/29/96: http://afpc1.tamu.edu/pubs/education/fb96text.htm.

Lash, S., and J. Urry. 1987. *The End of Organized Capitalism*. Madison: University of Wisconsin Press.

Lasley, P., E. Hoiberg, and G. Bultena. 1993. "Is sustainable Agriculture an Elixir for Rural Communities? *American Journal of Alternative Agriculture* 8(3):133–139.

Levi D., and S. Kocher. 1995. "The Spotted Owl Controversy and the Sustainability of Rural Communities in the Pacific Northwest." *Environment and Behavior* 27(5):631–649.

Lipietz, A. 1987. *Mirages and Miracles*. London: Verso.

Lobao, L. M. 1990. *Locality and Inequality: Farm Industry Structure and Socioeconomic Conditions*. Albany: State University of New York Press.

Lyson, T. A., and R. Welsh. 1993. "The Production Function, Crop Diversity, and the Debate Between Conventional and Sustainable Agriculture." *Rural Sociology* 58(3):424–439.

MacNeill, J. 1991. "Toward Global Action." In *Beyond Interdependence*. New York: Oxford University Press.

MacNeill, J., P. Winsemius, and T. Yakushiji. 1991. "Recasting Domestic Policy." In *Beyond Interdependence*. New York: Oxford University Press, 29–51.

MacRae, R. J., S. B. Hill , J. Henning, and A. J. Bentley. 1990. "Policies, Programs and Regulations to Support the Transition to Sustainable Agriculture in Canada." *American Journal of Alternative Agriculture* 5(2):76–92.

Metzner R. 1993. "The Emerging Ecological Worldview." In *Worldviews and Ecology*, edited by M. E. Tucker and J. Grim. Lewisburg: Bucknell University Press, 163–172.

Meyer, P. B., and M. Burayidi. 1991. "Is Value Conflict Inherent in Rural Economic Development? An Exploratory Examination of Unrecognized Choices." *Agriculture and Human Values* 7(3):10–18.

O'Riordan T. 1981. *Environmentalism*. London: Pion.

________. 1995. "Frameworks for Choice. Core Beliefs and the Environment." *Environment* (October):4–9.

Paarlberg, D. 1980. *Farm and Food Policy: Issues of the 1980s*. Lincoln: University of Nebraska Press.

PCSD (President's Council on Sustainable Development). 1996. *Sustainable America. A New Consensus for Prosperity, Opportunity, and Healthy Environment for the Future*. Washington, D.C.: Government Printing Office.

Pearce, D., A. Markandya, and E. B. Barbier. 1989. *Blueprint for a Green Economy*. London: Earthscan Publications, Ltd.

Popper D. E., and Popper F. 1987. "From Dust to Dust." *Planning* 53(12):12–18.

Rees, W. E. 1995. "Achieving Sustainability: Reform or Transformation?" *Journal of Planning Literature* 9:343–361.

Richards, D. J., B. R. Allenby, and R. A. Frosh, eds. 1994. *The Greening of Industrial Ecosystems*. Washington, D.C.: National Academy Press.

Rural Sociological Society Task Force on Persistent Rural Poverty. 1993 *Persistent Poverty in Rural America*. Boulder, Col.: Westview Press.

Schmidheiny, S. 1992. *Changing Course: A Global Business Perspective on Development and the Environment*. Cambridge, Mass.: MIT Press.

Sessions, G. 1993. "Deep Ecology: Introduction." In *Environmental Philosophy*, edited by M. Zimmerman. Englewood Cliffs, N. J: Prentice-Hall, 151–170.

Siedenburg, K. 1996. "Farm Bill Signed into Law. Impacts on Environment and Agriculture." *Agrarian Advocate*, 3/29/96 on Global Action and Information Network (http://www.scruznet.com/~gain/FarmBillLaw.html).

Summers, G. F., and F. Clemente. 1976. "Industrial Development, Income Distribution and Public Policy." *Rural Sociology* 41:248–268.

Swyngendouw, E. A. 1992. "The Mammon Quest. Globalisation, Interspatial Competition and The Monetary Order: The Construction of New Scales." In *Cities and Regions in the New Europe*, edited by M. Dunford and G. Kafkalas. London: Belhaven, 39–67.

Thurow, L. C. 1996. *The Future of Capitalism*. New York: William Morrow.

Turner, R. K., D. Pearce, and I. Bateman. 1993. *Environmental Economics and Elementary Introduction*. Baltimore: Johns Hopkins University Press.

UMC (United Methodist Church). 1996. "New Farm Bill a Mixed Bag, According to Church Executive. New Farm Bill Has Pluses, Minuses." Washington, D.C.: United Methodist Church, New Services, 4/9/96. (http://www.umc.org/jbill.html).

Warren, R. 1972. *The Community in America*. Chicago: Randall McNally.

WCED (World Commission on Environment and Development). 1987. *Our Common Future*. Oxford: Oxford University Press.

Wimberley, R. C. 1993. "Policy Perspectives on Social, Agricultural, and Rural Sustainability. *Rural Sociology* 58(1):1–29.

Two

A Tradition of Rural Sustainability: The Amish Portrayed

Mark Lapping

INTRODUCTION

The concept of *rurality* lacks precision. Though hardly without meaning, it is expressed by a term that apparently defies neat definition. The difficulty in reaching a consensus or an operational definition for *rurality* has led to something akin to a scholarly and political "gentleman's agreement" not to pursue the matter very much further. Increasingly, much the same can be said for the word *sustainability*. The term has become so overused that it is difficult to provide a useable definition with enough specificity for clear meaning and understanding. Like *rurality, sustainability* is filled with implication and nuance. With its roots fixed in ecological science, *sustainability* suggests renewability, identity, and system integrity all within a larger complex of environmental connection, resilience, adaptability, versatility, perpetuation into a fundamentally unknown and uncertain future, and intergenerational equity and justice.[1] Simply put, then, *sustainable rural development* concerns itself with the ability of communities, however defined, to create and utilize those conditions and structures necessary to promote a rural future. The nature of that rural future may not be fixed nor well understood, but sustainability implies that a rural sector should exist and that it be one in which resources are such that individuals, families, communities, and nature have the opportunity to flourish within rural settings well into the future.

RURAL MYTH, RURAL REALITY

If the meaning of rural sustainability may be uncertain, a fundamental and core belief in its desirability and utility is not. Although the contemporary reality is that of an increasingly metropolitan nation and society, with rural areas becoming ever more peripheral, Americans persist in describing rural areas and small towns as their ideal community types. Moreover, Americans continue to believe that those who live and work in rural places live the good, natural, and all-American life, a life that is fundamentally sustainable and worthy. As a recent Roper organization survey indicates, Americans see rural people as "friendlier, better citizens, more family oriented, and harder working than the average American" and that "rural people have the last chance to pursue the 'American dream'" (*The Philadelphia Inquirer*, 11 June 1992). The presumption of rural sustainability is, then, deeply rooted in and tied to a popular conceptualization of the national past and a worldview that continues to define the American project. Often described as the agrarian myth, rural fundamentalism, neo-Jeffersonianism, or the pastoral myth, this concept is pervasive, powerful, and critical to developing any understanding of how Americans conceive of sustainable rural development. Though admittedly "best understood in terms of a set of ideological motifs too complex to permit monolithic categorization" (Buell 1989, 5), the myth underlying many conceptions of rural life is so ubiquitous that it has become a singular cultural icon. Suggesting "the virtues of family, tradition, and the cherished values of hard work and self-reliance . . . individual autonomy, nurturant family life, and the spirit of community," the rural myth is both symbol and commodity in mass culture, no matter the "sharp disparity" between the image and the reality of contemporary rural life (Goldman and Dickens 1983). The larger point is that one cannot appreciate the potential for sustainable rural development without recognizing the longevity, pervasiveness, and power of the larger rural myth within American culture. And although much of the future of the debate over sustainable rural development will depend on the ability to move beyond the rural myth, confronting it becomes a clear and obvious necessity (Worster 1991).

The rural myth, which colors all discussions of sustainable rural development, has come down to us as embodying Jeffersonian egalitarianism through the agrarianism of the yeoman farmer,[2] the domesticity of the "middle landscape" that lies somewhere between the city and the wilderness (Marx 1964), the innocence of the Adamic view of the American experience (Lewis 1955), arcadian simplicity (see Schmitt 1969 and Shi 1985), and an antipathy for the market, class structure, and corporate capitalism.[3] No matter how sanitized, ahistorical, and flawed—we are, after all, speaking about a myth—these motifs have been woven into this elaborate construction that has traditionally defined much of the rural debate. Both the market and the federal state have been among the primary agents for the elaboration of the rural myth. Together, the powers of the market and the state have worked to commodify the rural myth while successfully integrating resource development and agriculture into export and surplus production, privatizing much of the national land base, creating the bureaucracy of modern agriculture, subsidizing technology transfer and human capital development in certain ways, and reducing the potential of conflict to generate fundamental change in the countryside (see Goldman and Dickens 1983 and Lapping 1992). Even some attempts to promote rural development, such as the Country Life movement at the turn of the twentieth century, actually led to accommodation with the larger urban-industrial need for the construction of a countryside that was "a socially and economically organized and efficient part of an increasingly interdependent nation," as historian David Danbom (1979, 139) has observed. The result has been that the reality of so much of rural America—persistent poverty over the generations; dilapidated and poor-quality housing and physical infrastructure; the lack of adequate medical, educational, nutritional, and social services; the erosion of extended family support systems in the face of increasing single parenthood and teen parenting; self-destructive behaviors; and the overall collapse of economic opportunity—is substantially masked and hidden by the pervasiveness of the rural myth of good and plenty across a verdant landscape.[4]

THE AMISH PORTRAYED

To be truly sustainable, rural development must reject such a fictitious characterization of rural life, for everywhere the culture of the countryside is collapsing or in atrophy, except in a few instances. A growing number of advocates for such an approach find in the Amish way of life the active metaphor and example for rural renewal along the lines of sustainable development. While the Amish speak to much that resonates as authentic and true within the rural myth, they have created a community and familial prosperity that confirms its viability, integrity, and sustainability. Primary among these advocates are Gene Logsdon, David Kline, and Wendell Berry, most especially.

A non-Amishman living in an area with a substantial Amish population—northeastern Ohio—Gene Logsdon operates a small family farm and has written a number of books and many essays and articles on rural, agricultural, and environmental themes. A recently published collection of essays (Logsdon 1994) includes the republication of his "Amish Economics," which first appeared in the *Whole Earth Review* in 1986.

Logsdon's view of the future of rural America is surprisingly optimistic. It has at its core a "new rural community," based on the small farmstead of less than 50 acres, which is highly integrated; uses "new biological methods"; is relatively debt-free because it is small and is managed under "fiscally conservative" assumptions; and, most important, is part of a larger, supportive community culture. Indeed, efforts to build an agriculture without a supportive cultural context, such as that provided by the Amish community, are doomed to fail because they are "essentially islands in an alien culture. . . . If you want to remake an agriculture that is technically correct for sustainability, we must make sure that it is also culturally correct, or the effort will not succeed" (99).

Amish agriculture "resists financial chaos and the decline that follows by fortifying individuals against their own frailty. The culture sanctifies the rural virtues that make good farming, or good work of any kind, possible: a prudent practice of ecology, moderation in financial and material ambitions, frugality, attention to detail, good work habits, interdependence (neighborliness), and common sense" (33). For Logsdon the matter comes down to the existence of a social and community fabric defined by cooperation, connection, and mutual responsibility. The Amish have been successful at creating a sustainable rural life and culture because "they cooperate to get the work done, forming a network of helping hands, each feeding into the other, turning money over again and again, keeping more of the ultimate consumer dollar in the community and forming a strong interdependent web of local self-reliance. This is perhaps the most important aspect of Amish economy" (145). This economy is then integrated back into society, with the result that "there is no distancing of work from family life that breeds the idea that what one does at work is not bound by the same moral good as what one does at home" (153).

The continuity between means of livelihood and manner of living, what Wendell Berry (1990a) calls the Amish negation of the division of "the life of the mind from the life of the body," is central to the writings of David Kline, an Amish farmer from Ohio. A frequent contributor to the Amish magazine *Family Life*, Kline's (1990) *Great Possessions* initially appears to be a naturalist's observation of a year in the life of a farm. His possessions are the gifts that nature and life bestow on one willing to take the time to observe and appreciate what the natural world brings. But this quiet book is much more than that. It describes Amish agriculture in detail and its continuing quest to place modern technology in perspective. "The Amish," he writes, "are not necessarily against modern technology. We have simply chosen not to be controlled by it" (xv). Logsdon (1995, 179) has called the Amish "past masters at using advanced technology to lower rather than raise the cost of machinery, the opposite of what modern agribusiness does." In using technology in this way, the Amish give us answers to each of the three critical questions that, Kline (1990, xix) argues, are facing agriculture: "Should we, instead of working the land traditionally, which requires the help of most family members, send our sons to work in factories to support Dad's farming habit? Should we be willing to relinquish a nonviolent way of farming that was developed in Europe and fine-tuned in America (by what Wendell Berry calls 'generations of experience')? Should we give up the kind of farming that has been proven to preserve communities and land and is ecologically and spiritually sound for a way that is culturally and environmentally harmful?"

For Kline, as well as for others, scale is the fundamental issue.[5] The Amish have placed limits on their lives, not just on things agricultural. The nuclear family and the larger extended community of believers and congregants define the borders of the rural community. The "proper scale" for a sustainable agriculture is one "that enable[s] each farm to be worked by a family" or remains viable by "staying with the horse. The horse has restricted unlimited expansion," thus enforcing limits that encourage the farm family to truly know its land, animals, machinery, and capacity (xxi). Scale has allowed the Amish to know their neighbors as well. Kline writes; "Probably the greatest difference between Amish farming and agribusiness is the supportive community life we have" (xxiii). "The assurance and comfort of having caring neighbors is one of the reasons we enjoy our way of farming so much" (xxiv).

Kline's metaphor for the overall Amish rejection of the secular American religion of plenty and limitlessness is the disappearance of fencerows in rural areas. "Sadly, fencerows have become unfashionable. They began disappearing when the bulldozer became affordable, farm size increased, and the 2-4 D brush killers were developed. Soon after the demise of the fencerow hunters began complaining about the scarcity of rabbits and pheasants. The blame was mistakenly put on the fox and the owl" (109). Here, as elsewhere, Kline makes the appropriate connections between scale, the petrochemical-based and expansionist character of contemporary agriculture, and ecological change and disruption. In another illustration one can see how the naturalist's sensitivity and the farmer's appreciation for the land come together in the Amish critique: "In many parts of the country the silk moths are disappearing. No one is sure of the reasons why, but they likely include destruction of habitat, the overuse of pesticides, and, as some experts insist, the effects of sodium and mercury lights around homes and along streets and highways. The large imperial moth, for instance, has a weakness for artificial light. It is attracted to the lights and often lingers into daylight when it is then eaten by birds. This moth is becoming rare where artificial lights are abundant" (127).

Kline's essays suggest a writer who knows the world in a genuinely worldly way and yet possesses a traditionalism initially born of faith, but rooted in much that has come down to us as the national rural myth: ties to the land and natural life through a reflective agrarianism; self-reliance and individual responsibility; the simple life; and the thorough engagement in life through work, family, and community.

No contemporary student of the Amish has so successfully and completely made the argument for sustainable rural communities based on Amish agrarian principles as has Wendell Berry. Perhaps this is the result of his own great popularity as a poet, essayist, and student of American culture; he is a prodigious and prolific writer with an enormous following. But, just as likely, the appeal of Berry's writing on Amish living is tied to his belief that a future for the rest of America, odd as it might initially appear, resides in the Amish example and concern with the nature and meaning of community. To many, this is an appealing idea. Berry does not ask us to become Amish, for he understands and appreciates the nature, complexity, and totality of the Amish confession. Rather, he implores us to learn from the Amish and to emulate many of the principles of Amish society that have been developed over the past several hundred years both in Europe and in North America. In "Seven Amish Farms," one of his many essays on the Amish, Berry (1981) argues that Amish agriculture is both aesthetically and socially pleasing. Amish farms truly reflect the best of what a working rural landscape is like: "The Amish neighborhoods are more thickly populated than most rural areas, and you see more people at work. And because the Amish are diversified farmers, their plowed croplands are interspersed with pastures and hayfields and often with woodlots. It is a varied, interesting, healthy looking farm country" (250). Such places are the physical manifestation of a society that integrates family, community, and livelihood through a practiced and sustainable relationship with the land (see Lapping 1982).

The history of Amish persecution, Berry (1981) tells us, has a good deal to do with their agricultural and cultural practices. Relegated to the poorest lands by the authorities, the core aspects of Amish agriculture—diversification, crop rotation (including the seeding of legumes), the use of horses for work, and manures for fertilizers—were developed out of necessity as well as principle. By keeping farms small and labor intensive rather than highly capitalized and mechanized, these practices have continued to serve Amish communal life well. Amish agricultural practices have kept the Amish on the periphery and have guaranteed a high degree of independence and freedom from power and authority. Thus the Amish have been permitted to remain both aloof and largely insignificant or marginal to authority. Their remaining something of a curiosity rather than a threat to the majority has given the Amish community the ability to sustain itself even in the face of official indifference or, sometimes, outright hostility. In addition, "intelligent planning, sound judgment, and hard work that good farming requires," which were also essential, have been great assets in cultural survival over the generations (252).

Central to the Amish critique is the recognition that the problem of scale as a defining element of sustainability has had to be addressed time and again. "What these Amish farms suggest, on the contrary, is that in farming there is inevitably a scale that is suitable both to the productive capacity of the land and to the abilities of the farmer; and that agricultural problems are to be properly solved, not in expansion, but in management, diversity, balance, order, responsible maintenance, good character, and in the sensible limitation of investment and overhead" (Berry 1981, 257). Amish farms become those fully capable of conserving both land and community. Amish agriculture is a "complicated structure that is at once biological and cultural, rather than industrial or economic" (258). Berry continues: "I do not think that we can make sense of Amish farming until we see it, until we become willing to see it, belonging essentially to the Amish practice of Christianity, which instructs that one's neighbors are to be loved as oneself. To farmers who give priority to the maintenance of their community, the economy of scale (that is, the economy of large scale, or 'growth') can make no sense, for it requires the ruination and displacement of neighbors. A farm cannot be increased except by the decrease of a neighborhood. What the interest of the community proposes is invariably an economy of *proper* scale" (261, emphasis added).

This is but part of a larger critique of rural America that Berry offers. In another essay, "The Work of Local Culture," Berry (1990b, 197) writes: "In Rural America, which is in many ways a colony of what the government and the corporations think of as the nation, most of us have experienced the losses that we have been talking about: the departure of young people, of soil and other so-called natural resources, and of local memory. We feel ourselves crowded more and more into a dimensionless present, in which the past is forgotten and the future, even in our most optimistic 'projections,' is forbidding and fearful. Who can desire a future that is determined entirely by the purposes of the most wealthy and the most powerful, and the capacities of machines?"

The Amish stand, then, in opposition to this trend. And their opposition is rooted both in belief and understanding: "an understanding and acceptance of the human place in the order of Creation—a proper humility" (Berry 1983), which comprehends and appreciates the nature of community, scale, restraint, the land, the proper place and role of technology, and simple faith (Berry 1977).

Berry (1987, 177–178) advocates a sustainable rural development based on several key Amish principles: the preservation of family, neighborhood, and community; the maintenance of the "domestic arts of kitchen and garden household and homestead"; placement of limitations on technology so that is does not displace or alienate labor; home teaching and service to the community; the fostering of esteem for farming and manual arts; and the imposition of limits in accord with notions of scale consistent with the virtue of "proper use" of people, the land, nature, and community.

Each of these writers argues in convincing ways that the Amish have solved some of the riddles facing contemporary rural America. For example, the Amish emphasis on living in community and with family helps to resolve some of the problems of rural depopulation. Rather than sending their children away or their elderly into nursing homes, the Amish subdivide farms to create still other farms for the young and build "Grandpa houses" near existing farmhouses to maintain patterns of family-based care and the intergenerational transfer of knowledge and tradition. Likewise, the Amish live and farm in ways that harness technology rather than allowing themselves to be harassed by it, economically and physically. Technological solutions are genuinely appropriate in terms of the issues of scale, purpose, need, and the dictates of the principle of "kindly use." Finally, Amish practices have provided a high level of self-sufficiency and autonomy within the larger society in which they live. Family, community, control of technology, and self-sufficiency—these are among the essences that Logsdon, Kline, and Berry identify in Amish living. How very similar these are to some of the most cherished elements of the American rural idyll!

AMISH IMPLICATIONS FOR RURAL SUSTAINABILITY

A commitment to Amish principles of rural sustainability does not merely mean the mass adoption of a new life-style. The Amish require, after all, that one live an all-encompassing way of life rooted in a particular religious tradition. It is, as Testa (1992, 176) has observed, a "preoccupation" built on faith. We must recognize that the contemporary portrayal of Amish life by Logsdon, Kline, Berry, and others is highly idealized. Soil erosion, water pollution, and land development are, unfortunately, part and parcel of Amish agricultural and land use practices (see Place 1993). Yet the fact remains that Amish living *is* the most profound example of rural sustainability in practice within contemporary America. If nothing else, its longevity as a model merits further discussion. With its emphasis on self-reliance, family farming, neighborliness and communitarianism, a shared community bio-ethic, and a "healthy" rejection of both expertise and scientific agriculture and the institutional framework that diffuses it—the USDA, the land-grant university system, and agribusiness—the Amish example is a characterization of rural life that connects with some of the most deeply held and cherished notions Americans have about themselves as a people (see Lapping 1987).

For a number of obvious reasons, not the least being the fascination they hold for those who do not identify with an active "faith community," the Amish remain a compelling subgroup. As we craft a definition of rural sustainability, the Amish stand as something of a test for other expressions of the concept. More specifically, we must determine whether the practical implementation of other forms of rural sustainability can meet Kline's (1990) test: that they be family-based, small-scale, environmentally benign and convivial, and supportive of communitarian values and structures. Is it possible to "secularize" the Amish way to provide us with principles for rural sustainability? Each in his own way—Logsdon, Kline, and Berry—appears to suggest that this may be the case. Callicot (1994) argues that it has already been done, that by 1980 the "stewardship" model of environmental ethics, which shares so much with Amish belief and practice, had been institutionalized through the work of Wes and Dana Jackson in Kansas, the Sustainable Agriculture Working Group, and the Land Stewardship Project, among others. Yet although each advances the argument and points the way to significant change, none of these examples makes of agricultural sustainability what Paul B. Thompson (1995) calls "a trait of a system."[6] Surely other conceptions of rural sustainability, such as the organic and community farming movements, the advocacy of local economic diversification, and a number of green initiatives like the new forestry, have a number of things in common with that of the Amish. Beyond placing farming at or near the center of the equation, the problems of proper human scale, building community, supporting families and connecting generations, working and living in tandem with the environment, and the control of technology all appear to be critical issues for the larger debate over the meaning of rural sustainability. Because Amish life is so complex and yet basic, and because other definitions of rural sustainability ought to take stock of and test themselves against it, the Amish way retains its relevance. Thus the Amish tradition will also continue to hold promise. It works; it moves us closer to a manner of living that is more generous and gentle and that can be sustained—precisely because the tradition supports and defines rural life in ways that are productive, that strive to be environmentally benign, and that are respectful of individuals, families, communities, and the great forces of nature.

NOTES

1. This brief discussion of semantic problems relative to the terms *rurality* and *sustainability* owes much to a larger and more complex analysis by Vail (1994).
2. The classic statement here is that of Smith (1950).
3. See the classic statement on the problem by Hofstadter (1955, 1962).
4. For the expanding literature on rural poverty, see, for example, Brown and Warner (1989), Deaver and Hoppe (1992), Fitchen (1981, 1991), Lichter and Eggebeen (1992), Lyson and Falk (1993), Porter (1991), Rural Sociology Society, Task Force on Persistent Poverty (1993), and Shapiro (1989).
5. For a rich discussion of the problem of scale in the American context, see Sale (1980).
6. Thompson (1995, 171), a foremost student of environmental and agricultural ethics, discusses and assesses concepts of sustainability in terms of philosophical holism. He argues, correctly, I believe, that most contemporary attempts at charting a sustainable future are "emicly expressible goals because they state objectives that we can pursue for a while, achieve or give up on. They imply a narrative temporality because the story begins when a goal is chosen, and ends when it is achieved or abandoned (or when the character dies trying). Sustainability as I have described it can't be understood as a goal that imposes a self-interpreting story line on one's life. It is understandable only as a trait of a system with a non-narrative, yet finite temporal horizon, a system intelligible only from a non-participant's perspective."

REFERENCES

Berry, W. 1977. *The Unsettling of America: Cultural and Agricultural.* San Francisco: Sierra Club Books.

———. 1981. *The Gift of Good Land: Further Essays Cultural and Agricultural.* San Francisco: North Point Press.

———. 1983. *Standing by Words.* San Francisco: North Point Press.

———. 1987. *Home Economics.* San Francisco: North Point Books.

———. 1990a. Preface to *Great Possessions: An Amish Farmer's Journal,* by D Klein. San Francisco: North Point Press.

———. 1990b. *What Are People For?* San Francisco: North Point Press.

Brown, D., and M. Warner. 1989. "Persistent Low Income Nonmetropolitan Areas in the United States: Some Conceptual Challenges." *Policy Studies Journal* 19:22–41.

Buell, L. 1989. "American Pastoral Ideology Reappraised." *American Literary History* 1:5.

Callicot, J. B. 1994. *Earth's Insights: A Survey of Ecological Ethics from the Mediterranean Basin to the Australian Outback.* Berkeley: University of California Press.

Danbom, D. B. 1979. *The Resisted Revolution: Urban America and the Industrialization of Agriculture, 1900–1930.* Ames: Iowa State University Press.

Deaver, K. L., and R. Hoope. 1992. "Overview of the Rural Poor in the 1980s." *Rural Poverty in America,* edited by C. M. Duncan. New York: Auburn House.

Fitchen, J. M. 1981. *Poverty in Rural America: A Case Study.* Boulder, Col.: Westview Press.

———. 1991. *Endangered Spaces, Enduring Places: Change, Identity, and Survival in Rural America.* Boulder, Col.: Westview Press.

Goldman, R., and D. Dickens. 1983. "The Selling of Rural America." *Rural Sociology* 48:585–606.

Hofstadter, R. 1955. *The Age of Reform: From Bryan to FDR.* New York: Random House.

———. 1962. *Anti-Intellectualism in American Life.* New York: Random House.

Kline, D. 1990. *Great Possessions: An Amish Farmer's Journal.* San Francisco: North Point Press.

Lapping, M. 1982. "Toward Working Rural Landscape." In *New England Prospects: Critical Choices in a Time of Change,* edited by C. H. Reidel. Hanover, New Hampshire University Press of New England.

———. 1987. "Peoples of Plenty: Agriculture as Metaphor and National Character in North America." *Journal of Canadian Studies* 22:121–128.

Lapping, M. B. 1992. "American Rural Planning, Development Policy, and the Centrality of the Federal State: An Interpretive History." *Rural History* 3:219–242.

Lewis, R. W. B. 1955. *The American Adam: Innocence, Tragedy, and Tradition in the Nineteenth Century.* Chicago: University of Chicago Press.

Lichter, D. T., and D. J. Eggebeen. 1992. "Child Poverty and the Changing Rural Family." *Rural Sociology* 57:152–172.

Logsdon, G. 1994. *At Nature's Pace: Farming and the American Dream.* New York: Pantheon.

———. 1995. *The Contrary Farmer.* White River Junction, Vt.: Chelsea Green Publishing.

Lyson, T., and W. Falk. 1993. *Forgotten Places: Uneven Development in Rural America.* Lawrence: University Press of Kansas.

Marx, L. 1964. *The Machine in the Garden: Technology and the Pastoral Ideal.* New York: Oxford University Press.

Place, E. 1993. "Land Use." In *The Amish and the State,* edited by D. Kraybill, Baltimore: Johns Hopkins University Press, 191–210.

Porter, K. 1991. *Poverty in Rural America.* Washington, D.C.: Center on Budget and Policy Priorities.

Rural Sociology Society, Task Force on Persistent Poverty. 1993. *Persistent Poverty in Rural America.* Boulder, Col.: Westview Press.

Sale, K. 1980. *Human Scale.* New York: Coward, McCann and Geoghegan.

Schmitt, P. J. 1969. *Back to Nature: The Arcadian Myth in Urban America.* New York: Oxford University Press.

Shapiro, I. 1989. *Laboring for Less: Working but Poor in Rural America.* Washington, D.C.: Center on Budget and Policy Priorities.

Shi, D. 1985. *The Simple Life: Plain Living and High Thinking in American Culture.* New York: Oxford University Press.

Smith, H. N. 1950. *Virgin Land: The American West as Symbol and Myth.* Cambridge, Mass.: Harvard University Press.

Testa, R. M. 1992. *After the Fire: The Destruction of the Lancaster County Amish.* Hanover, NH.: University Press of New England.

Thompson, P. B. 1995. *The Spirit of the Soil: Agriculture and Environmental Ethics.* London and New York: Routledge.

Vail, D. 1994. "Resilient Organizations and Sustainable Rural Development: Crisis and Response in Sweden." Paper presented at a meeting of the Rural Sociology Society, Portland, Oregon.

Worster, D. 1991. "Beyond the Agrarian Myth." In *Trails: Toward a New Western History,* edited by P. Limerick, C. Milner, and C. Rankin. Lawrence: University Press of Kansas, 3–25.

Three

Ecological Footprints and the Imperative of Rural Sustainability

William E. Rees

INTRODUCTION AND PURPOSE

We live in an increasingly urban society. Seventy-five% to 80% of the people in the so-called industrialized countries already live in towns and cities, and urbanization has become a global phenomenon in the last half century. The resultant mass movement of people from farms and rural villages everywhere constitutes the greatest human migration in history. Having first swept through Europe and North America with the industrial revolution, modernization and urbanization are now in the process of transforming Asia and the rest of the world. It seems likely that fully half of the human family will be city dwellers by the year 2000.

One major consequence of urbanization is that people are first spatially and then psychologically distanced from the land that supports them. Even our food, that most vital of basic needs, seems increasingly dissociated from its origins in the sun and the soil. Once the final product for market, today food at the farm gate is only raw material for an elaborate—and mostly urban—processing industry. Indeed, most wealth generated by the food sector is in value-added processing, packaging, transportation, and retailing, effectively trivializing farming as economic activity and a way of life. (Grain farmers see only a few cents from a $2.00 loaf of quality bread.) Little wonder that city folk lose all sense of personal ecological reality, of their connection to nature.

This alienation is further abetted by the fact that by 1988 in the United States, only three million individuals (2% of the population) owned farmland and 44% of these owners leased out rather than worked their land. The two million farms and actual farm operators represented only 1.3% of the population (U. S. Department of Agriculture 1994). Similarly, in Canada 390,575 farm operators constituted only 1.4% of the population in 1991. Farm operators plus household members totaled only 3% of Canadians, down from a peak of 31% in 1931 (Harrison and Cloutier 1995). These data explain how many urban North Americans can live their entire lives without meeting a farmer or anyone else engaged in primary agriculture.

Nor does it help that ours is a culture in which the short-term logic of the marketplace and the language of efficiency have come to override most other values and considerations in both public and private life. In these circumstances, far from being treated with special reverence, food lands have been thoroughly commodified. To many academic analysts and landowners alike, agricultural land is just another tradable good, and farming must compete for it with other uses. So it is that some of the world's finest farmland now earns higher economic returns as suburban parking lots and shopping centers. Thus, we have a paradox. Food is vital to survival, and the availability of cheap, high-quality food is taken for granted by all, yet agriculture is increasingly marginalized both in the modern consciousness and on the national landscape. A similar case could be made that modern society has an equally diminished appreciation of the true worth of our once magnificent forest lands.

All of this, of course, is only one reflection of the great cultural myth of the twentieth century: that humankind is now the master of nature. It is this myth that motivates the analysis presented in this chapter. Its purpose is to show that modern urban civilization is, in fact, still very much constrained by "the environment." The popular belief in human dominance is little more than a shared illusion sustained by brute technologies and conventional economic analyses that are blind to biophysical reality.

The issue of humanity's current place in the ecosphere is examined through the newly crafted lenses of ecological economics and modern reinterpretations of the second law of thermodynamics. This perspective reveals that far from achieving freedom from the natural world, human beings are more dependent than ever on goods and services provided by nature. We are therefore increasingly vulnerable to human-induced global ecological change. Thus, the analysis also shows that achieving rural sustainability is not only a desirable goal in its own right, but a necessary prerequisite for the sustainability of our citified high-income civilization. It is no small irony that the more urban we become, the greater our dependence on the functional integrity of the countryside.

FRAMING THE ANALYSIS

The late twentieth century marks a critical turning point in the ecological history of human civilization. For the first time since the dawn of agriculture and the possibility of geographically fixed settlements 12,000 years ago, the aggregate scale of human economic activity is capable of altering global biophysical systems and processes in ways that jeopardize both global ecological stability and geopolitical security.

Examples abound, many associated with agriculture: more artificial nitrate is now applied to the world's croplands than is fixed by microbial activity and other natural processes combined (Vitousek 1994); about 26% of global evapotranspiration now moves through human-dominated ecosystems; humanity currently appropriates 54% of accessible terrestrial fresh water runoff, with the largest share going to irrigation (Postel et al. 1996); partially as a consequence of excessive use of fertilizer, irrigation, and poor tillage practices, soil erosion and related degradation have forced abandonment of a third of the world's cropland in the past 40 years (Pimentel 1996). At the same time, deforestation has recently been claiming 11 million hectares per year of tropical forests alone. This contributes to an accelerating rate of human-induced species extinction which now approaches that caused by "the great natural catastrophes at the end of the Paleozoic and Mesozoic eras—in other words, [they are] the most extreme in the past 65 million years" (Wilson 1988, 11–12). Meanwhile, "residuals" discharged by industrial economies are depleting stratospheric ozone and altering the preindustrial composition of the atmosphere, and both these trends contribute to (among other things) the threat of climate change, itself the most potent popular symbol of widespread ecological dysfunction. Perhaps the most significant trend, from the perspective of long-term sustainability, is the evidence that human beings, one species among millions, already consume, divert, or otherwise appropriate for their own purposes at least 40% of the product of net terrestrial photosynthesis (Vitousek et al. 1986) and up to 35% of primary production from coastal shelves and upwellings, the most productive marine habitats (Pauly and Christensen 1995). Were it not for the fact that fish catches are in decline because of stock depletion, *both* these proportions would be steadily increasing.

All such empirical evidence suggests that the human economy has effectively converged with the ecosphere from within and that such "rural" activities as agriculture, forestry, and commercial fishing play a significant role in this uncomfortable merger. Humankind no longer merely affects local environments, but instead has effectively co-opted the ecosphere for its own purposes. A growing number of ecological economists interpret this convergence of the economy with the ecosphere as signaling a new constraint on material growth. Goodland (1991) makes the case that growth in the energy and material throughput of the world economy cannot be sustained; Daly (1991a, 29) argues that this reality marks "an historical turning point in economic development" that should force a transition from "empty-world to full-world economics." Any such economic transition would clearly require a parallel sociocultural transformation.

Official recognition that global ecological trends have serious implications for world security led to the creation of the United Nations World Commission on Environment and Development (WCED) (the Brundtland Commission) in 1983. The Commission's report succeeded in popularizing the concept of sustainable development and stimulated the continuing global debate on its meaning and implications.[1] The Brundtland Commission defined sustainable development as "development that meets the needs of the present without compromising the ability of future generations to meet their own needs" (WCED 1987). Although it communicated a basic idea, this definition was left sufficiently vague that academe, governments, industry, and environmental organizations everywhere were able to accept the concept and have been striving ever since to flesh it out, each in its own image (Rees 1995a).[2] Mainstream analysts stress the development side, whereas environmentalists emphasize the sustainability part. *Our Common Future* (WCED 1987) was therefore an important work, "not so much for what it says, but for the reaction it has generated. It has had a galvanizing effect on international development at a crucial time" (Brooks 1990, 24).

The major product of this "galvanization" has been Agenda 21, the far-reaching accord on sustainable development, arrived at during the United Nation's (UN) 1992 Earth Summit in Rio de Janeiro. The goal of Agenda 21 was to launch humanity into "a global partnership for sustainable development" designed to achieve "the fulfillment of basic needs, improved living standards for all, better protected and managed ecosystems, and a safer, more prosperous future" (UN 1992; Quarrie 1992, 47). Of course, this goal itself reflects the contradiction many see in the term *sustainable development* —how can we produce the growth deemed necessary to "[improve] living standards for all" and provide a "more prosperous future" while simultaneously protecting ecosystems, when historical patterns of material growth seem responsible for the present unsustainable levels of ecological disintegration (Rees 1995a, 144)?

Revisiting the "Second Law"

> In a sufficiently long time-frame, it becomes evident that the most important [potential] scarcity is of thermodynamic potential (Berry 1972).

Many analysts agree that Daly's (1991a) "full-world economics" will be economics based on principles of ecology and the second law of thermodynamics. The second law is arguably the ultimate governor of all economic activity, but is totally ignored in conventional analysis.

The second law formally states that the entropy of any isolated system always increases. That is, available energy spontaneously dissipates,[3] gradients disappear, and the system becomes increasingly unstructured and disordered in a inexorable slide toward thermodynamic equilibrium. This is a state in which "nothing happens or can happen" (Ayres 1994). What is often forgotten is that all systems, whether isolated or not, are subject to the same forces of entropic decay. In other words, any complex differentiated system has a natural tendency to erode, dissipate, and unravel. The reason that complex, self-producing systems, such as the human body, do not run down in this way is that they are able to import available energy and material (essergy, see note 3) from their host environments, which they use to maintain their internal integrity. Such systems also export the resultant entropy (waste and disorder) into their hosts. Modern formulations of the second law therefore suggest that all highly ordered systems develop and grow (increase their internal order) "at the expense of increasing the disorder at higher levels in the systems' hierarchy" (Schneider and Kay 1992, abs, 2). Because such systems continuously degrade and dissipate available energy and matter, they are called *dissipative structures*.

Of what relevance is this to the human condition? Most important, the economy is a prime example of a highly ordered, dynamic, far-from-equilibrium dissipative structure. At the same time, it is an open, growing, subsystem of the materially closed, nongrowing ecosphere (Daly 1992). It follows that the economy can grow and develop (i.e., remain in a dynamic state of nonequilibrium) only because it is able to extract available energy and material from the ecosphere and discharge its wastes back into it (Fig-

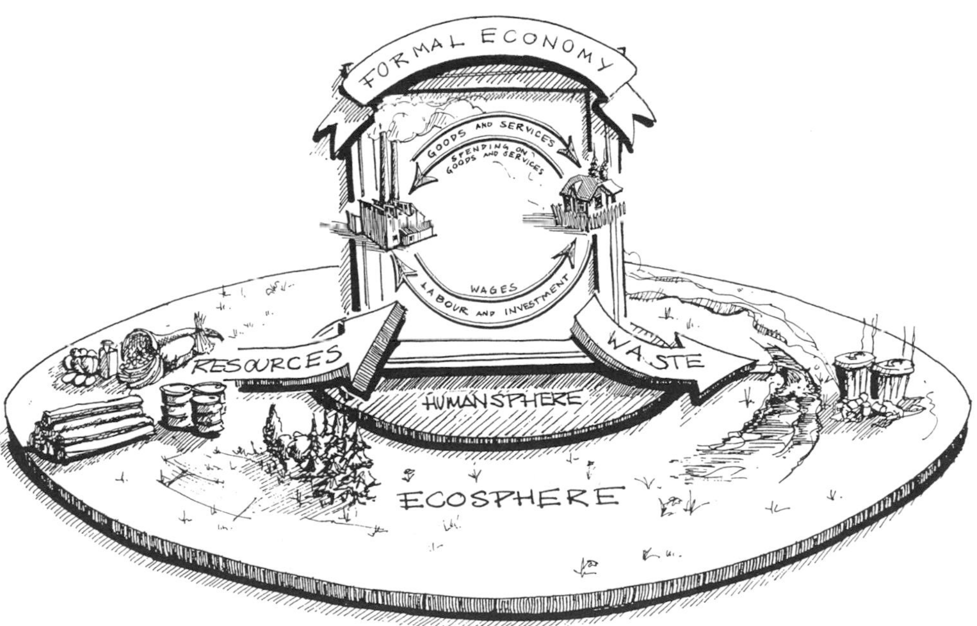

Figure 3.1 The ecosphere contains and sustains the economy. The linear throughput of useable energy and matter (essergy) produced in the ecosphere sustains the circular flows of exchange value (money flows) in the economy. This entropic throughout connects the human enterprise to the environment, yet is invisible to conventional economic analyses (from Wackernagel and Rees 1995).

ure 3.1). Unfortunately, the hierarchical nature of this relationship implies that beyond a certain point, *the continuous growth of the economy can be purchased only at the expense of increasing entropy or disorder in the ecosphere.*

Thermodynamic law thus suggests a new physical criterion for sustainable development. An economy or a community exists in a sustainable state *if there is no significant net increase in the entropy of its host ecosystem(s).* It is, otherwise, unsustainable, a condition that may be characterized (in the case of the global economy) by such trends as fisheries collapse, deforestation, desertification, soil degradation, reduced biodiversity, receding water tables, and waste accumulation (air/water/land pollution). These things are the stuff of daily headlines. We seem to be witnessing the destructuring and material dissipation of the ecosphere, a continuous increase in global net entropy. According to this criterion, society should acknowledge that the present global economy is unsustainably bankrupt. With the prevailing technology, it can grow and maintain itself only by simultaneously consuming and polluting its host environment.

Recognizing Natural Capital

The second law thus forces an uncomfortable reinterpretation of the nature of economic activity. In effect, it shows that what we normally think of as economic production is actually consumption—nature is the real producer. As Georgescu-Roegen (1993, 76) observed, "What goes into the economic process represents *valuable natural resources* and what is thrown out is *valueless waste.*" Moreover, because of the thermodynamic inefficiency of energy/material transformations, any economic activity necessarily consumes more essergy than is contained in the useful product. At best, therefore, the economic process can be said to extract economic goods and services—rather inefficiently—from energy and material produced elsewhere in the ecosphere.

Such considerations have forced ecological economists to recognize that ecosystems and other biophysical resources constitute a new class of natural capital and that our annual harvest of goods and life-support services from nature constitutes a (potentially) sustainable flow of "natural income" (see the box "On Natural Capital"). Without a reliable flow of natural income there can be no economy, no money income. This recognition elevates so-called natural capital to at least the same theoretical status as the more familiar manufactured or human-made capital (physical plant, machinery, communications infrastructure, etc.) and human/social capital (knowledge, skills, social infrastructure, etc.). Indeed, it is becoming clear to many analysts that maintenance of the functional integrity of the ecosphere is a necessary condition for sustainable development. This, in turn, extends the time horizon for economic policy and should elevate intergenerational equity to a place of prominence in developmental decision making. For the first time, environment becomes the independent variable, and economy the dependent one, in developmental decision making (Rees 1991).

SIDE BAR 1
On Natural Capital

The term *natural capital* refers to "a stock [of natural assets] that yields a flow of valuable goods and services into the future." For example, a forest or a fish stock can provide a flow or harvest that is potentially sustainable year after year. The stock that produces this flow is natural capital, and the sustainable flow is natural income. Natural capital also provides such services as waste assimilation, erosion and flood control, and protection from ultraviolet radiation (the ozone layer is a form of natural capital). These life-support services are also counted as natural income. Because the flow of services from ecosystems often requires that these function as intact self-producing systems, the integrity of such systems is an important attribute of natural capital.

There are three classes of natural capital. *Renewable* natural capital, such as living species and ecosystems, is self-producing and self-maintaining, using solar energy and photosynthesis. These forms can yield marketable goods, such as wood fiber, but may also provide unaccounted-for essential services when left in place (e.g., climate regulation). *Replenishable* natural capital (e.g., groundwater, agricultural soils, and the ozone layer) is nonliving but, like living forms, is often ultimately dependent on the solar engine for renewal. Finally, *nonrenewable* types of natural capital, such as fossil fuel and mineral ores, are like inventories in that any use implies liquidating part of the stock.

This chapter takes the position that because adequate stocks of self-producing and replenishable natural capital are essential for life support (and are generally nonsubstitutable), these classes of natural capital are individually more important to sustainability than are nonrenewable, more readily substitutable forms. Ironically (though predictably), technology has so far been able to maintain supplies of many nonrenewable commodities in the face of growing demand, albeit at great thermodynamic expense. This has enabled the human population and global economy to expand to the point where throughput now exceeds natural income, thus threatening the security of various forms of renewable natural capital. Both sources (e.g., fish stocks) and sinks (e.g., the atmosphere) are under siege.

Liberally adapted from Costanza and Daly (1992)

The Constant Capital Stocks Criterion

The implications of this new constraint on throughput growth are currently being explored through various interpretations of a "constant capital stock" condition for sustainability (Costanza and Daly 1992; Pearce et al. 1989,1990; Pezzey 1989). From the perspective of capital theory, no development path is sustainable if it depends on the depletion of productive assets (Pearce 1994; Solow 1986; Victor 1991). Thus society can be said to be economically sustainable only if it passes on an undiminished per capita stock of capital from one generation to the next. (This concept neatly addresses the intertemporal dimension of the WCED approach to sustainable development.) In the present context, the most relevant interpretation of the constant capital stock criterion is as follows: Each generation should inherit an adequate per capita stock of natural capital assets no less than the stock of such assets inherited by the previous generation.[4] Because of its emphasis on maintaining biophysical capital, this is a version of "strong sustainability" (Daly 1991b, 250).[5] The concept of strong sustainability best reflects known ecological principles. In particular, it acknowledges the multifunctionality of biological resources "including their role as life support systems" (Pearce et al. 1990).

Strong sustainability theory recognizes that manufactured and natural capital "are really not substitutes but complements in most production functions" (Daly 1991b, 250). In other words, many forms of biophysical capital perform critical functions that cannot be replaced by technology. For sustainability, a critical minimal amount of such capital must be conserved intact and in place. This will ensure that the ecosystems on which humans depend remain capable of continuous self-organization and self-production.

As observed by Pearce and associates (1990), the need to conserve natural capital may seem particularly relevant to developing countries in which socioeconomic stability is immediately and directly threatened by deforestation, desertification, soil erosion, falling water tables, and so forth. In these circumstances there can be little doubt that existing stocks of natural capital are well below bioeconomic optima and must actually be enhanced for survival, let alone sustainability. However, there is increasing agreement that further reductions of natural capital may impose significant risks on society, "even in countries where it might appear we can afford to [reduce stocks]" (p. 7). These risks lay in our imperfect knowledge of ecological functions, the fact that loss of such functions may be irreversible, and our inability to substitute for those functions once lost. In short, because of the unique and essential services provided by ecological capital, we cannot risk its depletion. "In the face of uncertainty and irreversibility, conserving what there is could be a sound risk-averse strategy" (p. 7).

This last point naturally poses the question as to how to specify an adequate—or better—optimal stock of biophysical capital. If we assume that total natural stocks decline as development advances, then the optimum stock in economic terms might correspond to that theoretical point at which the quantifiable marginal costs of further stock liquidation (e.g., the value of lost ecological services) just exceed the marginal economic benefits (e.g., gains in jobs and income). However, it is extremely difficult to quantify, let alone price, many ecosystem products and functions. This virtually disqualifies conventional cost-benefit approaches to the optimal stocks problem.

Indeed, monetary analyses generally are suspect from the ecological perspective. For example, economists may interpret *constant capital stock* to mean the constant money value of stocks (see Pearce and Atkinson [1993] for an example involving the weak-sustainability criterion). This would allow declining physical stocks with rising prices over time. Alternately, it might mean constant resource income over time in situations where intensified exploration or technological substitution is able to compensate for increasing absolute scarcity. The problem here is that the market prices of resource commodities do not reflect the value of any associated life-support services. These are lost as the rescue base declines. It seems that the only ecologically meaningful interpretation of constant stocks, particularly for renewable resources, is in terms of constant physical stocks. This idea is explored in the following section.

HUMAN ECOLOGICAL FOOTPRINTS

The relationship between an animal population and its physical demands on its habitat is often described in terms of carrying capacity. *Carrying capacity* is defined as the maximum population of a given species that can be supported indefinitely in a defined habitat without permanently impairing the productivity of that habitat. However, because of our seeming ability to increase our own carrying capacity by eliminating competing species, by importing locally scarce resources, and through technology, this definition seems irrelevant to humans. Indeed, trade and technology are often cited by economists and technological optimists as reasons for rejecting out of hand the concept of human carrying capacity (Rees 1996).

This is a conceptual error. Trade does not free humans from their obligate dependence on the ecosphere, it merely displaces the entropic load of import-dependent populations onto distant exporting regions. Meanwhile, technology often *reduces* carrying capacity while creating the illusion of increasing it! We often use technology to increase the short-term energy and material flux through exploited ecosystems. This seems to enhance systems productivity while actually eroding permanently the resource base (Rees 1996).

We can account for these realities and meet the economists' objections to human carrying capacity simply by inverting the standard carrying capacity ratio. Rather than asking how many people a particular region can support, the question becomes: How much region is needed to support a given population?

My students and I have developed a novel approach to this question that uses land or an ecosystem area as a proxy for natural capital. Our method, ecological footprint analysis, is based on the fact that many forms of material and energy consumption can be converted to a land/water area equivalent. We estimate the area of productive land and water (agricultural land, forest land, etc.) required to produce the material and energy flows and to assimilate the wastes associated with the production of a typical shopping basket of ecologically significant consumption items used by the subject population. Summing the ecosystem areas required for each major resource and waste flow provides an estimate of the total area functionally appropriated by that population to maintain its current life-style. In short, the true ecological footprint of a defined population is the total area of land/water required on a continuous basis to provide the resources consumed and to assimilate the wastes produced by that population, wherever on Earth that land is located (for details, see Rees 1996; Wackernagel and Rees 1995; Rees and Wackernagel 1994).

Our analyses show that the average North American needs 4 to 5 hectares (approximately 10 to 12 acres) to support his or her consumer life-style, assuming the land is being used sustainably. This is three times his or her "fair earthshare" of 1.5 hectares.[6] For example, Canadians each appropriate the ecological services of almost 1.3 hectares to produce their food alone. Our high-protein diets require about 0.9 hectares of crop and pasture land. Most of the rest of the agricultural footprint represents the carbon sink needed to assimilate carbon dioxide emissions from fossil fuel use (Wackernagel and Rees 1995).

Footprinting Cities . . .

The findings discussed thus far are first applied to my home city of Vancouver in British Columbia, Canada (Rees 1996). In 1991, Vancouver proper had a population of 472,000 and an area of 114 square kilometers (11,400 hectares). Assuming a per capita land consumption rate of 4.3 hectares, the 472,000 people living in Vancouver require, conservatively, 2.03 million hectares of land for their exclusive use to maintain their current consumption patterns (assuming such land is being managed sustainably). However, the area of the city is only about 11,400 hectares. This means that the city population appropriates the productive output of a land area *nearly 178 times larger than its political area* to support its present consumer life-style.

We can also estimate the marine footprint of the city's population based on fish consumption. Available data suggest a maximum sustainable yield from the oceans of about 100 million metric tons of fish per year.[7] First we divide the global fish catch by total productive ocean area. About 96% of the world's fish catch is produced in shallow coastal and continental shelf areas that constitute only 8.2% of the world's oceans (about 29.7 million square kilometers). Average annual production is therefore about 32.3 kilograms of fish per productive hectare (0.03 hectare per kilogram of fish). Because Canadians consume an average of 23.4 kilograms of marine fish annually, their marine footprint is about 0.7 hectares each. If we add this per capita marine footprint to the terrestrial footprint, the total area of Earth needed to support Vancouver's population is 2.36 million hectares (5.83 million acres), or more than 200 times the geographic area of the city (Figure 3.2).

Other researchers have obtained similar results for their cities. Using our methods, British researchers have estimated London's ecological footprint for food, forest products, and carbon assimilation to be 120 times the surface area of the city proper (International Institute for Environment and Development 1995). Folke and Associates (1994) report that the aggregate consumption of wood, paper, fiber, and food (including seafood) by the inhabitants of 29 cities in the Baltic Sea drainage basin appropriates an ecosystem area 200 times larger than the area of the cities themselves. (Although this study includes a marine component for seafood production, it has no energy land component.) Cities may be among the brightest stars in the constellation of human achievement, but in material terms they more resemble entropic black holes.

. . . And Urban Regions

This analysis can be extended to the entire Lower Fraser Basin region within which Vancouver is located (the 1991 population was 1.78 million; the area is approximately 5,500 square kilometers). Even though only 18% of the region is dominated by urban land use (i.e., most of the area is rural agricultural or forested land), consumption by its human population appropriates through trade and biogeochemical flows the ecological output and services of almost 7.7 million hectares of terrestrial ecosystem. In other words, the true ecological footprint of the region is 14 times larger than the geographic basin it physically occupies. The people of the Lower Fraser Basin, in enjoying their consumer life-styles, have overshot the terrestrial carrying capacity of their geographic home territory by a factor of 14 (Table 3.1).

Putting this in terms of natural capital, it is clear that the consumer life-style of the region's human population could not nearly be sustained on the internal natural capital endowment of the region. Material consumption exceeds local natural income nearly fourteenfold. However healthy the region's economy may be in monetary terms, the Lower Fraser

Figure 3.2 Urban ecological footprints. The ecological footprints of urban populations and high income cities are typically at least two orders of magnitude larger than the geographic areas they occupy. Through commercial trade and natural biogechemical cycles, cities thrive on the ecological output of rural regions all over the world (from Wackernagel and Rees 1995).

TABLE 3.1 Ecological Footprints of Vancouver and The Lower Fraser Basin

Geographic Unit	*Population*	*Land Area (ha)*	*Ecol. Ftprnt (ha)*	*Overshoot Factor*
Vancouver City	472,000	11,400	2,029,600	178.0
L. Fraser Basin	1,780,000	555,000	7,654,000	13.8

Basin is running a massive ecological deficit with the rest of Canada and the world.

The Global Context: Larger Illusions

The situation of the Lower Fraser Basin is typical of so-called industrialized (or high-income) regions and even entire countries. Our data show that many industrialized countries run an ecological deficit of an order of magnitude larger than the sustainable natural income generated by the ecologically productive land within their political territories (Rees 1996; Wackernagel and Rees 1995). For example, with an area of 34,000 square kilometers and a population of 15 million, the population density of the Netherlands is 4.4 people per hectare. Although the average Netherlander consumes fewer resources and has a smaller ecological footprint (3.3 hectare) than the average North American, Holland still uses more than 15 times more land than lies within the country's own political boundaries. The Dutch use approximately 5,400 square kilometers of built-up area, 100,000 square kilometers for food production for domestic consumption, 70,000 square kilometers for forestry products, and 320,000 square kilometers for carbon dioxide absorption (Wackernagel and Rees 1995).

The ecological economics of the Netherlands' agriculture and food sectors are particularly interesting. Holland is a regular net exporter of food products in monetary terms. This is often seen as a remarkable achievement for a small, overpopulated country. However, Dutch government data show that just to grow fodder for its livestock (including that used to produce processed food products for export), the Netherlands appropriates 100,000 to 140,000 square kilometers of arable land from North America and certain developing countries (National Institute for Public Health and Environmental Protection 1992). This is five to seven times the area of agricultural land in the entire country. In short, Holland's "remarkable" dollar trade surplus is sustained by a large hidden ecological deficit with the rest of the world. What the Netherlands actually exports is relatively high-cost value-added processing. The ecologically critical components of Dutch food exports (energy and matter) are imported from elsewhere.

These data pose a serious problem for current world development models. For example, the Netherlands and Japan boast positive trade and current account balances measured in monetary terms, and their populations are among the most prosperous on earth. Densely populated yet relatively resource (natural capital) poor, these countries are regarded as stellar economic successes and held up as models for emulation by the developing world. At the same time, the Netherlands' 3.3 hectares per capita and Japan's 2.5 hectares per capita give these countries national ecological footprints about 15 and 8 times larger, respectively, than their total domestic territories. In short, the monetary accounts of such economic successes do not reveal the whole material story. Ecological deficits are real deficits, and global sustainability cannot be deficit financed. Simple physics dictates that not all countries or regions can be net importers of biophysical capacity. The levels of material consumption currently enjoyed in high-income countries simply cannot, therefore, be extended sustainably to the entire human population using anything like existing prevailing technology. (This would require the equivalent of two additional Planet Earths.) What does this say to a world in which the principal sustainable development strategy is economic growth?

Urban Sustainability: Standing on Rural Sustainability

The first lesson to take from the foregoing analysis is that high-income cities and urban regions have their ecological feet firmly planted in the global rural hinterland. In tangible terms, urban populations get their food, water, fresh air, natural fiber, and most other material and energy from the extraurban environment. They also use the latter as a carbon sink and general waste dump.

The implicit dependence of the city on the countryside is total. Imagine what would happen to a typical modern city, as defined by its political boundaries or built-up area, if it were enclosed in an impermeable glass or plastic hemisphere. Clearly, this city would cease to function and its inhabitants would perish within a few days. The population and the economy contained by the capsule would be cut off from vital resources and essential waste sinks, leaving it both to starve and to suffocate at the same time. In other words, the ecosystems contained within our imaginary human terrarium would have insufficient carrying capacity to service the ecological load imposed by the contained human population (Rees 1996).[8]

Many commentators have interpreted the urbanization of the human population as reflecting a decreasing dependence of humans on nature and its resources. This is false reasoning—economic accounting systems are simply blind to physical reality. Even the foregoing simple mental experiment shows that cities everywhere are utterly dependent on the sustainability of rural landscapes. In fact, rising incomes that accompany urbanization imply increasing material consumption and, therefore, increasing per capita demand for the goods and services of nature. From an ecological and thermodynamic perspective, then, cities are simply increasingly concentrated nodes of consumption within the increasingly human-dominated global landscape. The rural landscape is the ecologically productive and certainly co-equal component of the system. In fact, even urban cash income can often be traced to rural natural income (as in the Dutch example). Whereas the countryside could be quite viable without the city, there could be no city without the countryside.

THE EXPANDING FOOTPRINT OF AGRICULTURE

The preceding discussion suggests that if we wish to maintain the ecological security of urban populations, we must maintain the viability and productivity of the rural hinterland. It is unsettling, therefore, that the prevailing patterns of human exploitation of rural landscapes are themselves inherently unsustainable. Indeed, industrial forestry and agricultural practices use some of humankind's most ecologically brutal technologies.

Ironically, these same technologies heighten the illusion that human ambition has been freed from environmental constraints. For example, cheap food, declining farm populations, and the apparently endlessly rising land and labor efficiency of high-tech crop production has lulled the developed world into thinking that we have abolished forever the Malthusian threat of widespread food shortages and famine. In contrast, the present analysis argues that conventional agriculture may actually contribute to future global food crises. In our preoccupation with increasing productivity, we have ignored the entropic costs and hidden land requirements of modern high-input farming.

Agriculture has long been one of the primary means by which humans appropriate essergy from the ecosphere. In short, through photosynthesis, food crops concentrate a tiny portion of the solar energy striking the earth, and the food they produce is the source of usable energy that most directly supports people. In ideal circumstances, using low-input, low-tillage agriculture and minimal processing of produce, we should be able to maximize the entropic efficiency of the food production-consumption cycle. That is, agrophotosynthesis *could* feed a substantial population of humans with minimal degradation of input energy and matter—soil, fossil fuel, and so on—and minimal pollution. Indeed, on a finite planet, agriculture should strive for sufficiency with as little contribution as possible to the increase in net global entropy.

This is clearly not the case today. High-input conventional agriculture consumes much larger quantities of useful energy and material than it produces. Insecticides, herbicides, and inorganic fertilizers (produced from fossil-energy feedstocks), irrigation, heavy equipment, and so forth, all require large inputs of nonrenewable material and energy, which are degraded in use and subsequently dissipated into the ecosphere as pollutants. The high cost of chemicals and machinery and the perceived need for farmers' incomes to keep pace with those of their urban compatriots have driven a steady increase in the economic size of farms and working land units. This has led to the destruction of hedgerows, the drainage of farm ponds, and the spread of uniform monoculture over the countryside, contributing to the declining biodiversity and aesthetic impoverishment of rural areas.

Ironically, the near-complete mechanization of agriculture has meant the replacement of renewable nonpolluting solar-based essergy—human and animal labor—with depletable polluting terrestrial sources of essergy. In effect, the vast increase in land productivity in this century represents the conversion of petroleum to food, the virtual substitution of oil for sunlight and land. This substitution has produced more reliable and abundant food supplies in the short term, but with steadily decreasing output/input efficiency and at a great cost in ecological degradation. It has also enabled the massive expansion of the human population whose continuing growth (currently 90 million per year) is dependent on further increasing these artificial levels of production.

According to Giampietro and Pimentel (1994), today's high-tech monocultures produce 350 times as much corn per hour of labor as could be raised by Native Americans with traditional agriculture. However, the greater labor and land productivity of modern agriculture comes at a great cost of natural capital depletion: 10 calories of exosomatic energy (mostly nonrenewable fossil fuel) for every calorie of food delivered to the consumer;[9] surface and ground water pollution; toxic contamination of food webs; accelerated erosion, compaction, oxidation, and mineralization of prime soils; greenhouse gas enhancement; mining of groundwater, and so on. Because long-term productivity of agroecosystems depends on the sustainability of natural income flows, it is clear that today's agriculture production is on a collision course with ecological reality. In short, "at present and projected world population levels, the current pattern of human development is not ecologically sustainable. The world economic system is based on depleting, as fast as possible, the very natural resources upon which human survival depends" (Giampietro and Pimentel 1994, 61).

The "Invisible Foot" of High-Tech Agriculture

One of our recent ecological footprint studies dramatically illustrates the hidden ecological costs of high-input food production. Wada (1993, slightly revised here) compared the effective land requirements of mechanized field cropping and heated hydroponic greenhouse production of tomatoes in southern British Columbia, Canada. The high-tech greenhouses in

Wada's study required only 2.0 to 2.3. hectares of actual growing area to produce 1,000 metric tons of tomatoes, as compared with 12 to 18 hectares per 1,000 metric tons for the high-input field farms. By this measure, greenhouse production appears to be five to nine times more land-efficient than field farming. Moreover, revenue and profits per hectare were as much as 13 and 9 times higher, respectively, for the greenhouses than for the field farms. Such data as these are sometimes used to illustrate the superiority of superindustrial agriculture over ordinary field cultivation and to show the capacity of human ingenuity to substitute for nature. Both points seem to weaken concerns about food production and arguments for the preservation of arable land.

Ecological footprint analysis turns this argument on its head. When the land equivalents of the major material and energy inputs to both types of production agriculture were considered (fertilizer, pesticides, wood chips as soil medium, fossil energy, building materials, etc.) the greenhouse operations were found to appropriate the ecological services of 613 to 760 hectares per 1,000 metric tons of tomatoes, as compared with only 37 to 49 hectares per 1,000 metric tons required for the field farms (Figure 3.3). In other words, the ecological footprint of a greenhouse tomato is 12 to 20 times that of a field tomato. Because of greatly increased energy and material inputs, the ecological impact of greenhouse agriculture is of a far greater order of magnitude per unit output than that of field farming. Keep in mind that the ecological footprint of the high-input field farms examined here is already three times their nominal growing area.

These data can be interpreted in terms of natural capital, the law of entropy and sustainability. The true ecological footprint of crop production includes the land previously unaccounted for, needed to maintain the entropic balance of the total system. Wada's (1993) numbers show that an area of productive land vastly larger than the nominal growing area is required to produce the inputs and assimilate the wastes (mainly carbon dioxide) associated with high-tech tomato production. In effect, the greater the material and energy throughput in agriculture, the more ecoproductivity elsewhere is needed to repair the damage. Such full (ecological) cost accounting for high-input production shows that, contrary to conventional wisdom, the land efficiency of agriculture is actually decreasing, not increasing. We had better not give up our arable land to nonagricultural uses just yet—there is simply not enough ecological space on the planet to support the invisible foot of hydroponic greenhouses or other forms of high-throughput agriculture.

Global Change: Added Uncertainty and Risk[10]

If the sustainability of our urban civilization is already threatened by the direct abuse of rural lands, how much more vulnerable does it become

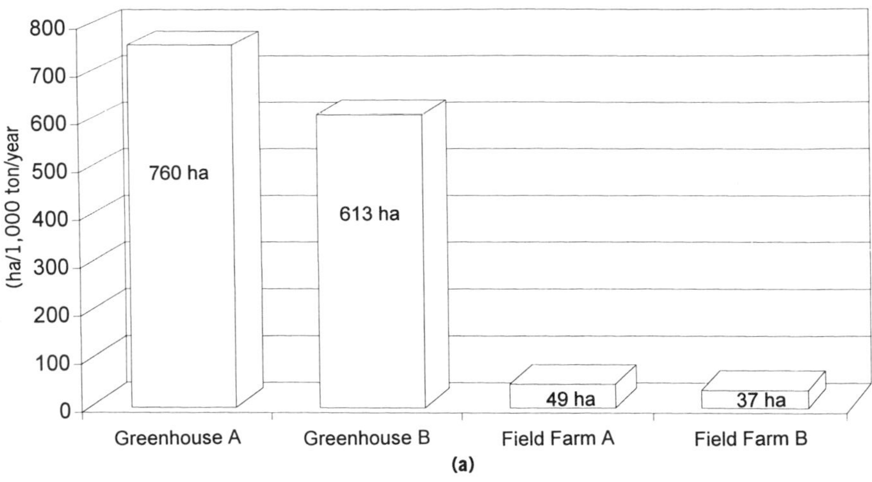

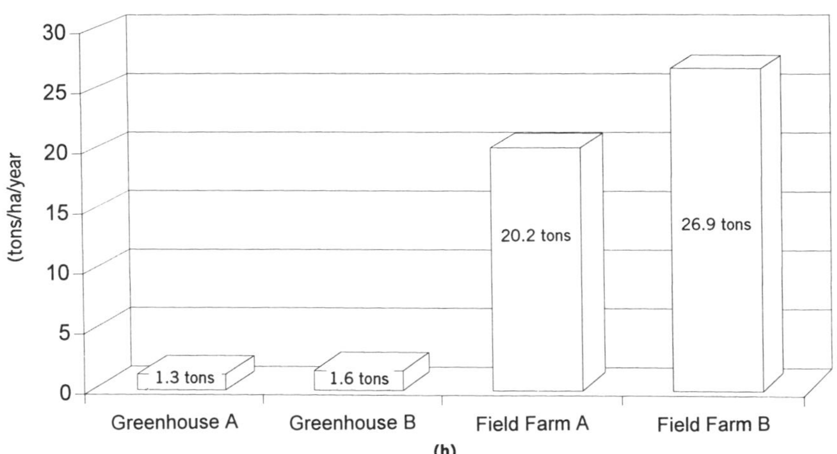

Figure 3.3 The relative ecological footprints of greenhouse and field farm production, (a) ecological footprints of greenhouses and field farms producing 1,000 metric tons of tomatoes; (b) productivities per hectare based on total land approprtations by greenhouses and field farms. (from Wada 1993).

with the risk of global ecological change? For example, in the last 100 years human activity has increased atmospheric carbon dioxide 28% above preindustrial norms. At 350 parts per million, current carbon dioxide levels are "higher than at any time in the past 150,000 years" (Flavin 1996, 22). This makes anthropogenic carbon dioxide the largest contributor to anticipated greenhouse enhancement and the attendant climate change that may threaten food security.

We can only speculate on the climatic effects of enhanced greenhouse forcing. However, computer simulation models indicate that significant global warming will be accompanied by shifts in the amount and distribution of rainfall, the frequency and distribution of heat waves, and the length of growing seasons. Increasing extremes of climate are also likely in many regions (Intergovernmental Panel on Climate Change 1995). The effects of climate change on agriculture will certainly be mixed, depending on crop types and the geographic distribution of impacts (Smith and Tirpak 1988; Parry 1989).

Although estimating the net impact is difficult, higher temperatures and reduced precipitation will likely result in lower soil moisture in two of the world's major food-producing regions, the North American heartland and the grain-growing areas of the former Soviet Union. An indication of the possible result was provided by the North American drought of 1987 and 1988. While we cannot claim that this drought was caused by greenhouse enhancement, the general warming trend with which it was associated in the 1980s is consistent with expectations of greenhouse effects (Wigley et al. 1985) and with various models predicting increasing frequency and severity of drought (Williams et al. 1988). (see text boxes on Global Ecological Constraints on Agriculture and Ecological Constraints on Global Food Production)

At a minimum, these trends will affect the quantity, quality, and price of food in the global marketplace. At worst, difficulties in sustaining recent gains in global food production may contribute to civil disorder and greater geopolitical instability. Furthermore, shifting to sustainable practices may require more rather than less arable land in the future. The potential impact of global ecological change on future food security suggests a very conservative strategy for the planning, allocation, and use of productive rural lands. Whether or not we presently acknowledge it, the sustainability of the entire human enterprise depends on maintaining the viability of all extraurban ecosystems.

TOWARD RURAL SUSTAINABILITY

If there is little doubt that global change will affect the quality and productivity of rural landscapes, it is generally less appreciated that rehabilitating natural landscapes can mitigate global change. Indeed, recent history and capital theory suggest that stabilizing and then rebuilding terrestrial natural capital is at the heart of ecological sustainability.[11]

Such primarily rural activities as agriculture and forestry are major factors in the entropic mining of the ecosphere. Agriculture, overgrazing, and deforestation have already degraded 1.96 billion hectares of soils, about 22% of the total arable and forested land on earth (WRI 1992). Deforestation and the accelerated oxidation of agricultural soils resulting from mech-

SIDE BAR 2
Global Ecological Constraints on Agriculture

Agriculture everywhere is increasingly constrained by ecological trends, including loss of topsoil, excessive runoff, waterlogging and salting of soil by irrigation, falling water tables, farmland conversion, and now, possibly, climate change (Brown 1988; Brown et al. 1989).

Possible Climate Change Effects on Harvests

- Between 1950 and 1990, three U.S. corn harvests were severely reduced by drought. Each reduced harvest occurred during one of the six years since 1980, described as the warmest in the last century.
- From the beginning of the 1987 harvest to the beginning of the 1989 harvest, "world reserves of grain [more than half of all human caloric intake . . . dropped from the highest level ever to the lowest since [just after] WW II" (Brown 1988, 5). Carryover stocks fell from 104 to less than 60 days of consumption and have remained below this level since (Brown 1988, 1995a, b).
- The 1988 drought was, by many criteria, the most severe ever to afflict the United States: "For the first time in decades . . . the U.S. ran up a 10-million-ton food deficit and had to dig into its food reserves" (Bertin 1989, B1–B2).
- The year 1995 was the hottest in the instrumental record, and the 10 hottest years in the last century have all occurred since 1980 (several sources cited in Flavin 1996).

  It seems likely that another severe drought in the United States and Canada in 1989 would have "reduce[d] exports to a trickle, creating a world food emergency" (Brown 1988, 5). Keep in mind, however, that there is a considerable buffer in the system. More than a billion people could be fed with the grain and soybeans currently consumed by U.S. livestock alone.

Loss of Top-Soil

- The best grain-growing soils of Canada have lost more than half of their natural nutrients and organic matter since the turn of the century, resulting in significant productivity losses. Similarly, states with some of the best soils in the U.S. report losses of one half their topsoil in just over 100 years of farming (Pimentel et al. 1987, 1989).
- Until recently, average topsoil erosion from cultivated land in North America has been estimated at 18 metric tons/hectare/year, 18 times the typical regeneration rates. Erosion rates in many less-developed countries are even greater.

If data were adequate to assess the world as a whole, they would undoubtedly show that food consumption is already running well ahead of sustainable production (Brown 1988). Consider the following:

- World population continues to grow by 90 million per year.
- After outpacing population growth for 50 years, grain production has now fallen behind. Annual output peaked at about 345 Kilograms per capita in the early 1980s, but has dropped erratically to less than 300 kg per capita in 1995 (USDA data, cited in Gardner 1996).
- Although technology has yet to be fully exploited in some areas, "the backlog of unused agricultural technologies that farmers can draw upon in some countries is dwindling" (Brown 1988, 5). For example, increasing fertilizer use has been the chief cause of the explosive growth of food output since the 1950s, but crop response in most areas has slowed dramatically and fertilizer use, globally, itself peaked in 1989, dropping by 15% in the ensuing four years.
- Even as the strategy of substituting energy for land begins to stumble, and arable land itself is increasingly disordered by industrial agriculture, the amount of grain land per capita has dropped by 50% to just over 0.1 hectare per capita since 1950.
- Stratospheric ozone depletion threatens to exacerbate the problem. The loss of ozone exposes plants to the damaging effects of enhanced ultraviolet b radiation (UV-b). Of 200 species of terrestrial plants tested, 70% were found to be sensitive to increases in UV-b. In some cases, productivity losses of 1% to 2% for each 1% reduction in ozone seem possible (NASA 1988).
- Increased UV-b may have serious negative effects on primary production and therefore, on wildlife, agriculture, forestry, and ecosystem structure and function. Significant reductions in net primary production (which sequesters carbon dioxide in biomass) would also result in increased atmospheric carbon dioxide, thereby "feeding back" to accelerate greenhouse warming.

anized cultivation accounted for as much as 50% of atmospheric carbon dioxide accumulation by midcentury and may still contribute as much as 20% (burning fossil fuels contributes an increasing proportion) (Bolin 1986). However, the world's remaining vegetation and soils still store three times as much carbon as is resident in the atmosphere. It follows that modified agricultural practices and large-scale reforestation might significantly slow atmospheric change. The rehabilitation of rural landscapes will therefore necessarily be an essential component of any sustainability revolution.

And a revolution it must be. Achieving sustainability will require reducing society's entropic throughput and moving toward a steady-state equilibrium. This will certainly involve a shift to known ecologically sound agricultural practices and a massive increase in the technological efficiency of virtually everything we do. Indeed, there is a growing consensus that a 90% reduction in the material and energy content of goods and services may be necessary in the developed countries by 2040 to create the ecological space for necessary economic growth in the developing world (Ekins and Jacobs 1994; Ekins 1993; RMNO 1994; BCSD 1993).

There is also increasing acceptance that market forces alone are incapable of driving the needed efficiency revolution; government intervention in the form of ecological fiscal reform may well be essential. Ecological and economic sustainability requires the restructuring of economic incentives and taxation policy to encourage energy and material conservation. Existing subsidies in the energy and agricultural sectors, for example, must be replaced by a system of accelerating resource use, input, and depletion taxes and marketable quotas. The increase in government revenues (and consumer prices) should be offset by reductions in payroll, value added, and income taxes. By raising energy and resource prices closer to the full social costs of the goods and services they provide, resource taxes would create the incentive needed for industry to reduce material throughput. At the same time, lower labor costs would further increase workers' relative advantage over materials and capital in the production process (von Weizs-Säcker 1994; Rees 1995b).

REDUCING THE FOOTPRINT OF AGRICULTURE

As hard technology begins to lose ground to population growth and global environmental change, the search is on for more sustainable forms of agriculture (Edwards et al. 1990; National Research Counsel 1990). The objective is to maintain productivity while enhancing rather than depleting the resource base. The ecological footprint of agriculture must be reduced as closely as possible to the actual growing area. Existing agricultural subsidies encourage unsustainable practices, such as overusing manufactured fertilizers and biocides, squandering irrigation water, overproducing certain crops, and specialized monocropping. The removal of price supports and a shift to full-cost pricing of energy and material inputs to production agriculture (through, for example, a stiff depletion tax on fossil fuel and license fees for the use of artificial fertilizers and pesticides) would greatly accelerate the adoption of more ecologically benign approaches to farming.

Agriculture in the twenty-first century will use fewer man-made inputs and rediscover how to put nature to work. Indeed, a primary objective of sustainable agriculture is to replace brute technological force with ecological intelligence. Conventional high-input agriculture is characterized by monoculture over vast areas, heavy machinery, extensive use of synthetic chemicals (pesticides and fertilizers), and deep tillage, all of which contribute to the slow but certain destruction of both soils and supporting ecosystems. Alternative practices include mixed rotation cropping with stubble mulching and the return of crop residues and manure to the soil; biological pest control or integrated pest management to minimize the use of pesticides; smaller fields and the reestablishment of hedgerows to protect fields from wind and water erosion, to aid in the accumulation of snow (vital for soil moisture), and to provide habitat for bees and other native pollinators, and for birds, which help control insect pests; and the retention of ponds to help stabilize soil moisture and provide insurance against drought. All such nature-based technologies reduce the use of fossil energy, machinery, chemicals, and other man-made inputs to agriculture.

First and foremost, sustainable agriculture requires maintenance of the soil. As noted, deep plowing techniques, heavy equipment, and synthetic chemicals accelerate soil erosion, compaction, and other forms of soil degradation. In contrast, various forms of conservation tillage (ridge tillage, chisel plowing, deep subsoiling, etc.) reduce soil erosion and oxidation, facilitate pest management, and maintain a natural soil structure with improved drainage and water retention (Edwards 1990).

The economics of alternative agriculture are still disputed, but increasing evidence suggests that both productivity and profitability can be maintained. For example, a recent World Resources Institute study compared conventional and organic farming practices in Pennsylvania and Nebraska. In Pennsylvania, conservation cultivation of corn and corn-soybean production eliminated the use of chemical fertilizer and pesticides, cut costs by 25%, reduced erosion by 50%, and actually increased yields over conventional norms after five years. Off-farm damages were estimated to have been reduced by $75 per hectare of farmland, and a 30-year income loss (present value $306 per hectare) was avoided by preventing a 17% loss in soil fertility. All things considered, the resource-conserving practices outperformed conventional approaches in economic value per hectare by a two-to-one margin. In flatland Nebraska, where the costs of erosion are lower, low-input cultivation was slightly less financially competitive than the prevailing high-input corn-bean rotation, but was found to be environmentally superior overall (Faeth et al. 1991, reported in WRI 1992).

Ecologically necessary and economically feasible, sustainable agriculture may also be socially desirable for rural North America. The realistic pricing of resources, attention to the ecology of land, and ecotechnology implies a return to smaller farms and more labor and information-intensive practices. The countryside might therefore regain population as human labor and ingenuity once more become an important (and renewable) factor in primary food production. Thus, sustainable agriculture would help restore a historical cultural landscape through salvation of the family farm and revitalization of dependent communities. Achieving sustainability also raises long-dormant questions about optimal city size and the appropriate distribution of population. In any event, urban society would reap special dividends with the restoration of ecological diversity and beauty to the rural landscape and through reduced air, water, soil pollution, and other off-farm impacts. We might even enjoy more wholesome food.

Ideally, the move to sustainable agriculture will be international if not universal. Until this happens, the terms of trade may have to be adjusted to protect participating domestic producers from unfair competition from offshore producers who are still willing to write off the environment for short-term gain. In the long run, reduced subsidies and realistic prices for food, combined with reduced capital and operating costs, should help restore profitability to agriculture.

Carbon Sink Forests: A Second Key to Rural Sustainability[12]

Agriculture exemplifies the dependence of urban populations on the rural hinterland for essential resources. But what about the second major component of the urban ecological footprint—the rural land required for waste assimilation, particularly carbon dioxide? We can address this issue too through an integrated approach to sustainability-oriented fiscal reform. Fossil fuel taxes could become the centerpiece of a program to reduce carbon emissions through the partial integration of energy, forest, and conservation policies. As a major by-product, this scheme could provide a second significant boost to rural ecological and economic sustainability.

Carbon is the building block of life on earth as well as the chemical basis of the fossil fuels that provide 86% of the commercial energy used by industrial economies. As the central element of organic life, carbon takes part in one of the great biogeochemical cycles by which all the elements essential to life circulate within the ecosphere. As part of the carbon cycle (and the global entropy cycle), respiration by plants and animals, bacterial decomposition, and natural fires release tens of billions of metric tons of carbon as carbon dioxide into the atmosphere each year. Meanwhile, green plants assimilate an equivalent quantity of carbon dioxide from the atmosphere (or from water, in the case of aquatic ecosystems), fixing it in their tissues in the form of carbohydrates, fats, and proteins. This photosynthesis by green plants makes the carbon (and chemical essergy) available once again for consumption by animals, bacteria, fungi, and fire, thus completing the cycle.

Burning hydrocarbons (fossil fuels) to heat the buildings, power the vehicles, and drive the machinery of urban-industrial economies constitutes a form of industrial metabolism that dissipates about six billion additional metric tons of carbon into the atmosphere annually. This fossil carbon is being taken from a relatively inactive pool, where it has been concentrated and stored for millions of years, and is being injected into active circulation in the ecosphere at rates far in excess of assimilation rates. Human-caused deforestation adds about another 1.6 billion metric tons to the fossil carbon loading, further overloading the contemporary carbon cycle. Anthropogenic carbon releases have thus significantly and rapidly increased the carbon emission:assimilation ratio in what had been a relatively balanced natural entropic cycle. Indeed, human activity is apparently responsible for the more than three billion metric tons of carbon accumulating annually in the atmosphere. This is equivalent to 11 billion metric tons of carbon dioxide, the gas responsible for 60% of the extra greenhouse effect so far. The global carbon cycle is one waste sink clearly filled to overflowing.

Humankind has never explicitly accounted for the impact of such artificial releases of carbon in energy policy or pricing. An opportunity exists, therefore, to internalize the rising costs of atmospheric and possible climatic change by integrating energy policy with a national policy for forest ecosystem rehabilitation.

The Forest Connection

Forests are one type of terrestrial habitat that humans have considerable capacity to manipulate. Rapidly growing young-to-middle-aged forests can accumulate up to six metric tons of carbon per hectare per year, raising the possibility of managing large tracts of presently understocked forest lands as explicit carbon-sink forests. This could be done by implementing a policy of taxing fossil fuel consumption in proportion to carbon content and dedicating a significant share of the revenues to national programs of ecosystem rehabilitation and restoration forestry. Such an approach could lead to increased economic efficiency, enhanced family and community stability in woods-based communities in much of the United States and Canada, and would simultaneously address an array of ecological issues central to sustainability. Most important, it would represent a significant investment in several forms of valuable natural capital. Because the constant natural capital criterion applies on a per capita basis, natural capital stocks must be enhanced as the population expands.

Several coal-burning electrical utilities have established, or are contemplating the establishment of, tropical carbon-sink forests to offset their carbon emissions. Although forests in temperate climates may be less area-efficient than tropical forests, there are many compensating reasons to extend the logic to temperate U.S. and Canadian forests.

Consider the forests of Canada first. Because of a relatively low population (28 million in 1991) and large forest area, this country has the potential to balance the carbon cycle and approach entropic neutrality. Canada's trade-corrected fossil carbon emission rate of 4.1 metric tons per capita per year (15 metric tons CO_2) is among the highest in the world (Canada 1993). To extract this carbon from the atmosphere, the equivalent of less than one hectare of managed tropical carbon-sink forest or one to several hectares of Canadian forest (depending on latitude, management practices, and other factors affecting productivity) would be required for every Canadian (population, 28 million). Assuming an average carbon assimilation rate of 1.5 metric tons per hectare per year (ha^{-1} yr^{-1}), a total of 76.5 million hectares of dedicated carbon-sink forest would have to be established in Canada to offset current fossil carbon emissions. This represents approximately 20% of the country's total closed forest and woodland area. The 115 million metric tons of carbon that would be sequestered annually is only 0.05% of the estimated 226 billion metric tons of carbon currently stored in Canada's forest ecosystems.

The United States' situation is rather different. The country's 250 million people have an average annual carbon emission rate of about 5 metric tons per person. At a carbon assimilation rate of 1.8 metric tons per hectare per year (ha^{-1} yr^{-1}) (the global average), the United States would require a carbon-sink forest of about 694 million hectares. This is 235% of the entire forest and woodland area in the nation. These data show that although the United States has massive carbon-sink potential, it is grossly inadequate to compensate for the enormous entropic load generated by fossil energy consumption in the this country. The energy footprint of the United States clearly extends far beyond the nation's borders. It would overwhelm even the entire Canadian surplus, assuming that the entire forested area of the two countries were managed solely as a carbon sink.

This is, of course, all the more reason to make carbon-sink forests in the United States a major component of integrated energy and environmental policy. The country has an obligation to show leadership in reducing its ecological footprint on the global commons, and dedicated carbon-sink forests provide an effective, if partial, means to do so. The remaining sustainability gap would have to be addressed by an absolute reduction in consumption, or at least in the energy intensity of consumption. In the meantime, the sustainability of rural America would be significantly enhanced by myriad socioeconomic and ecological spin-offs of the carbon-sink program. These effects are described in the following paragraphs using Canadian data, but the general principles apply equally to rural forest-based communities throughout the United States.

Having 57,000 Jobs and the Environment Too

Taxed at a rate of $100 per metric ton of carbon, current average fossil fuel use by 28 million Canadians would raise about $11.5 billion annually. If only half of this amount were made available for the proposed carbon-sink program, it could create 57,500 jobs in a whole new woods-based industry, assuming an average job-creation rate of 10 jobs per million dollars invested.[13] Ideally, these new jobs would be created in locally owned, rural community-based ecosystem management firms, which would undertake the work on contract from an appropriate federal agency.

It should be emphasized that the new industry should be oriented not only to the carbon-sink function, but also to comprehensive forest-related ecosystems restoration—fish and wildlife, soil management, biodiversity, erosion control, outdoor recreation, and tourism values would all benefit as spin-offs of the program, and with this would come additional sustainable employment in rural regions.

The extractive forest industry would be subject to its own revised management regime involving steep depletion taxes or marketable quotas on the available sustainable cut. This keeps the ecological accounts for goods production by the forest separate from the accounts for its carbon dioxide assimilation services. It would also enhance job security in the traditional woods industries, as labor, reversing historical trends, would become more attractive relative to energy and capital. (While the harvest steadily increased, timber jobs in British Columbia alone declined by 25% during the 1980s, partially as a result of mechanization.)

Ecological Benefits

A carbon-sink forest closes the loop in the industrially amplified carbon cycle, returning carbon previously stored in fossil fuels back to storage in contemporary biomass. This slows the accumulation of carbon dioxide in the atmosphere and helps to restore entropic balance by creating the biomass equivalent of the fossil-fuel consumed. A global system of carbon-sink forests would reduce the potential risk of climate change for several decades, thus avoiding some of the associated damage costs (e.g., of sea level rise) and buying time for the development of alternative energy technologies.

Rehabilitated forests throughout rural North America could provide other significant ecological functions and services with economic implications, such as the regulation of water supply and maintenance of water quality, flood protection, erosion control and soils maintenance, wildlife and fisheries habitat, outdoor recreational potential, and so on. There are also less tangible ecological benefits; for example, national carbon-sink forest networks could be managed to maximize biodiversity across the continent, thus addressing one of the most pressing and widely acknowledged problems of unsustainability.

Further Economic and Employment Benefits

By allowing energy prices to "tell the truth," a phased fossil fuel depletion tax would acknowledge the external costs of hydrocarbon consumption.

By internalizing these costs, the country would be implementing a form of full-cost pricing and the user-pays principle—fossil fuel users would be paying to repair the damage caused by this use—thus enhancing overall economic efficiency. The tax would also provide a general incentive for energy conservation, thereby reducing both demand for primary fuel and carbon dioxide emissions into the atmosphere.

There are several additional structural benefits that would accrue to both urban and rural areas. The increased cost of energy-intensive manufacturing and related economic activities would stimulate the search for more energy and material-efficient production technologies and associated skills. Higher prices for hydrocarbons would also encourage research and development of alternative energy sources, such as hydrogen fuel cells and photovoltaic and other solar energy sources, thus creating employment in science, engineering, and innovative technologies.

The increased transportation costs for many products and commodities would shift the competitive advantage away from imports to locally produced goods, thereby stimulating local employment. This would particularly assist farmers to recapture a higher share of their local—or even national—markets. Additional employment benefits may be realized as the increasing cost of energy-intensive capital relative to labor gradually reduces the incentive to substitute capital for labor.

Linking energy and forest policies could also revolutionize forest economics. A fossil fuel depletion tax could provide the economic means to develop a new, creative, forest-based industry in rural regions, involving a specialized set of skills in ecology, silviculture, soil science, wildlife biology, hydrology, and so forth. Forest rehabilitation is more labor-intensive than the traditional forest-based industries. Thus, national programs of forest ecosystem restoration would employ more workers in logging communities than have been displaced from their jobs by mechanization (capitalization) in the forest sector. Meanwhile (and somewhat ironically) it would help to ensure an adequate resource base for the full range of extractive forest-based industries in the future, thus contributing to maintaining employment in the traditional forestry sector.[14] Finally, restoring our forests would enhance the resource base for tourism, sport fishing, wildlife-related activities, and outdoor recreation generally, creating increased employment in all associated businesses.

Social Benefits

The ecological integration of energy and forest policy can also improve the quality of life in rural and resource-based communities. The creation of employment in restoration forestry and related ecosystem rehabilitation could more than compensate for the loss of jobs resulting from mechanization in the extractive sector, thereby reducing individual and family stress and enhancing community sustainability. In addition, by eliminating the perceived conflict between conservation and jobs, it would also alleviate the present debilitating tension between conservationists and loggers (a critical issue in the Pacific Northwest and in British Columbia) and thus help restore unity to forest communities. Meanwhile, the increased viability of sport fishing, ecotourism, and outdoor recreation generally would help diversify the permanent economic base of many rural regions, further enhancing long-term sustainability.

SUMMARY AND CONCLUSIONS: KEYS TO SUSTAINABILITY

In the twilight of the twentieth century, short-term economic efficiency has become the central pillar of sociopolitical discourse. Indeed, narrow economic rationality would permit our best agricultural land to be converted to almost any more "productive" use, at least until food prices make commercial farming competitive with other economic activities. It seems that even the biological necessity for food holds no special sway in the global marketplace. In the minds of many analysts, the food problem has been handled—technology, not land or people, produces our food. Agriculture, particularly near cities, is further undermined by cheap food imports and the short-term calculus of the marketplace.

This conventional economic perspective is an integral part of the prevailing cultural myth of human mastery of nature. Unfortunately, mainstream economic models are dangerously simplistic. Devoid of ecological content, they say nothing about the structural and functional complexity of ecosystems and completely ignore the entropy law. This means that the theory driving development policy today is blind to the physical flows (the appropriated natural income) associated with rising money incomes and mute on the possibility of consequent systems destabilization. Ecological footprint analysis reveals starkly that, contrary to conventional wisdom, technology and trade have not freed humanity from ecological constraints. In fact, the high levels of consumption and waste generation made possible by modern technology and international trade have made individuals and nations *more dependent than ever* on the products and processes of nature, even as they undermine the functional integrity of nature's production system.

As rural populations shrink and the relative contribution of agriculture and forestry to gross domestic product declines, cities are increasingly seen as the engines of economic growth and sustainability. This too is at least partial illusion. It is true that human capital and knowledge-based enterprise are increasingly the source of GDP (gross domestic product) growth and prosperity in high-income countries. However, money is not real wealth. It merely confers access to natural income flows, or their products, without which money would be worthless.

Unfortunately, the natural capital producing these physical income flows and other life-support functions is being depleted. It is no small irony that this destructuring of the ecosphere is both driven and obscured by rising money incomes. The globalization of the economy increases demand, while the expansion of world trade gives the rich access to remaining stocks of natural wealth anywhere on earth. This system keeps prices down while insulating wealthy consumers from the negative consequences of overpopulation, overconsumption, and natural capital depletion in their own countries.

Conventional agriculture is a major factor in global ecological decline, according to the second law of thermodynamics and natural capital criteria for sustainability. Mechanized high-input cropping consumes more available energy and material than is represented by the useable crop and, in some cases, appropriates the ecological output of several times more land than is represented by the physical growing area. If preindustrial agriculture had a minimal impact on global entropy, industrial agriculture is a virtual entropy pump. A similar case could be made against industrial forestry.

Ironically, then, even as land and labor efficiency appears to reach new heights, we may be moving ever closer to a sustainability crisis in agriculture and a global food emergency. Even as governments leave ever more resource allocation decisions to the short-term, entropy-blind logic of the marketplace, interventionist policies to protect agricultural land for farming may be more important nows than at any time in history.

In the final analysis, nature remains the ultimate source of material wealth. It also maintains the life-support functions of the ecosphere. These functions are critically undervalued—indeed, virtually ignored—by our current economic system. Market prices for food, forest products, and most other consumer goods do not reflect the value of nonmarket life-support functions, the depletion of natural capital, the loss of future natural income, nor the costs of pollution damage associated with their production. Worse, there are no markets nor any chance of establishing reasonable dollar prices for the majority of essential services provided by nature.

The constant, natural capital stocks condition necessary for sustainability therefore places boundaries on market economics. Our strong sustainability variant of the constant stocks criterion requires the preservation of adequate per capita stocks of essential self-producing natural capital. This approach is fully supported by second law considerations that recognize natural capital as the source of essergy for the economy and the absolute need to arrest the entropic decline of the ecosphere. Of course, what constitutes adequate natural capital is subject to debate and is dependent, in part, on the prevailing (or desired) material standard of living. However, this uncertainty should not present a barrier to the general implementation of the concept. The accelerating global change is sufficient evidence that remaining natural assets are inadequate to support even present material demands. Indeed, the richest quarter of the human population alone has appropriated the entire productive capacity of the planet (Wackernagel and Rees 1995). This means that natural capital stocks must be enhanced just to satisfy the basic needs of the presently poor, whatever the final definition of adequacy. It also challenges the presently rich to reduce their ecological footprints to free up needed ecological space for others. A sustainable world will be a more equitable world. (The need to arrest population growth is self-evident.)

Together, these considerations suggest a development path based, at a minimum, on no net loss of natural wealth. In fact, in the medium term, with a growing human population and rising material expectations, sustainability depends on large-scale investment in the rehabilitation of ecosystems and in the rebuilding of natural capital stocks. Herein lies the ecological key to rural sustainability. Indeed, the security of our entire global urban civilization depends on our capacity to maintain and increase the natural income from ecoagricultural, forest, and other rural ecosystems.

In the sociocultural sphere, various nongovernment environmental organizations, motivated by powerful emergent philosophies ranging from deep-ecology through ecofeminism to bioregionalism, are coming together in a spontaneous coalescence of insight and energy for change. They see the key to sustainability not so much in technical enhancement of the ecosphere, but rather in fundamental changes to the deeply rooted values and behaviors that spawned our global consumer society. To them, the presumption of dominance (mostly male) of humans over humans and of humankind over nature is at the root of the problem. In the present context, this means that we must come to see ourselves as living *in* nature as part *of* nature, demanding no more than nature can provide.

Living within our ecological means and sharing the available natural income may seem sensible enough to the ecologically and socially aware, but appears impossibly idealistic to the modern self-indulgent technoexpansionist consciousness. Nevertheless, closer to the mainstream, ecological fiscal reform is being advocated as one means to alter popular perceptions and induce some of the behavioral changes required to begin the transition to sustainability. In theory, as more people come to recognize their absolute dependency on nature, personal and cultural values should evolve accordingly.

What is missing, however, is the political key. Society at large has been lulled into a false sense of security by a truncated economic paradigm and entrenched technological arrogance; scientific controversy over the details and implications of human-induced global change obscure the consensus that such change poses a genuine threat to global security; the swing to the neoconservative right undermines our sense of "the public interest" and denies the existence of common social purpose. In these circumstances, ordinary people find it difficult enough to identify with their immediate neighbors, let alone the global community, and do not know what to believe about the so-called environmental crisis. This makes it easy for powerful vested interests to seize the political agenda and divert attention from the sustainable development agenda.

In sum, there is both good news and bad in our ongoing quest for sustainability. The good news is that our conceptual and analytic tools for understanding core issues are constantly improving, there is a growing consensus on the urgency of the problem, and there are no insurmountable technical barriers. Even more encouraging, a grass-roots movement to redefine humanity's place in the natural order of things is gathering momentum. The bad news is that in the absence of broad public understanding of the sustainability imperative, there is little political will to implement the changes in government policy necessary to knock global society off its present perilous course. Thus, even if the analysis in this chapter is correct, the status quo may well prevail for the foreseeable future. This is no reason to despair but, rather, a challenge to the converted to redouble their efforts to seize center stage in the political debate.

ACKNOWLEDGEMENTS

The author's and his students' work on ecological footprinting is supported by a Canadian Tri-Council EcoResearch Grant to the University of British Columbia, in which the author is a coinvestigator. Special thanks to Mathis Wackernagel and Yoshi Wada for their dedication and hard work. Phil Testamale prepared the illustrations for Figures 3.1 and 3.2.

NOTES

1. Sustainable development has deep roots in early twentieth-century theory of renewable resource management and was later advanced as a more fully integrated approach to conservation and development by the International Union for the Conservation of Nature (IUCN 1980).
2. Daly and Cobb (1989) suggest that the Commission's reluctance to refine the concept was intentional and politically astute. It facilitated general acceptance of the general concept and "guaranteed eventual discussion of its radical implications."
3. Available or unbound energy (referred to here as *essergy*) represents the thermodynamic potential to do work. For example, some of the energy (essergy) in a lump of coal can be used to heat a kettle of water. In contrast, unavailable or bound energy cannot be put to use. There is vastly more heat in the water of a swimming pool at room temperature than in a lump of coal, but in the absence of a thermal gradient it cannot be used. The heat in the pool water is low-grade dissipated energy.

4. The term *natural assets* encompasses not only material resources (e.g., petroleum, the ozone layer, forests, soils) but also process resources (e.g., waste assimilation, photosynthesis, soils formation). It includes renewable as well as exhaustible forms of natural capital. Our primary interest here is in essential renewable and replenishable forms. Note that the depletion of nonrenewables could be compensated through investment in renewable natural capital.
5. The prevailing alternative interpretation refers to an aggregate stock of manmade and natural assets. This version of the constant capital stocks criterion reflects the neoclassical premise that manufactured capital can substitute for natural capital, and is referred to as weak sustainability (Daly 1990; Pearce and Atkinson 1993; Victor et al. 1995).
6. There are only about 8.9 billion hectares of ecologically productive land on earth, 1.5 hectares for each of the 5.8 billion people in 1996. This approach recognizes that photosynthesis and dependent ecological processes are the major means by which the ecosphere compensates for the disorder resulting from economic production/consumption processes. For sustainability, essergy produced in the ecosphere must at least offset the entropy pumped out by the economy.
7. Recent catches (including discards) have slightly exceeded this amount but have been steady or in decline since 1989, despite increased fishing effort.
8. Assume the city were surrounded by typical terrestrial waterscape and landscape types. If the glass hemisphere were then expanded until it encompassed sufficient land and water (ecosystem) area to support the internal population's consumption and waste generation indefinitely, then the area so contained would be equivalent to that city's de facto ecological footprint on the earth.
9. As compared with four calories of exosomatic energy (in the form of renewable biomass) per calorie of food delivered by subsistence agriculture.
10. Abstracted and revised in part from Rees (1990).
11. As noted, massive increases in energy use have enabled rising rates of material throughput in virtually all sectors of the economy throughout this century. The result has been accelerated resource depletion and increasing net global entropy—deforestation, soils degradation, fisheries collapse, species extinctions, and so forth. One aggregate measure of the dissipative destructuring of terrestrial ecosystems with the spread of industrial society is the net release of 120 billion metric tons of carbon from biomass and soils into the ecosphere between 1850 and 1950. This is an average loss of 1,100 metric tons of carbon per kilometer (km^{-1}) or 11 metric tons per hectare. (Houghton and Skole 1990).
12. Extended from Rees (1995c).
13. Investment on this scale should be possible, because billions of dollars would also be saved in the form of recovered hydrocarbon subsidies and related tax expenditures, unrequired welfare payments and unemployment compensation, and so forth.
14. Mature carbon-sink forests must be continuously harvested and the products kept as wood or fiber in houses, furniture, and so on, to maintain the net sink function. Allowing the accumulated biomass to burn or decompose would return the sequestered carbon to the atmosphere. Unharvested old growth may not be accumulating net carbon.

REFERENCES

Ayres, R. U. 1994. *Information, Entropy and Progress: A New Evolutionary Paradigm.* Woodbury, N.Y.: AIP Press.

BCSD. 1993. *Getting Eco-Efficient.* Report of the BCSD First Antwerp Eco-Efficiency Workshop, November 1993. Geneva: Business Council l for Sustainable Development.

Berry, R. S. 1972. "Recycling, Thermodynamics, and Environmental Thrift." *Bulletin of the Atomic Scientists* 28:8–15.

Bertin, O. 1989. "Bleak Forecast for World Grain Stock." *Globe and Mail* (Toronto, 26 January).

Bolin, B. 1986. "How Much Carbon Dioxide Will Remain in the Atmosphere?" In *The Greenhouse Effect, Climatic Change, and Ecosystems* (SCOPE 29), edited by B. Bolin, B. Doos, J. Jaeger, and R. Warrick. New York: John Wiley & Sons.

Brooks, D. 1990. "Beyond Catch Phrases: What Does Sustainable Development Really Mean?" *IDRC Reports.* Ottawa: International Development Research Centre.

Brown, L. 1988. *The Changing World Food Prospect: The Nineties and Beyond.* Worldwatch Paper 85. Washington, D.C.: The Worldwatch Institute.

_______. 1995a. "Natures Limits." In *State of the World 1995*, edited by L. Brown, D. Dennison, C. Flavin, H. French, H. Kane, N. Lessen, M. Renner, D. Roodman, M. Ryan, A. Sachs, L. Starke, P. Weber, and J. Young. Washington, D.C.: Worldwatch Institute, 3–20.

_______. 1995b. *Who Will Feed China?* New York: W.W. Norton.

Brown, L., A. Durning, C. Flavin, L. Heise, J. Jacobson, S. Postel, M. Renner, C. P. Shea, and L. Starke. 1989. *State of the World 1989.* Washington, D.C.: Worldwatch Institute.

Canada. 1993. *Environmental Perspectives 1993.* Ottawa: Statistics Canada, National Accounts and Environment Division.

Costanza, R., and H. Daly. 1992. "Natural Capital and Sustainable Development." *Conservation Biology* 1:37–45.

Daly, H. 1990. "Sustainable Development: From Concept and Theory Towards Operational Principles." In *Steady State Economics*, 2d ed., edited by H. E. Daly. Washington, D.C.: Island Press.

_______. 1991a. "From Empty World Economics to Full World Economics: Recognizing an Historic Turning Point in Economic Development." In *Environmentally Sustainable Economic Development: Building on Brundtland*, edited by R. Goodland, H. Daly, S. El Serafy, and B. von Droste. Paris, France: UNESCO.

_______. 1991b. "Sustainable Development: From Concept and Theory Towards Operational Principles."In *Steady-State Economics*, 2d ed. Washington, D.C.: Island Press, 241–260.

_______. 1992. "The Concept of a Steady-State Economy." In *Steady-State Economics*, 2d ed. Washington, D.C.: Island Press, 14–19.

Daly, H., and J. Cobb. 1989. *For the Common Good: Redirecting the Economy Toward Community, the Environment, and a Sustainable Future.* Boston: Beacon Press.

Edwards, C. 1990. "The Importance of Integration in Sustainable Agricultural Systems." In *Sustainable Agricultural Systems*, edited by C. Edwards, R. Lal, P. Madden, R. Miller, and G. House. Ankeny, Iowa: Soil and Water Conservation Society.

Edwards, C., R. Lal, P. Madden, R. Miller, and G. House, eds. 1990. *Sustainable Agricultural Systems.* Ankeny, Iowa: Soil and Water Conservation Society. (See review by D. Pimentel, *BioScience* 41:[1]48–49.)

Ekins, P. 1993. "'Limits to Growth' and 'Sustainable Development': Grappling with Ecological Realities." *Ecological Economics* 8:269–288.

Ekins, P., and M. Jacobs. 1994. "Are Environmental Sustainability and Economic Growth Compatible?" In *Energy-Environment-Economy Modelling Discussion Paper No. 7.* Cambridge, UK: University of Cambridge, Department of Applied Economics, 1–18.

Faeth, P., R. Repetto, K. Kroll, et al. 1991. *Paying the Bill: US Agricultural Policy and the Transition to Sustainable Agriculture.* Washington, D.C.: World Resources Institute.

Flavin, C. 1996. "Facing Up to the Risks of Climate Change." *In State of the World 1996,* edited by L. Brown, J. Abramovitz, C. Bright, C. Flavin, G. Gardner, H. Kane, A. Platt, S Postel, D. Roodman, A. Sachs, and L. Starke. New York: W. W. Norton, 21–39..

Folke, C., J. Larsson, and J. Sweitzer. 1994. "Renewable Resource Appropriation by Cities." Paper presented at the Third International Meeting of the International Society for Ecological Economics, San Jose, Costa Rica, 24–28 October.

Gardner, G. 1996. "Preserving Agricultural Resources." In *State of the World 1996,* edited by L. Brown, J. Abramovitz, C. Bright, C. Flavin, G. Gardner, H. Kane, A. Platt, S. Postel, D. Roodman, A. Sachs, and L. Starke. New York: W. W. Norton, 78–94.

Georgescu-Roegen, N. 1993. "The Entropy Law and the Economic Problem." In *Valuing the Earth: Economics, Ecology, Ethics,* edited by K. Townsend and H. Daly. Cambridge, Mass.: MIT Press, 75–88.

Giampietro, M., and D. Pimentel. 1994. "The Tightening Conflict: Population, Energy Use, and the Ecology of Agriculture." *Focus* 4:51, 57–62.

Goodland, R. 1991. "The Case That the World Has Reached Limits." In *Environmentally Sustainable Economic Development: Building on Brundtland,* edited by R. Goodland, H. Daly, S. El Serafy, and B. Von Droste. Paris, France: UNESCO.

Harrison, R., and S. Cloutier. 1995. *People in Canadian Agriculture.*Statistics Canada Cat.# 21-523E Occasional. Ottawa: Ministry of Industry and Science.

Houghton, R., and D. Skole. 1990. "Carbon." In *The Earth as Transformed by Human Action,* edited by B. Turner II, W. C. Clark, R. W. Kates, J. F. Richards, J. T. Mathews, and W. D. Meyer. New York: Cambridge University Press, 393–408.

Intergovernmental Panel on Climate Change. 1995. "The IPCC Assessment of Knowledge Relevant to the Interpretation of Article Two of the UN Framework Convention on Climate Change: A Synthesis Report" (draft). Geneva.

International Institute for Environment and Development. 1995. *Citizen Action to Lighten Britain's Ecological Footprint.* A report prepared for the UK Department of the Environment. London.

IUCN. 1980. *The World Conservation Strategy.* Gland, Switzerland: International Union for the Conservation of Nature.

NASA. 1988. *Executive Summary of the Ozone Trends Panel.* Washington, D.C.: National Aeronautics and Space Administration.

National Institute for Public Health and Environmental Protection. 1992. *National Environmental Outlook 2, 1990–2010.* Bilthoven, Netherlands.

National Research Council. 1990. *Alternative Agriculture*. Washington, D.C. (See review by D. Hodges, *The Ecologist* 20:5:198–199.)

Parry, M. 1989. "Impact of Warming on Agriculture," *Forum for Applied Research and Public Policy* 4:4:43–46.

Pauly, D., and V. Christensen. 1995. "Primary Production Required to Sustain Global Fisheries." *Nature* 374:255–257.

Pearce, D. 1994. *Valuing the Environment: Past Practice, Future Prospect*. CSERGE Working Paper PA 94-02. London: University College Centre for Social and Economic Research on the Global Environment.

Pearce, D., and G. Atkinson. 1993. "Capital Theory and the Measurement of Sustainable Development: An Indicator of Weak Sustainability." *Ecological Economics* 8:103–108.

Pearce, D., E. Barbier, and A. Markandya. 1990. *Sustainable Development: Economics and Environment in the Third World*. Hants, England: Edward Elgar Publishing.

Pearce, D., A. Markandya, and E. Barbier. 1989. *Blueprint for a Green Economy*. London: Earthscan Publications.

Pezzey, J. 1989. *Economic Analysis of Sustainable Growth and Sustainable Development*. Environment Department Working Paper No. 15. Washington, D.C.: The World Bank.

Pimentel, D. 1996. "The Environment: How Many Humans Will It Support." Paper presented to the scientific session of the AAAS Annual Meeting, Baltimore, Maryland 8–13 February.

Pimentel, D. et al. 1989. "Low-Input Sustainable Agriculture Using Ecological Management Practices." *Agriculture, Ecosystems, and Environment* 27:3–24.

Pimentel, D., J. Allen, A. Beers, L. Guinand, R. Linder, P. McLaughlin, B. Meer, D. Musonda, D. Perdue, S. Poisson, S. Siebert, K. Stoner, R. Salazar, and A. Hawkins. 1987. "World Agriculture and Soil Erosion." *BioScience* 37(4):277–283.

Postel, S., G. Daily, and P. Ehrlich. 1996. "Human Appropriation of Renewable Fresh Water." *Science* 271:785–788.

Quarrie, J., ed. 1992. *Earth Summit '92*. London: Regency Press.

Rees, W. E. 1990. "Atmospheric Change: Human Ecology in Disequilibrium." *International Journal of Environmental Studies* 36:103–123.

———. 1991. "Conserving Natural Capital: The Key to Sustainable Landscapes." *International Journal of Canadian Studies* 4:7–27.

———. 1992. "Ecological Footprints and Appropriated Carrying Capacity: What Urban Economics Leaves Out." *Environment and Urbanization* 4:121–130.

———. 1995a. "Achieving Sustainability: Reform or Transformation?" *Journal of Planning Literature* 9:343–361.

———. 1995b. "More Jobs, Less Damage: A Framework for Sustainability, Growth, and Employment." *Alternatives* 21:24–30.

———. 1995c. "Taxing Combustion and Rehabilitating Forests: Achieving Sustainability, Growth, and Employment Through Energy Policy." *Alternatives* 21:31–35.

———. 1996. "Revisiting Carrying Capacity: Area-Based Indicators of Sustainability." *Population and Environment*. 17(3):195–215.

Rees, W. E., and M. Wackernagel. 1994. "Ecological Footprints and Appropriated Carrying Capacity: Measuring the Natural Capital Requirements of the Human Economy." In *Investing in Natural Capital: The Ecological Economics Approach to Sus-*

tainability, edited by A-M. Jansson, M. Hammer, C. Folke, and R. Costanza. Washington D.C.: Island Press.

RMNO. 1994. *Sustainable Resource Management and Resource Use: Policy Questions and Research Needs.* Publication No. 97. Rijswijk, The Netherlands: Advisory Council for Research on Nature and Environment (RMNO).

Schneider, E., and J. Kay. 1992. "Life as a Manifestation of the Second Law of Thermodynamics." Working Paper Series. (Preprint from *Advances in Mathemetics and Computers in Medicine.*) Waterloo, Ont.: University of Waterloo Faculty of Environmental Studies.

Smith, J., and D. Tirpak, eds. 1988. "The Potential Effects of Global Climate Change on the United States" (Executive Summary, Draft Report to Congress). Washington, D.C.: United States Environmental Protection Agency.

Solow, R. 1986. "On the Intergenerational Allocation of Natural Resources." *Scandinavian Journal of Economics* 88:1.

United Nations. 1992. *Agenda 21.* New York.

U. S. Department of Agriculture. 1988. *World Grain Situation and Outlook.* Washington, D.C.: U. S. Department of Agriculture, Foreign Aid Service.

———. 1994 *Agricultural Resources and Environmental Indicators.* Agricultural Handbook No. 705. Washington, D.C.: United States Department of Agriculture, Economic Research Service.

Victor, P., E. Hanna, and A. Kubursi. 1995. "How Strong Is Weak Sustainability?" *Economie Applique* XLVIII: 75–94.

Victor, P. A. 1991. "Indicators of Sustainable Development: Some Lessons from Capital Theory." *Ecological Economics* 4:191–213.

Vitousek, P. 1994. "Beyond Global Warming: Ecology and Global Change. *Ecology* 75:1861–1876.

Vitousek, P. , P. Ehrlich, A. Ehrlich, and P. Matson. 1986. "Human Appropriation of the Products of Photosynthesis." *BioScience* 36:368–374.

von WeizsSäcker, E. U. 1994. *Earth Politics.* London: Zed Books (see Chapter 11).

Wackernagel, M., and W. Rees. 1995. *Our Ecological Footprint: Reducing Human Impact on the Earth.* Gabriola Island, B.C., and Philadelphia, Pa: New Society Publishers.

Wada, Y. 1993. *The Appropriated Carrying Capacity of Tomato Production.* Master's thesis, University of British Columbia, School of Community and Regional Planning.

WCED. 1987. *Our Common Future.* Oxford: Oxford University Press.

Wigley, T., J. Angell, and P. Jones. 1985. "Analysis of the Temperature Record." In *The Potential Climatic Effects of Increasing Carbon Dioxide,* edited by M. MacCracken and F. Luther. DOE/ER-0237. Washington, D.C.: DOE/OER.

Williams, G., R. Fautley, K. Jones, R. Stewart, and E. Wheaton. 1988. "Estimating Effects of Climatic Change on Agriculture in Saskatchewan." *Climate Change Digest* CCD 88–06. Ottawa, Ontario: Supply and Services Canada.

Wilson, E. O. 1988. "The Current State of Biological Diversity." In *Biodiversity,* edited by E.O. Wilson. Washington, D.C: National Academy Press.

WRI. 1992. *World Resources 1992–93.* Oxford and New York: Oxford University Press (for the World Resources Institute).

Four

Sustainability and Rural Revitalization: Two Alternative Visions

Richard M. Clugston

INTRODUCTION

It is well known that rural America—those areas where major income is based on farming—has undergone an extensive process of depopulation for the past 100 years. The farm population has dropped from 30 million in 1940 to 5.4 million in 1985 and is still declining (Staten 1987, 28). Technological innovation has reduced the amount of human labor necessary to produce a unit output of food or fiber, and the number of farmers and farm workers needed to produce a bushel of wheat or a pound of beef has dropped dramatically. On the other hand, the amount of energy necessary to produce this same bushel or pound has risen dramatically: it now takes about 13 calories of fossil fuels to produce 1 calorie of food. Petroleum (as fuel and as agrichemicals) has replaced human and animal labor, and this replacement is financed by large amounts of capital. Capital has been substituted for labor. Farmers, in the words of Earl Butz, a former secretary of agriculture, have gotten big or gotten out (Kirschenmann 1994).

Much of the country's rural population has migrated to the cities. With rural depopulation has come the consolidation and regionalization of services. Small town, main street, ma-and-pa operations shut down. Large, chain-store–dominated malls open at regional centers. Hospitals close, schools consolidate. Many rural small towns become retirement centers with few young people.

Even with its ebbs and flows—such as the agricultural boom of the late 1970s and the bust of the 1980s—the overall trend has been remarkably persistent and reflective of the transition from locally self-reliant economies to an international economy dominated by huge, vertically integrated corporations. Once largely self-sufficient, local rural economies have become more and more dependent on external inputs ranging from pesticides, fertilizers, tractors, and televisions to state subsidies for schools and health care. The prosperity of rural areas has come to depend less and less on mutual assistance and self-sufficiency and ever more on producing high volumes of export commodities.

TWO SCENARIOS OF RURAL REVITALIZATION

Rural revitalization is usually a synonym for the economic growth (or at minimum, the stabilization) of the small town and its surrounding farm economy. It sometimes connotes recognition of the richness of rural culture and the need to preserve it. Thus it includes job and wage growth, excellent schools, and the preservation and enhancement of rural ways of life. Those striving to revitalize rural America focus primarily on developing new agricultural products or attracting new enterprises. To a lesser extent, they focus on achieving greater efficiencies within the rural economy. Such efforts have met with limited success, particularly in areas that are remote from urban centers.

Almost by definition, the revitalization of rural areas will depend largely on improving the farm economy, supplemented, when possible, by improvements in the small town business climate and efficiencies in a range of supportive industries. Thus any conception of rural revitalization must clarify how the farm economy can become sufficiently robust to support small towns. The following are two scenarios for a vital farm economy—the first based on agribusiness industrialization and the second requiring a shift to sustainable agriculture, utilizing intermediate technology.

Industrial Agribusiness

Farms continue to grow in size and to become increasingly mechanized and automated to produce a maximum of outputs (food units) per unit of cost. The following are examples of technological innovation that may be integrated into the high-tech megafarm:

- Increasing use of biotechnology. This includes using plant and animal growth regulators (e.g., brassia) and genetic engineering to increase resistance, yields, and plant varieties that tolerate high levels of salinity, herbicides, pesticides, and other forms of stress.
- New feedstuffs based on animal and plant waste, plastics, and existing by-products of agricultural production.
- Mechanized technologies, particularly computer-assisted planting, animal confinement systems, and automated harvesting systems, which do some initial processing and eliminate hand-harvesting of vegetables and fruits (Staten 1987, 193–202; Production Credit Associations 1983).

In this scenario, capital continues to be substituted for labor. The number of farm owners and farm workers continues to decline to some bottom limit (about one-fourth the current number) determined by the feasibility of automation and the mobility of the rural poor. Small towns continue to disappear while regional hubs flourish. These hubs contain malls with major retail outlets, state universities, community colleges, and governmental centers and have populations of 25,000 to 50,000. The new highly educated farm workers who operate and program the hardware and software of the megafarms commute to the farm factories. Within the regional hubs, these workers/managers live in refurbished and picturesque small towns or on hobby farms (farmhouses on land that is part of the megafarm) (Staten 1987).

In this industrialized best-case scenario, rural revitalization consists of a highly paid, highly educated farm manager/technician work force. Although much of the wealth moves to remote profit centers, enough remains to support a well developed service economy at the regional hub (where perhaps one fourth of those displaced from farms and small towns can relocate).

In a worse case, the industrialization of rural America does not work. While megafarms increase in number, they do so in the midst of a deteriorating resource base and infrastructure, increasing rural poverty and unemployment and the destruction and loss of a sense of place and culture. Given this outcome, farmers and small town dwellers are displaced, but because there are no decent or adequate jobs in either the big cities or the regional centers, they become welfare dependent (returning to or remaining in rural areas). Petroleum-intensive, monoculture-based, mechanized agriculture continues to deplete and degrade the environment. Soil erodes, water is used up, crops and animals are more subject to disease and stress. Environmental damage increases. Both the quantity and quality of food and fiber drop. As energy gets more expensive, megafarms become less profitable and cannot support vital regional centers. Nothing trickles down except displacement and the degradation of community and environment (Fritsch 1989).

Sustainable Agriculture

In the alternative scenario, previously dominant trends to increase capital and higher external inputs are reversed. Low input, sustainable agriculture, which relies on intermediate technology, more farmers and laborers, and ecologically sound practices, results in rural resettlement.

"Humane Sustainable Agriculture (HSA) produces adequate amounts of safe, wholesome food in a manner that is ecologically sound, economically viable, equitable, and humane. HSA meets farm animals' basic physical and behavioral requirements for health and well-being through a food and agricultural system that respects all of nature—humans, soil, water, plants, and animals, wild as well as domestic" (Gips et al. 1993).

Major elements of sustainability in agriculture include a responsible reliance on local resources that does not require the major alteration or control of the land and preserves biological and cultural diversity (Gliessman 1990, 380). Sustainable agriculture depends on the substitution of labor—particularly as a harmonious working with the natural forces in the form of intermediate technology-based, low-input farming—for capital. The average farm is smaller, more diversified, and more self-reliant. Off-farm inputs (chemical pesticides and fertilizers, energy, and nonindigenous building materials) are minimized. These are replaced by ecologically sound practices such as the use of allelopathic crops, living mulches, crop rotations, biological pest control, organic matter inputs, and many others (Edwards 1990). Equipment reflects a more intermediate technology, and the systems of planting and cultivation create, as much as possible, a self-maintaining crop and animal system (e.g., permaculture).

Sustainable agriculture requires the farmer to be more intimately acquainted with the land and with the individual farm animals. It is this immediate knowledge of the needs and potential of a particular crop, terrain, pig, or cow that enables the farmer to design and adjust his or her action to maximize yield. Uniformity, so necessary to the automated megafarm, is replaced by the cultivation of diversity, based on the farmer's knowledge of and responsiveness to each ridge or gully, pig or cow. The technologies and techniques used—such as integrated pest management, permaculture design, mixed planting—are sophisticated, yet require a relatively modest capital investment. More people are needed to work the farm.

In the sustainability scenario, energy is conserved not only in food and fiber production, but also in the design of dwellings and small towns. Local resources (building materials, fuel supplies) are developed. Fewer inputs are needed from outside to support the local community. Rural development based on sustainable agriculture would halt depopulation and may even repopulate various rural areas with family farms, thus providing an adequate population base to support more main street ma-and-pa business and local schools and churches.

The risk in supporting sustainable agriculture and development is real. As families, farms, and small towns become more self-reliant, they rely less on money; instead, they barter goods and services and purchase less from retail centers. They use more of what they have on their farms, they recycle, and they buy to last. The wealth generated by overconsumption of soon-obsolete consumer items is replaced by a different form of prosperity—one that is more satisfying and sustainable, yes, but which may also be less profitable for large corporate interests and less taxable for governments. Food prices may rise. Both large corporations and large cities may realize less support by rural areas. The combination of such forces could bring about a significant recession (or even a depression), which would be particularly hard for those sectors of the economy that are dependent on economic growth.

RURAL REVITALIZATION THROUGH SUSTAINABLE AGRICULTURE AND DEVELOPMENT

The policies of the United States have favored the agribusiness megafarm as they have the same approach to wealth generation in agriculture (e.g., consolidation, vertical integration, and large corporate control) as in other sectors (Sale 1986). This mode of economic development, however effective it may be in benefiting stockholders, is increasingly viewed as destructive to the environment and to local communities (Daly and Cobb 1989). It exploits for short-term gain and displaces other modes of valuation with money, consumption, cheapness, and convenience.

Although the cultural and economic system that supports true rural revitalization will be based on a mix of large-scale, export-oriented agribusiness and diversified sustainable agriculture, planners, policymakers, and consumers must, at this moment, act to promote sustainable agriculture and development. Given the scenarios discussed here, it is more likely that the rapid industrialization of agriculture as the predominant mode of food and fiber production will result in a deterioration in the quality of life for current rural (and, increasingly, all global) citizens. This system depletes, erodes, and pollutes the land and environment and undermines cultural and biological richness and diversity, and its true cost will be paid primarily by future generations.

Rural revitalization can be best achieved through a comprehensive shift to sustainability. This involves not only shifts in farming, but in the design and operations of the farm and the rural community. The same principles that call for minimizing dependence on external, purchased inputs and using locally available and renewable resources apply to the development of housing, energy, and waste management systems on farms and in small towns.

Various agencies and communities, such as the Alternative Energy Resources Organization (AERO) in Helena, Montana, have been pursuing sustainable development in agriculture, energy, and shelter design and have made significant contributions to the self-sufficiency and viability of local communities. AERO's (n.d.) brochure articulates well the interconnected dimensions of rural sustainable development, and these are used in the following paragraphs to organize a brief discussion of the components of a sustainable strategy for rural revitalization.

Agriculture

> AERO sponsors farm tours, on-farm research, and farm improvement clubs. We've identified farmers and ranchers who are reducing their dependence on farm chemicals and expanding the variety of crops they grow, while building the soil, breaking weed and other pest cycles, and increasing the flexibility of their operations (AERO n.d.).

Many organizations, such as land stewardship projects and sustainability centers, provide technical assistance to farmers who wish to make the transition to sustainable agriculture. *The Humane Producer and Consumer Guide* (Gips et al. 1993) lists more than 1,400 farms, research and education organizations, training programs, and others promoting humane, sustainable agriculture.

Food

Support for sustainability requires a wholesale and retail infrastructure to promote and distribute farm products. AERO builds the links necessary for the creation of locally and regionally based food systems that can provide consumers with fresher food, support local growers and their communities, and reduce the amount of energy used in packaging and transportation (AERO n.d.). State and county officials have developed "Buy Local" programs and supported farmers markets. Food cooperatives often are committed to supporting sustainable agriculture and provide an intermediate level of organization that supports local products and recirculates profits at the local level. In addition, more farmers are becoming linked directly with consumer groups through community-supported agriculture. In this structure consumers purchase shares, providing the farmers with a guaranteed income up front, and receive a percentage of the produce on a weekly basis throughout the season.

Communities

AERO and other organizations promote efforts to meet local needs with local resources. An example of an economic development approach based on sustainability is provided by the Center for Maximum Potential Building Systems in Austin, Texas. This organization has created a strategy for development based on the local resource base and sustainable planning methods. Central to its strategy is the use of indigenous building materials and passive climate design (architecture stressing solar energy and the use of shade and wind). Local wastes are recycled as raw materials. In addition, this center offers a community development process, in which members of the business community are brought together to discuss their needs and to determine how resources might be developed locally to meet these needs (Center for Maximum Potential Building Systems 1988).

The Sustainable Urban Rural Enterprise Project (SURE) (see Chapter 16), initiated by the U.S. Department of Housing and Urban Development, also seeks to mobilize sustainable local development. The SURE experience in Wayne County, Indiana, illustrates the benefits of consciously constructing a more self-reliant economy. Members of the local Chamber of Commerce identified goods and services they could receive from one another and agreed to buy them, even if cheaper products became available elsewhere. Although prices and supplies of local materials were not as competitive or reliable at times, the multiplier effect of recycling money locally more than compensated for such difficulties (Euston 1995).

Energy

AERO promotes the renewable and efficient use of energy, recommending the use of passive solar energy, super-insulation, small-scale greenhouses, and methanol. Although smaller farms may also be energy-intensive, they too can shift away from oil, both by substituting human and animal labor and by substituting renewable and solar energy for fossil fuel inputs. Sale (1986, 235) points out:

> A series of tests by the Small Farm Energy Project in Nebraska has shown that retrofitting and insulation, wind and biogas systems, and solar heat and hot water collectors can be installed cheaply in single-family farms and that the savings they produce are almost immediate. At least a dozen farms in the U.S. get their full energy needs from methane systems powered by the manure of their animals.

Research centers, such as the Land Institute in Salinas, Kansas, have developed the range of intermediate technologies needed to promote self-sufficient, ecologically sound farming operations (Preston 1989, 92). The Rocky Mountain Institute in Snowmass, Colorado, also has a sustainable economic development initiative in which small towns engage in energy conservation through utility-sponsored weatherization programs and switching to more efficient lighting systems.

The efforts of the Clinch Powell Sustainable Development Forum (see Chapter 19) illustrate a comprehensive strategy by a rural area to begin this shift to sustainability. The Clinch Powell initiative supports the development of sustainable livelihoods in the Appalachian region of southwest Virginia and northeast Tennessee, an area that the Nature Conservancy has described as one of America's "Last Great Places" (Austin 1996). Austin, a native of Appalachia, environmental theologian, and local farmer, describes the Forum as being organized to "(1) create quality jobs through an economic life greatly diversified and locally controlled; (2) create and support ecologically sensitive businesses; and (3) build skills and promote entrepreneurial innovation" (120).

The Clinch Powell Sustainable Development Forum, organized in 1993 by regional citizen-action and environmental leaders, used some state support to bring together public and private efforts to explore strategies for sustainable development. Its priorities are sustainable agriculture, sustainable wood products, sustainable home construction, nature tourism, and the establishment of a regional information bank. The wood products pilot project called **eco•log** includes selective logging that adheres to rigorous environmental standards and provides for training additional loggers, sawing at a family-run sawmill, curing wood in a state-of-the-art solar kiln, and finish milling and sizing in high school workshops. To demonstrate the environmental quality of the wood and the development of sustainable livelihoods, high school students produced a videotape documenting the entire process to accompany the wood being sold. Crafts people in the region who work with this wood may add to the video a section demonstrating their own techniques and market the video with their wood products. As a second phase of the project, there are plans to further develop commercial enterprises from **eco•log**.

In both forestry and agriculture initiatives, the goal is to make a decent return by reducing off-farm inputs (heavy machinery, fertilizer, pesticides, electricity) and adding value by recapturing the finishing and marketing tasks. In both, the farmer is substituting his or her wits for repetitive, capital-intensive, high-volume commodity production, as well as marketing more directly to consumers who value environmentally sound products. Austin further argues that such development strategies affirm the broad diversity of life and culture, providing broader returns than short-term economic goals. He states, "The goal of this strategy is renewing the community of life: more people on the land and in surrounding towns, productively employed serving each other; healthier soil; more diverse crops for regional markets; more livestock raised on the farm under humane conditions; more wildlife, wild edges, and wild places; more small businesses, home crafts, and entrepreneurial activities; greater community sufficiency" (Austin 1994, 10).

The various initiatives described here illustrate the diversity of approaches being developed to promote sustainable agriculture and communities. They share a common emphasis on strengthening local self-reliance within vibrant ecological and social systems, and these movements are growing (DeMuth 1995; Euston 1995).

As sustainable agriculture advocacy and practices have spread, various efforts have been made to investigate empirically what the impact of shifting to sustainable farm practices has been on farm income and, more indirectly, on rural economies. For example, Dobbs and Cole looked at the effects of shifting to sustainable practices (both organic and integrated pest management). They concluded, "Overall rural economy effects were negative in three of five South Dakota case farm comparisons when organic premiums were included, and in four of five when they were excluded" (Dobbs and Cole 1992, 78).

In contrast, a six-year Northwest Area Foundation project comparison of sustainable and conventional farming in Minnesota, Iowa, North Dakota, South Dakota, Montana, Oregon, and Washington concluded that sustainable practices did maintain or increase farm income, particularly in periods of hardship (drought). This extensive study clearly demonstrates the contribution of sustainable agriculture to rural revitalization (Bird et al. 1994).

Although these results are mixed, it is generally good news for the sustainability scenario for two main reasons: (1) sustainability does not necessarily endeavor to increase income, but rather to increase quality of life. Thus, in a more locally-based sustainable system, the exchange of many goods and services are not monetized and traditional income measures are less reliable indicators of prosperity; and (2) some efforts are successful despite the fact that the deck is still stacked against sustainability, given federal policies and subsidies.

SUPPORTING SUSTAINABILITY

To support a shift to the sustainable agriculture scenario, we must shift the system of federal regulation, incentives, and subsidies that for the last 40 years has exclusively favored growth of the industrial agribusiness complex. A range of policies have supported agribusiness, including price supports, soil-bank arrangements, direct payments, export controls, research-and development funds, disaster-assistance payments, marketing agreements, and tax write-offs (Sale 1986, 232). The industrial agribusiness scenario will prevail as long as current oil and agricultural policies remain unchanged. Daly and Cobb (1989, 271–272) eloquently present the policy choice that must be made:

> Policies have been based on the belief that there are economies of scale in bigness, and the collapse of so many smaller farms is taken as evidence that this is true. But on closer examination it turns out that it is government policy that has given the advantage to the largest farmers. Study after study has shown that small family operations are in fact more productive per acre. Though cash income may be small, they can support a family. It is when they are drawn into increasing their size or into excessive borrowing for "modern-

ization" that they are sucked into the downward currents that lead to bankruptcy.

The only measure by which the large scale farms are better is that of productivity of labor. On these large farms enormous quantities of energy are substituted for labor. But if productivity is measured in other ways, such as production per acre or per unit of energy or amount of capital input, it is the small farm that always excels. It is federal policy that has destroyed family farms in so much of the country, not any inherent weakness in the family farm system. The cessation of federal interference is the first requirement for the recovery of healthy rural life.

Of course, helpful governmental policies are possible. The most important would be taxing oil to include all externalities. The true price of oil-based agriculture would then become apparent, and monocultural production on huge acreages orientated to supplying distant markets would be at a disadvantage in relation to small farms raising a variety of foods for nearby consumers. Almost all regions of the country can produce most of the food they need close to population centers. Growing food locally is a practical goal, though it will have to struggle against well-entrenched interests in the present complex global system of production, processing, and distribution.

POLICIES, PLANS, CONSUMER CHOICES

Rural revitalization requires the development of more self-sufficient, sustainable farms and communities. This means putting into place regulations, incentives, and consumer education that encourages the development of local and sustainable food, energy sources, and building materials. Based on a series of forums and conferences throughout the United States, the Global Tomorrow Coalition in Washington, D.C., drafted an agenda for shifting policies to support sustainable agriculture and communities. Their recommendations, which illustrate aspects of the emerging sustainability policy agenda, include the following.

- Gradually increase Low-Input Sustainable Agriculture (LISA) funding for research and extension to 5% of the Department of Agriculture budget to make American farming more profitable and more sustainable.
- Pass a conservation-oriented farm bill providing more flexibility and incentives for farmers shifting to sustainable methods.
- Provide technological and financial assistance for agroecology, on-farm nutrient cycling, integrated pest management, soil improvement practices, crop rotation, windbreaks, efficient livestock/crop interaction, water conservation, and alternative energy sources.

- Reduce dependence on agrochemicals by 75% and remove the worst toxins from the market.
- Shift tax incentives to encourage local food production and more small farms.
- Require environmental impact reports as a component of farm policy.
- Include experience with nature and gardening in curricula at levels from kindergarten through secondary school.
- Promote the concept of partially self-reliant new communities by incorporating trees, gardens, alternative energy, alternative transportation, local job opportunities, etc. into community planning (Lesh and Lowrie 1990, 76–77).

In addition to federal policy changes, the agenda requires local and regional planning authorities, in cooperation with local land owners, to develop and use plans that consider the local ecosystem. As Barrett and associates (1990, 633) have pointed out, "sustainable landscapes can best be created by management at the watershed or regional scale because the flux of materials and organisms at these scales significantly influences management decisions for individual fields." Kirschenmann, in his testimony to the Sustainable Agriculture Scoping Group of the President's Council on Sustainable Development, pointed out that fewer than 6% of current farmers are under age 34. Unless we are preparing the next generation of farmers to steward the land, consolidation is inevitable and sustainability impossible. He states,

> We must maintain a sufficient base of [these] human resources [i.e., farmers] on the land to effectively manage the soil and water resources on which agriculture depends. Conservation biologists are now telling us that due to the ever changing and site specific character of natural ecosystems, there is no way that we can properly manage natural resources without populating local ecosystems with people who have been acquainted with those systems long enough and intimately enough to know how to manage them. Natural *agro*-ecosystems are no exception!" (Kirschenmann 1994, 3).

Any of these changes are unlikely in an antiregulatory and "property rights" climate. Such changes must have strong local governmental and consumer support to succeed. A value shift, from the deeply American notion that land is just a commodity to a recognition that land is primarily a community of living beings and a common trust, will be required. So too must our choices as consumers express our commitment to sustainability. Rural revitalization requires changing the food we eat; we must "just say no" to the products of chemically dependent, exploitative agriculture and buy that which is organically, humanely, and locally raised, even if it is initially more costly and inconvenient. The effort we make and expense we assume to create a humane and sustainable diet is a direct indication of the concern we have for rural communities and the wider life community of which we are a part. To claim concern for the diversified family farmer, and then fail to bury and eat local and organic food because it is too expensive or inconvenient, is hypocrisy. We must create demand for sustainability, even when the current market system makes this difficult.

CONCLUSION

Lasley and associates (1993) describe the challenge facing proponents of sustainable agriculture and sustainable communities:

> To present a convincing argument that sustainable agriculture will strengthen rural communities, it will be necessary to prove that sustainable farmers are stronger proponents of public good or community philosophy. . . . [Proponents ask] whether the adoption of sustainable agriculture will contribute to more local [quality] employment opportunities, or simply bring a shift in employment patterns. . . . A major weakness in the sustainable farming argument is how processing and marketing jobs will be captured for local community development . . . [proponents need to show] how decentralization of the food processing and marketing sector can be achieved to create jobs in local communities. A cleaner and more healthful environment, a more vibrant rural community, more local jobs and less dependence on foreign energy sources clearly are public benefits, but by themselves may be insufficient to persuade farmers to adopt sustainable practices (137–138).

The rural crisis is a symptom of the moral and spiritual crisis of America. Decision making is driven by finding the lowest price, the greatest short-term financial return, and the greatest increase in productivity. This emphasis on bigness and capital in rural development, though it enriches some, is destructive to the social and natural capital on which vital communities, and even the market, ultimately depend. Instead, efforts must be geared toward supporting rural revitalization through the strategies for sustainability and self-reliance discussed in this chapter.

Making the necessary policy shifts (e.g., taxing oil to include all externalities, basing land-use decisions on a recognition of the intrinsic value of natural and rural communities, and revising building codes to favor passive solar energy and indigenous building materials) goes against the grain of policy and planning that have been operative for decades.

Sustainability requires that we, in our personal life-style choices, political advocacy, and neighborly behavior, reduce our dependence on capital-

and energy intensive consumption, and refuse to use products and services that cause harm to others. Without embodying this sustainable way of living, we cannot escape deep complicity in an exploitative system. Making a difference for life is primarily a matter of producing and consuming in ways that revitalize the soil, preserve biodiversity, treat animals well, enhance local self-reliance, and create genuine options for the poor.

As we grope toward sustainability, we discover how entrenched in our culture the old paradigm is—we fear recession, don't want to pay more for organic food, hope to earn more and more money. Yet only by creating policies, plans, and consumer choices that consciously value ecological soundness, social justice, and humaneness—even if they sacrifice some short-term economic gain—can we create truly vital rural communities. Although such a change in the bottom line will prompt considerable resistance by the current agribusiness sector, it is necessary to ensure wholesome food for the future, as well as the moral fiber necessary for true prosperity.

REFERENCES

AERO. n.d. Information brochure. Helena, Montana.

Austin, R. C. 1994. "The Spiritual Crisis of Modern Agriculture." *Christian Social Action.* 10:4–11.

———. 1996. "Clinch Powell Sustainable Development Forum." *Towards Sustainable Livelihoods.* Rome, Italy: Society for International Development.

Barrett, G. W. 1990. "Role of Sustainable Agriculture in Rural Landscapes." In *Sustainable Agriculture Systems,* edited by C. A. Edwards. Ankeny, Iowa: Soil and Water Conservation Society.

Bird, A., G. Bultena, and J. Gardner. 1994. *Planting the Future: Developing an Agriculture That Sustains Land and Community.* Ames: Iowa State University Press.

Center for Maximum Potential Building Systems, Inc. 1988. *Mobilizing Human and Natural Resources in Rural Communities.* Informational brochure. Austin, Texas.

Daly, H. E., and J. B. Cobb. 1989. *For the Common Good.* Boston: Beacon Press.

DeMuth, S. 1995. *Sustainable Agriculture in Print: Current Periodicals.* Beltsville, Md.: National Agricultural Library.

Dobbs T. L., and J. D. Cole. 1992. "Potential Effects on Rural Economies of Conversion to Sustainable Farming Systems." *American Journal of Alternative Agriculture* 7(1, 2):70–80.

Edwards, C. A., ed. 1990. *Sustainable Agricultural Systems.* Ankeny, Iowa: Soil and Water Conservation Society.

Euston, A. 1995. "Community Sustainability: Agendas for Choice-Making and Action." Draft guide. World Bank, Third Annual Conference on Environmentally Sustainable Development.

Fritsch, A. J. 1989. *Communities at Risk—Environmental Dangers in Rural America.* Washington, D.C.: Renew America.

Gips, T., M. Adcock, R. Clugston, and E. Truong. 1993. *The Humane Consumer and Producer Guide: Buying and Producing Farm Animal Products for a Humane Sustain-*

able Agriculture. Washington D.C.: The Humane Society of the United States and The International Alliance for Sustainable Agriculture.

Gliessman, S. R. 1990. "Understanding the Basis of Sustainability for Agriculture in the Tropics: Experiences in Latin America." In *Sustainable Agricultural Systems*. edited by C. A. Edwards. Ankeny, Iowa: Soil and Water Conservation Society.

Kirschenmann, F. 1994. Testimony. The Sustainable Agriculture Scoping Group of the President's Council on Sustainable Development. Presidents Council on Sustainable Development, Washington, D.C.

Lasley, P., E. Hoiberg, and G. Bultena. 1993. "Is Sustainable Agriculture an Elixir for Rural Communities?" *American Journal of Alternative Agriculture* 8:133–139.

Lesh, D. R., and D. G. Lowrie. 1990. *Sustainable Development: A New Path for Progress*. Washington, D.C: The Global Tomorrow Coalition.

Preston, D. 1989. *Healthy Harvest III: A Directory of Sustainable Agriculture and Horticulture Organizations 1989–1990*. Washington, D.C.: Potomac Valley Press.

Production Credit Associations. 1983. *Agriculture 2000—A Look at the Future*. Columbus, Ohio: Battelle Press.

Sale, K. 1986. *Human Scale*. New York: Cowan, McCowan, and Gesgheyon.

Staten, J. 1987. *The Embattled Farmer*. Golden, Col.: Fulcrum.

Five

Exploration of a Framework for Evaluating Sustainable Development in a Rural Agricultural Context

Erik Davies and George Penfold

INTRODUCTION

Sustainable development, as presented in *Our Common Future* (World Commission on the Environment 1987), challenges classical approaches to understanding and evaluating development. These approaches have failed to adequately account for and protect the biophysical environment and/or make a meaningful change in the quality of life of most of the world's poor. The integration of human development with ecological objectives requires that socioeconomic development frameworks be replaced with those that recognize the ubiquitous role of the biophysical environment in our lives. This, in turn, has implications for the way we monitor change and evaluate data used to assess development in rural areas.

This chapter uses a definition of sustainable development based on one proposed by Rees (1989). Using this definition, four development frameworks are compared to assess the extent to which they address an ecological basis of sustainability. The lessons learned from these models are then integrated into a proposed sustainable development framework that may be used in the generation of indicators of sustainable development within a rural context. This framework is then applied to a case study of agricultural change in southern Ontario. The implications for analysis of sustainable rural development are discussed in the final section.

SUSTAINABLE DEVELOPMENT

Sustainability expands the concept of development by recognizing the ecological limits imposed on achieving a given set of development objectives (Davies 1990, 29). According to Rees (1989, 3), "Sustainable development is positive socioeconomic change that does not undermine the ecological and social systems upon which communities and social systems are dependent." Implementation requires integrated policy, planning, and social learning processes; its political viability depends on the full support of the people it affects through their governments, their social institutions, and their private activities (Rees 1989, 3).

What Rees has not adequately accounted for in his definition of sustainable development is the innate value of ecological systems, which goes beyond those values directly related to the support of human life. Thus, for the purpose of this paper, Rees's definition is modified to read as follows: "Sustainable development is positive socioeconomic change that does not undermine the ecological systems upon which humans and all living organisms are dependent, nor the social systems on which human communities are dependent." This definition reinforces the need for an integrated approach to planning and development that recognizes the dependency of human systems on the biophysical environment. A sustainable development framework should reflect this integration.

LIMITATIONS OF EXISTING PROCESS AND CONTENT DEVELOPMENT FRAMEWORKS

Development frameworks are implicitly or explicitly based on some conceptualization of reality and change. They include measurable elements of the development context that are used to characterize the type of change or development taking place over time. Some frameworks place importance on the process of change, attempting to illustrate the relationship between components. For example, Weitz (1986) designed a regional development framework that deals with time and space. This framework indicates, at a very general level, that ecological and social factors interact with macro-, micro-, and sectoral planning at a regional level.

A resource management framework by O'Riordan (1971) conceptualizes how economics, science and technology, philosophy, ethics, and aesthetics interact to affect resource management decisions.

A problem common to both of these frameworks is that, at a macrolevel, links between various sectors indicate general relationships rather than social, economic, and ecological interactions and therefore do not assist in microlevel management decisions. Often interactions between various sectors are ignored. For example, how the harvesting of natural resources would impact ecological systems is not apparent in Weitz's framework. The lack of descriptive elements for each sector also means that the interpretation of these frameworks is highly subjective.

More complex models of interactions tend to be sectoral, based on disciplines that more easily lend themselves to exact analysis (e.g., field ecology, financial models). Although these models may be useful subcomponents of a larger framework, they are not adequate for planning rural sustainable development because they explicitly emphasize one set of factors (e.g., economic) over others (e.g., social, ecological).

Content-oriented frameworks list elements of systems that can be measured to characterize the type of change taking place. For example, in the field of social impact assessment, Finsterbusch (1981) proposes a framework based on five components: economic, political, social, cultural, and environmental. Each component contains discrete elements that are well defined and measurable. There are two interrelated levels of analysis: (1) the household and community and (2) organizations and groups, and societal institutions and systems.

A second example of a content-oriented framework can be found in social indicators research, which has developed a social economic accounts system (SEAS) based on a comprehensive set of community goals (Fitzsimmons and Lavey 1975, 389).

Fifteen sectors were developed primarily from social and economic literature that derived from efforts to monitor large-scale structural change in the United States in the late 1960s and the early 1970s. They are useful in that they are conceptually distinctive aspects of community that are immediately identifiable by theoreticians, researchers, evaluators, and makers of public policy. Social indicators include education, health, economic base, social services, employment and income, recreation and leisure, welfare, housing and neighborhood, government operations and services, transportation, communications, law and justice, religious life, environment, and family life.

There are a number of problems related to these content-oriented frameworks. First, the conceptual separation of the (biophysical) environment as a discrete entity ignores the omnipresent role of ecological systems in our economy and in our social systems. Second, these frameworks do not consider the interaction between discrete components. Acknowledgment of these interactions is one of the premises of sustainability analysis. Third, they do not take into account the regulatory function of sociopolitical institutions in the process of development. Finally, social indicators such as leisure and neighborhood proposed by SEAS are culturally defined and conceptual in nature. They do not readily lend themselves to measurement or evaluation.

In summary, when biophysical systems are considered in development frameworks they are usually poorly defined and tend to be included as discrete components in the development equation with little or no interaction with other factors of development. Development frameworks that stress important qualitative aspects of human life often fail to account adequately for biophysical factors that are required in any monitoring framework for sustainable development. Moreover, social indicators often use a mix of quantitative and qualitative measures that attempt to approximate complex cultural relationships, some of which social sciences are still far from fully understanding.

Ideally, a development framework should reflect both process and content. However, existing frameworks place little emphasis on the interactions between systems (i.e., social and biophysical) or even discrete subcomponents within these systems. This is, no doubt, a reflection of our limited understanding of these interactions, the complexity and unpredictability of feedback loops, and the interdisciplinary complexity of the task—issues that continue to plague the development of general systems theories. Although it may be possible only in a heuristic sense, to indicate the full range of interactions possible within and between sectors and scales is particularly important in considering the role of ecological systems and how human activities affect their well-being.

PROPOSED RURAL DEVELOPMENT FRAMEWORK

Systems theorists have come closer to the mark in attempting to include both content and process. The concept of the "man-nature" system presented by Chadwick (1978) and the land management model presented by Odum (1975) both recognize the importance of the biophysical environment, but do not give the attention to social and economic systems that sustainability demands.

The rural development framework presented in Figure 5.1 attempts to build on these systems models based on a proposed definition of sustainable development: positive socioeconomic change that does not undermine the sociological systems on which all living organisms are dependent, nor the social systems on which human communities are dependent. This framework is applicable to all levels of planning, from household and communities to provincial and national levels.

In comparison to the development frameworks previously discussed, this framework expands the classic, static, sectoral role of the biophysical environment and recognizes the dynamic interactive relationship it has with other systems. It also explicitly recognizes the biophysical environment's role as a waste sink.

The biophysical environment encompasses all ecosystem components (e.g., flora, fauna, substrate) for which market values do not exist. It also includes the built environment and all natural resources and human activities. Conceptually, however, the latter aspects have been isolated from the biophysical environment in Figure 5.1 inasmuch as they are "managed" and have values placed on them in financial markets.

Natural resources are components of the biophysical environment that have a value in the marketplace. They include people, certain species of animals, commercial forests, and mineral resources, to name a few. The state of natural resources is well monitored both by government agencies

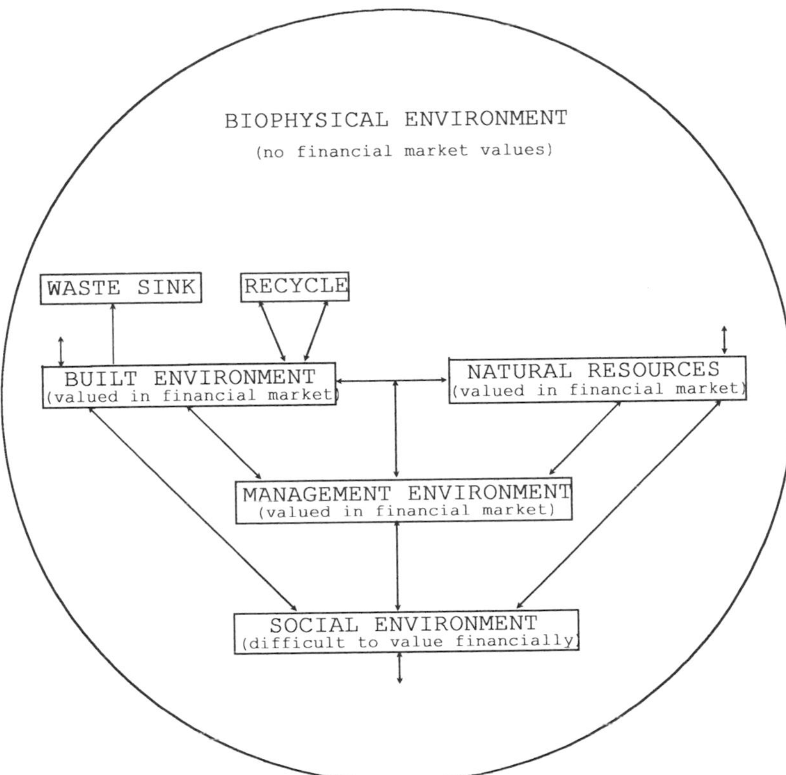

Figure 5.1 Proposed sustainable development framework.

(e.g., Ministry of Natural Resources in Ontario, Canada) and by private industries using those resources. Because natural resources are part of the biophysical environment, the interactive role they play in ecological systems is often completely overlooked. For instance, what function does a commercial forest have in supporting local wildlife species, processing carbon dioxide, and producing oxygen? There are no databases that can provide this type of information.

The built environment is also an integral part of the biophysical environment. As defined here, the built environment includes all buildings, transportation corridors, and consumption and production items manufactured from natural resources. Material used to manufacture these items is often returned to the environment as waste, yet sometimes it is recycled.

Although humans are included as part of the biophysical environment, the concept of sustainable development is still anthropocentric. Thus, social and management environments are acknowledged in Figure 5.1 as integral parts of the framework.

Qualitative aspects of human life, such as culture, religion, and the sense of well-being, are found and measured within the social environment. Values used to assess development and guide decision making are a product of the social environment and are, therefore, temporal and culturally defined. For this reason the social and management environments are integrally linked.

The management environment includes financial markets, policies, regulations, standards, and legislation (laws). These are used to regulate (1) the transformation of natural resources into components of the built environment, (2) processes occurring within both of these sectors, and (3) human interactions within the social environment. The management environment is a tool by which humans effect change to achieve development.

In summary, Figure 5.1 depicts the relationship between human production activities and the biophysical environment. It acknowledges the potential for the management of change through policies and law, and the intimate link between the values generated by society and the effect this has on how physical systems are managed.

AGRICULTURAL CHANGE IN SOUTHERN ONTARIO: APPLYING THE FRAMEWORK

From a planning and development perspective, it is essential that any sustainable development framework be developed to the point that it can be used to organize information. The purpose of this section is to explore some of the challenges involved in the application of the proposed framework in terms of generating information relevant to assessing the sustainability of rural development and change.

The Physical, social, natural resource, and environmental management aspects of change in the farming system in southern Ontario are used as the basis for this exploration. In addition, the potential limits of alternative agricultural systems as sustainable alternatives for agriculture are briefly discussed.

Farming in Southern Ontario

The Farming systems in Ontario vary regionally, depending on climate and soil quality. The soils and climate of the southern region support intensive cash crop production (corn, beans, wheat, and cereal grains). A few exceptional areas have microclimates and soils that support tobacco, fruit, and vegetable production. In the central and northern areas, where soil quality is generally poorer and microclimates are colder, livestock systems based on cereal grain and grass production predominate.

Continued economic restructuring of the farming system is a significant issue throughout Ontario. The southern part of the province also contains the major urban centers, and competition between agriculture and urban uses is a significant issue (Penfold et al. 1990).

Changes in Farms

Between 1971 and 1991, the area in census farms in Ontario declined from 15,963,045 acres to 13,470,653 acres.[1] This loss of approximately 2,500,000 acres provided the rationale for provincial and local land preservation policy. During this 20-year period the area of improved farmland declined, by 878,302 acres, to 9,551,950 acres. The actual reduction in productive land is approximately one third of the reduction in the area of farms. The difference of 1,614,090 acres is unimproved land, which may be assumed to have low production capability (see Table 5.1).

The area of land rented increased by 716,391 acres, and the total area rented in 1991 was equivalent to approximately 42% of the area of land in crops. The total number of census farms declined by 26,089 to 68,663. The volume of food produced increased by more than 30%. Between 1981 and 1986 the area of land converted to urban use in Ontario centers with populations of 25,000 or more was equivalent to approximately 5% of the loss of area in census farms. In southern Ontario, 94% of the land converted had class 1-3 CLI (Canadian Land Inventory) capability (Environment Canada 1989).

Since 1971 the census farm population has decreased from 391,713 to 226,755, and in 1991 represented 2.2% of the provincial population of 10,084,885. The farm population in 1991 represented 12.4% of the rural population of 1,831,045 (see Table 5.2).

Built Environment

Currently, information about agriculture is organized on the basis of the farm as an economic unit. Farms are described primarily according to their

Table 5.1 Change in Farms in Southern Ontario, 1971–1991

Year	*No. of Farms*	*Area in Farms*	*Improved Area*
1971	94,722	15,963,045 ac	10,430,252 ac
1981	82,448	14,923,280 ac	10,756,610 ac
1991	68,663	13,470,653 ac	9,551,950 ac

Source: Statistics Canada, Cat. 95-356.

Table 5.2 Population in Ontario, 1971–1991

Year	*Farm*	*Rural*	*Urban*	*Total*
1971	391,713	1,359,475	6,343,631	7,703,106
1981	279,806	1,578,705	7,046,402	8,625,107
1991	220,330	1,831,045	8,253,840	10,084,885

Source: Statistics Canada, Census of Population, various years.

functional production and financial characteristics. This information is useful in describing the dynamic of the industry and in developing economic policy. However, the biophysical basis of the proposed framework suggests that a unit of analysis, or of aggregation of information, should be an ecological unit such as a bioregion or subregion.

Farms are economic constructs that have social, environmental, and economic meanings, but how they function within a broader ecological context is not clear. For example, although the total area in farms has declined, the area in crops has increased by almost 600,000 acres and significant intensification in row crops, such as corn, has occurred. We have no specific information on the location or past or present uses of these additional croplands, or on the ecological significance of this change in use. Just as important from an ecological perspective is the change in use of the almost 2,500,000 acres no longer included in census farms. The focus on farm business units allows these unimproved lands to fall out of the information system. What is clear is that only a small proportion of this loss (less than 5% between 1981 and 1986) can be attributed to urban expansion. We have no information on past or present uses of the remaining 95% of land removed from census farms or its current ecological significance.

Although information is available on the location of current agricultural activities and the relative change in the context of administrative units, there are no ecological reference points, in either spatial or descriptive terms, to which to link these changes. A system of ecologically derived spatial units and descriptive measures are needed if sustainability is to be assessed. Given the relationship between farming, soil quality, and water quality, watershed areas may be an appropriate area for data collection and analysis.[2]

Natural Resources

The Canada Land Inventory provides a general inventory of the agricultural land resource and its capability for producing cereal crops. This inventory gives no information on the actual agricultural use of land or on the management systems employed. The inventory primarily considers the agricultural production potential of the soils. The fact that soils of similar type and/or agricultural capability may exist in different locations (e.g., near a watercourse or at higher elevation) and could support different flora and fauna or have different ecological consequences if used for agricultural purposes, is not considered as part of the agricultural inventory.

Current patterns of activities suggest that agricultural uses are clustered in those areas with better quality soils and climate. Intensive concentration of agricultural activity in these areas, as current land use policy encourages (Ontario Ministry of Agriculture and Food 1978), could result in significant ecological impact in those regions. Evidence of increased erosion and increased levels of chemical residue in soils and water, resulting from surplus or waste material from commercial production processes, are indicators of this trend.

Evaluative information on the ecological significance of resource units is needed. If land use change is to be understood from the perspective of sustainability, the ecological impacts resulting from different types of uses within resource units must be described and assessed.

Social Environment

Farming is an activity that not only uses natural resources to produce food for human and domestic livestock populations, but also generates a lifestyle and a context for addressing the needs and wants of those engaged in this endeavor. The quality of life of farmers and farm families is usually measured in terms of net farm income. The aggregate net income of Canadian farmers in 1971 was approximately the same as in 1989 in constant dollar terms, but was approximately half the net income generated in the mid-1970s. The proportion of government assistance increased between 1971 and 1986 from approximately 7% to approximately 67% of net farm income (*Statistics Canada Catalogue*, 21-603E, December 1990, quoted in Brinkman 1990). Thus, although the aggregate economic well-being of farm units appears not to have depreciated as compared with its status in 1971, dependency on government assistance has increased significantly and the economic well-being of the industry has declined significantly as compared with 1974.

From a human perspective, questions concerning well-being center on the family or the household as the unit of analysis. Almost all farm households generate additional off-farm income. What is not clear is the relationship between farm income and other sources of household income. Aggregation of household income to the community level is possible, but the boundaries of communities in the current information system (census subdivisions) do not fit with ecological boundaries, such as watersheds or subwatersheds.

In addition, nonincome aspects of well-being are not consistently documented. Other significant aspects of well-being, such as physiological and psychological health, safety, leisure, participation, skill/knowledge development, and life satisfaction, are not consistently monitored for farm families or for rural areas in general. The relationship between income and the ability of a family to access these aspects of well-being is important to sustainability, but is not well understood.

An analysis of household activity could also capture data on flows of material and energy and, therefore, information on waste returned to the ecosystem in various forms, ranging from solid waste to carbon dioxide and other emissions into the air.

The decline in the number of farms and farmers represents a significant change in the rural social environment and rural culture, which is currently not valued in the market system. Indicators of well-being that can capture these aspects of the social reality of agriculture are not well developed.

Management Environment

Increasing ownership of agricultural land by individuals whose main interest is not long-term farming appears to result in rental arrangements and management characteristics that depreciate the land resource and production potential (Culver and Seecharan 1986). Lands deemed not to be of production value are being excluded from the farming system. The increased economic value of these lands in the marketplace may result in their conversion to more intensive rural residential and recreational uses. The ecological consequences of this conversion resulting from new roads and septic tanks and the loss of wetlands and habitat are of concern, but current rural information systems focus on agriculture and do not adequately address other rural uses.

Emphasis on reorganization of local government to better manage the problems associated with growth (e.g., waste management) may assist in positive system changes (e.g., recycling). The boundaries and jurisdictions of these new management systems appear to have a service orientation, and whether they have appropriate jurisdiction to respond to ecological priorities adequately is not clear. The emphasis on watershed boundaries for water management may provide a more appropriate beginning point than a focus on service areas. Linking these two systems is a key management challenge.

SUSTAINABLE AGRICULTURE OPTIONS

Authors who have considered the conventional agricultural system and the changes it is experiencing (e.g., Gips 1988; Kneen 1989) suggest that the current system of agriculture is not sustainable. In their view, the current system is ecologically degrading, inhumane, unjust, and not economically viable. Organic, biodynamic, regenerative, and permaculture[3] methods are suggested as alternatives that are more compatible with principles of sustainability than are current farming systems.

Such interpretations appear to be based on several assumptions, which can be assessed within the proposed framework. The first assumption is that the appropriate unit of analysis is the farm. This suggests that a farm-level sustainable ecological microsystem can be created with little reference to the larger ecological unit on which it depends. Another assumption is that agriculture is a land use activity that can be considered independently of other uses of land and the characteristics of the ecosystem within which it occurs. However, agriculture is never the only land use or activity within an ecological unit. Transportation and communication systems, agriculturally related commerce and industry, and residential and recreational uses are common even in those few areas where agricultural production may be the dominant land use. Hence, the sustainability of the unit must consider other uses/activities in the system.

The implicit assumption appears to be that alternative agricultural systems are environmentally benign. Agriculture, even in organic form, represents a significant modification to the ecology of a region. There is also evidence that nitrogen from composted materials and animal waste is evident in groundwater and surface water. If the current system were to be converted to an organic system, it is not clear whether the present levels of production could be retained without additional land being added to the production system. From the perspective of an ecological unit, it is not clear that more land in organic production is better than less land in conventional agriculture, particularly if the residual land supports an ecologically important use, for example, forest cover or wetland. Indicators such as ratios of cropland and other developed rural lands to natural areas, combined with ecological sensitivity indicators, are needed as a basic starting point in evaluating sectoral changes within an ecosystem.

Although it may be true that alternative systems have characteristics that display relatively greater sustainability than current systems, they do not provide a comprehensive, normative concept of sustainability of human interaction with the environment. Comparing conventional and organic systems in a general way does not inform us as to the relative impact of either system in a given ecological unit. It may also be true that in many areas alternative systems would be more ecologically sustainable, but the difference between alternate and current systems may be rather small within some ecological units (e.g., moderate climate, coarse soils). Depending on the nature of a particular ecological unit, any form of agriculture may not be feasible without significant ecological consequences. If sustainability is to be implemented, the relative comparison might better be made with standards or indicators that are reflective of the ecology of the unit within which the activity occurs.

CONCLUSION

We explore the meaning of sustainable development within the context of a proposed monitoring framework. Rather than looking at classic economic constructs such as agriculture and industry, it is suggested that the elements of these activities be examined with reference to their interactions with the biophysical environment, as well as with each other. To do this, it is necessary to increase our knowledge of and information base on the state and functioning of biophysical, economic, and human systems. The utilization of this information then requires that the evaluation of rural development and change take place within ecologically based spatial units, such as watershed areas.

The framework suggested is exploratory, a starting place. Further efforts to define how sustainable development can be measured will force us to reexamine the values we use to assess socioeconomic change in relation to the biophysical environment. Until such a value shift occurs, a clear understanding of sustainable development or sustainable agriculture will be difficult to achieve. This shift could also result in a closer integration of rural and urban planning to produce a more holistic view of development and change.

NOTES

1. A census farm is defined as a farm that sells $50 or more worth of agricultural products, including crops, livestock, poultry, animal products, greenhouse or nursery products, mushroom, sod, honey, and maple syrup products (*Catalogue 95–629* Statistics Canada).
2. There are currently 17 subwatershed plans in various stages of development in Ontario. These pilot projects cover catchment areas in the range of 15,000 to 40,000 acres. Their purpose is to develop land allocation and management strategies to reduce the ecological impacts of development.
3. There are various definitions for these terms. The key common characteristic of these methods relevant to this discussion is their reliance on reduced commercial chemical and energy inputs, as compared with conventional agriculture.

REFERENCES

Brinkman, G. 1990. Course Notes 95–629: *Agricultural Land Use Policy*. Guelph:University of Guelph.

Chadwick, G. 1978. A Systems View of Planning: Towards a Theory of the Urban and Regional Planning Process. 2nd edition. New York: Permagon Press.

Culver, D., and R. Seecharan. 1986. "Factors That Influence the Adoption of Soil Conservation Strategies." *Canadian Farm Economics 20(2)*:9–13.

Davies, E. C. 1990. *Sustainable Development: The Case Study of Pond Aquaculture in South Sulawesi, Indonesia.* Master's thesis. University School of Rural Planning and Development, Guelph:University of Guelph.

Environment Canada. 1989. "Urbanization of Rural Land in Canada, 1981-86." *State of the Environment Fact Sheet,* No. 89–1.

Finsterbusch, K. 1981. "The Potential Role of Social Impact Assessments in Instituting Public Policies." In *Methodology of Social Impact Assessment. 2nd ed.,*edited by K. Finsterbusch and C. P. Wolf,. Pa: Hutchison Ross.

Fitzsimmons S. J., and W. G. Lavey. 1975. "Social Economic Accounts System: Toward a Comprehensive, Community Level Assessment Procedure." *Social Indicators Research vol. 2, 1976.*

Gips, T. 1988. "What is Sustainable Agriculture?" In *Global Perspectives on Agroecology and Sustainable Agricultural Systems. Proceedings of the Sixth International Scientific Conference of the International Federation of Organic Agriculture Movements.* Santa Cruz: Agroecology Program, University of California.

Goulet, D., 1977. *The Cruel Choice: A New Concept in the Theory of Development,* New York: Atheneum.

Kneen, B. 1989. *From Land to Mouth: Understanding the Food System.* Toronto: NC Press Limited.

Odum, E. P. 1975. *Ecology: The Link Between the Natural and the Social Sciences.* New York: Holt, Rinehart and Winston.

Ontario Ministry of Agriculture and Food. 1978. *Ontario Foodland Guidelines.* Toronto: Ontario Ministry of Agriculture and Food.

O'Riordan T. 1971. *Perspectives on Resource Managemen*t. London: Pion.

Penfold, G., Davies, E., Patry, M., Buysman, J. and J. Schnurr, 1990. *A Framework To Explore Development Trends and Issues in Rural Ontario.* Guelph: University School of Rural Planning and Development, University of Guelph.

Rees, W. 1989. "Sustainable Development: Myths and Realities."in *Environment and Economic Partners for the Future.* Conference proceedings, Winnipeg: Government of Manitoba.

Weitz R. 1986. *New Roads To Development.* New York: Greenwood Press.

World Commission on the Environment. 1987. *Our Common Future.* New York: Oxford University Press.

Six

Sustainable Community Development: A Systems Approach

Daniel D. Chiras and Julie Herman

INTRODUCTION

Throughout the world, intense interest is growing in a concept known as *sustainable development.* This spiraling interest permeates all levels of our society. At the international level, the World Bank and the United Nations Development Program have both embraced the concept. At the national level, the U.S. Agency for International Development has adopted it. Soon after taking office, President Clinton established the President's Council on Sustainable Development, designed to offer recommendations for achieving a sustainable future. At least 17 states have adopted some type of sustainable development mandate or legislation, and the local level, numerous communities throughout the United States have embarked on sustainable development initiatives.

Our experience with many such entities, especially rural communities in the West, suggests the need for a much deeper understanding of sustainable development. Without this foundation, many efforts will likely fall short of their full potential.

In an effort to help various entities, especially communities, understand sustainable development, our organization, the Sustainable Futures Society, has reviewed numerous books and articles on the subject.[1] From this work, we have assembled a working definition of sustainable community development (Chiras 1994a). In addition, in our review of this information and in discussions with theoreticians and practitioners of sustainable development, we have attempted to discern the underlying principles of sustainable development—its ideological foundation (Chiras 1994a; Chiras 1995). The definition and principles are presented in this chapter in conjunction with a new approach developed by the Sustainable Futures Society. This technique is designed to provide a means of operationalizing many of the principles of sustainability, thus assisting communities as they go about the difficult task of restructuring to create a humane, sustainable future.

WHAT IS SUSTAINABLE COMMUNITY DEVELOPMENT?

Countless definitions of sustainable development have been offered by activists, scientists, policymakers, and others over the past decade. One of the most widely quoted definitions was published by the World Commission on Environment and Development (WCED), established by the United Nations to examine environment and development and suggest ways to make the two much more compatible. In its seminal work, *Our Common Future*, the WCED (1987) defined sustainable development as "development that meets the needs of the present without compromising the ability of future generations to meet their needs."

We find the dictionary a useful source on the subject, too. According to *Webster's New World Dictionary*, *to sustain* means "to keep in existence, to maintain, and endure." *Development* refers to "a step or a stage in advancement or improvement." Sustainable community development, then, is a way of improving or advancing communities in ways that can be maintained over the long run.

Sustainable development offers an appealing alternative to our current unsustainable path. Although something of a newcomer on the political scene, sustainable development is not a new idea at all. In fact, 85 years ago President Theodore Roosevelt alluded to the concept in his annual message to Congress, a speech that has since become known as the State of the Union address. Roosevelt noted, "To waste our natural resources, to skin and exhaust the land instead of using it [so] as to increase its usefulness, will . . ." undermine the prospects "of our children" (cited in Reilly 1992). He went on to talk about the "the very prosperity which we ought by right to hand down to" future generations "amplified and developed." In other words, we ought to meet our needs in ways that ensure an equally rich natural resource base for future generations.

Nor was Roosevelt the first to champion this philosophy. Native American cultures and indigenous peoples the world over have espoused a similar view and, more important, lived according to sustainable principles thousands of years before Roosevelt's time.

Robert Costanza (1992) writes that sustainable development is a means of creating a "relationship between dynamic human economic systems and larger dynamic . . . ecological systems in which: (1) human life can continue indefinitely; (2) human individuals can flourish; (3) human cultures can develop; but in which (4) the effects of human activities remain within bounds, so as not to destroy the diversity, complexity, and function of the ecological support system." In truth, though, sustainable development is not primarily an economic challenge, as suggested by this definition. Although economics and economic systems are central to the pursuit of sustainability, sustainable development is much broader. That is, it concerns itself with finding sustainable ways of meeting a wide range of human needs such as food, shelter, water, and clothing. It also seeks ways to provide health care, education, employment, political freedom, a guarantee of human rights, freedom from violence, and much more—all while protecting, even enhancing, the diversity, complexity, and function of the ecological support system. In essence, it is concerned with creating a socioeconomic system—an economy and a way of life—that is ecologically sustainable.

It is important to understand that sustainable development is not, as some industry skeptics think, a means of promoting an environmental agenda. Nor does it advance purely economic or social goals. Rather, it is a genuine attempt to advance all three simultaneously, for all three are essential for human cultures to prosper and develop.

New human-environment relationships will not evolve from simply considering environmental concerns in economic and social decisions, or vice versa, as is sometimes asserted. Nor will they be achieved by redoubling old environment protection methods that often pitted environmental concerns against economics, placing society and environment in a precarious and largely unnecessary conflict. Rather, sustainable development requires an integration of social, economic, and environmental goals.

This integrated approach differs markedly from traditional government policy-making in which the social, economic, and environmental needs are often considered separately. In all too many instances, environmental policy has stipulated actions that have had adverse economic and social impacts. But, to be fair, economic policy has also traditionally ignored legitimate environmental needs. In contrast, sustainable development calls for solutions that make sense from all three perspectives concurrently.

We hasten to note that sustainable solutions are not merely compromises that partially satisfy human needs for a safe, healthy environment and social and economic well-being. Rather, sustainable solutions seek to achieve simultaneously social, economic, and environmental conditions vital to creating an enduring human presence. Several renewable energy technologies, such as wind generators and solar thermal electricity, exemplify the solutions that fit within this realm. As several studies have shown, modern wind farms produce electricity at a fraction of the cost of nuclear power plants and at the same cost as coal-fired power plants (Renner 1992). However, they employ far more people than either of the traditional technologies, thus providing great social benefit at a fraction of the environmental impact. In short, they create solutions that benefit people, the economy, and the environment.

Sustainable community development differs from traditional community development in several important ways. First, as noted earlier, it is an integrated approach. That is, it seeks strategies that satisfy the triple bottom line: social, economic, and environmental. Second, it seeks policies and actions that create long-term solutions. Although stopgap measures are often required to address immediate problems, sustainable solutions are long-term and preventive in nature. Third, sustainable community development seeks local solutions that solve local, regional, and even global issues. Although most people's interests are local or regional, local policies and practices have global effects. Local decisions that lead to loss of forests, for instance, will decrease the planet's ability to absorb carbon dioxide, further aggravating global climate change. To develop in a truly sustainable fashion, communities must be mindful of global repercussions, however small or remote they may seem. Buckminster Fuller said it best when he advised us to "think globally, act locally."

PRINCIPLES OF SUSTAINABLE DEVELOPMENT

The concept of sustainable development is built on a foundation of principles from five basic and sometimes overlapping realms: ecology, sociology, ethics, politics, and economics (Chiras 1995). This section summarizes twelve of these principles.

Ecological Principles of Sustainable Development

Principle 1: Dependence

Perhaps one of the most important of all principles is environmental dependence. This refers to a simple but widely unappreciated fact: Humans depend on ecosystems for a wide array of goods and services. Forests provide timber, but also protect watersheds and help maintain water quality. In addition, they provide oxygen, which is essential to life.

Global ecosystems also contribute significantly to the economy. Robert Goodland (1992) put it best when he wrote that the environment is the source of all material inputs of the economic system and is the sink of all our wastes. In essence, then, the earth is the biological infrastructure—or infra-infrastructure—of modern society. The environment makes possible all of the physical infrastructure: buildings, dams, ports, highways, and so on.

Recognizing our dependence on the earth renders acts of protection more than an exercise in enhancing the aesthetics of our world. It makes environmental protection and environmentally sustainable development a form of self-protection—an insurance policy for long-term human survival and economic prosperity. Planet care is truly the ultimate form of self-care.

Principle 2: Limits

The earth and its ecosystems provide a wide assortment of renewable and nonrenewable resources. Both types of resources have limits. There are, for example, limits to the amount of oil, a nonrenewable resource, that can be extracted from the earth's crust. And there are limits to how many fish, a renewable resource, that can be harvested without depleting a fishery.

Numerous technologies permit humankind to stretch the limits of resource supplies—for example, to extract more oil from a deposit or to catch more fish. Even so, there are very real upper limits to the resources the planet can supply. A growing body of evidence clearly shows that humankind has transcended certain biological and physical limits and stands perilously close to others (Meadows et al. 1992). Nowhere is this more evident than in rural communities. In the Western United States for example, many towns have encountered limits to the amount of water that can be withdrawn from local aquifers and rivers, limits to the amount of pollution that can be absorbed by the local airshed or the number of cattle that can be grazed on range land, even limits to the number of campers that can visit wilderness areas without causing irreparable damage.

In sum, then, much of the writing on sustainable development recognizes the presence, as well as the proximity, of limits and seeks ways to live within those limits, the third principle of sustainable development.

Principle 3: Living Within Limits

In the natural world, populations are sustained because they live within the carrying capacity of their habitats. Carrying capacity is defined as the number of organisms an environment can support indefinitely and is determined by two factors: resource supplies and the capacity of the environment to absorb, assimilate, and/or detoxify wastes. Proponents of sustainable development call for policies and practices that temper human activities so as to live within the limits of the environment. That is, they promote strategies for meeting human needs while honoring the biophysical confines—for example, limits to the amount of arable land, fish in the sea, and oil in the earth's crust. On a community level, development is best achieved by guiding it in accordance with local carrying capacity.

Principle 4: Interdependence

Human society has grown enormously, both in size and in technological power, in the past 200 years, so much so that the fate of the biosphere, the living skin of the planet Earth, is in our hands. We humans have become custodians of an entire planet. Nowhere is this more evident than in our own communities, where the effects of local actions are most visible. Therefore, although humans depend on the earth, the fate of the environment also depends on us. What we do and what we fail to do will have a profound effect on the biosphere.

Clearly, a precarious interdependence exists, for what befalls the earth befalls humanity. Soil erosion, water pollution, and groundwater overdraft in rural communities bode poorly for the environment, people, and rural economies.

Social/Ethical Principles of Sustainable Development

In our review of numerous books, articles, and speeches on sustainable development, we encountered three key social or ethical principles of great importance: intergenerational equity, intragenerational equity, and ecological justice.

Principle 5: Intergenerational Equity

The term *sustainable development* embodies a powerful idea: the notion of **intergenerational equity** or "fairness to future generations." Intergenerational equity refers to the responsibility of each generation to ensure that the next one receives undiminished natural resources and economic opportunity. The doctrine of intergenerational equity maintains that all generations hold the earth in common (Weiss 1990.). In other words, all generations are stakeholders in the planet's well-being. However, intergenerational equity maintains that all generations are endowed with certain rights and obligations. For example, whereas members of the present generation have the right to enjoy, even profit from, the earth, they are also obligated to bequeath the earth to their descendants in as good or better condition than it was received.

Principle 6: Intragenerational Equity

In the international arena, debates over global environmental protection have drawn attention to another important equity principle, notably the notion of *intra*generational equity. The doctrine of intragenerational equity calls on us to act in ways that satisfy our needs while safeguarding the welfare of all others who are alive today (Chiras 1995). Our actions are tempered by a concern for impacts on other people—our neighbors, our fellow citizens, and global citizens.

Intragenerational equity is based on a realization that all members of the world community share the same air and water and that abuse of these resources is often felt far from the site of initial impact. Soil erosion on a Kansas farm, for instance, may cause siltation in a downstream reservoir in Missouri and reduce its life span. It may also increase water treatment costs and impair fish life in the river. Impacts of human actions can even extend globally. Ozone depletion caused by the release of CFCs from refrigerators in a rural community in Canada, for instance, may cause skin cancer in Andean peasants who have never seen a refrigerator.

The lines of impact are reciprocal. That is, the developed nations can be profoundly distressed by resource abuse in the developing nations. Deforestation of tropical rain forests in Africa to accommodate population growth, for example, may have profound impact on the climate of European nations by altering the movement of moisture-laden air masses that affect precipitation patterns often hundreds of miles northward.

In sustainable community development, intragenerational equity compels us to consider the impacts of our actions on our planetary neighbors. In our experience in the western United States at least, intragenerational equity is becoming a potent driving force in rural sustainable development, encouraging regional planning to avoid problems that occur when communities develop without regard to impacts beyond their boundaries.

Principle 7: Ecological Justice

A small but growing number of proponents of sustainable development support yet another ethical principle, the doctrine of ecological justice (Regenstein 1991). One of the most controversial of all principles, ecological justice maintains that the earth is the rightful property of all species, not just humans. According to this view, humans have an obligation not only to other people, but to all living things, present and future. Our challenge, then, is to find ways to meet human needs while protecting and preserving biodiversity.

Political Principles of Sustainable Development

Principle 8: Participation and Cooperation

A humane, sustainable future will require more than new laws and regulations designed to promote the principles, policies, and practices of sustainability. It will require participation by individuals in all sectors of society. It will require actions by businesses, governments, not-for-profit organizations, and citizens.

Participation is especially important in rural communities. Fortunately, stakeholders are often easier to identify in rural communities than in urban communities. Friendships, business relationships, and an affinity (even love) for a particular place often unite communities and permit them to forge progressive plans.

Participation, essential at all stages of community development, from visioning to implementation of sustainable policies, not only taps into local passions and expertise, but fosters ownership. Ultimately, what is needed is a top-down and bottom-up approach, to ensure that policymakers, citizens, business owners, and all others take ownership of the process and help produce desirable outcomes.

Participation itself is not sufficient, however. We also need cooperation, people working together toward common goals. Cooperation is needed between not-for-profit organizations, government, citizens, and business—sectors of our society that are often at odds with one another. Regional cooperation is increasingly recognized as a vital need in the rural areas where communities often share airsheds, watersheds, labor forces, and other resources.

Principle 9: New Leadership

Cooperation of the magnitude required to build a sustainable society will not occur without a new kind of leadership: men and women who can help their communities create a vision of a sustainable future, bridge the differences that may exist between citizens, and forge shared concerns into strong working relationships (Gardner 1990). In short, we need leaders with the capacity to inspire cooperation and participation, rather than seed bitterness, distrust, and conflict—the divisive fuel of American politics.

The transition to sustainability also clearly hinges on the emergence of leaders who understand the big picture, as well as the problems, and are willing to search for lasting solutions rather than continue the ineffective patchwork that only passes problems on to future generations. Also needed are leaders who can show individuals that their contributions do count. In the words of Robert Gilman (1990), we need leaders that show that individuals' drops can actually fill up the bucket. In short, we need leaders who can empower citizens to take action and sustain the vision.

Principle 10: Addressing the Root Causes

One emerging realization in business and government is that many policies have focused on solutions that address the symptoms of environmental problems while overlooking root causes. Thus, environmental solutions enacted over the past two to three decades have been palliative in nature (Chiras 1992).

Treating symptoms is something of a human trait, not just in environmental protection, but in virtually all areas of human endeavor—from medicine to crime determent, from transportation policy to educational reform. In the arena of environmental protection, the examples of our shortsighted approach are many: smokestack scrubbers, catalytic converters on automobiles, hazardous waste incinerators, sewage treatment plants, and so on (Chiras 1993). Over the years, it has become painfully clear to many observers that these and other end-of-pipe measures are entirely inadequate, because they address pollution and hazardous waste too late—after they are produced. A pollutant attacked at the point of origin can be eliminated, but after it is produced it is too late. Once a pollutant has been generated, it becomes a social, economic, and environmental liability.

Fortunately, a growing number of individuals are beginning to realize that lasting solutions require policies and practices that confront root causes. Even prominent business leaders recognize the importance of root-level solutions, such as energy efficiency as means to comply with, even exceed, government environmental performance standards (Schmidheiny 1992). And many recognize that such an approach is beneficial to the bottom line (Postel 1990).

Addressing the root causes is essential to sustainable development at all levels of our society. It is needed not only to solve environmental problems, but also to solve the social and economic ills that plague communities. The last section of this chapter explains one approach to addressing root causes and operationalizing many principles of sustainability.

Economic Principles of Sustainable Development

Principle 11: Meeting the Needs of People

To be sustainable, development efforts in communities are most successful in the long run if they focus first and foremost on real human needs and seek ways to meet them, inexpensively and equitably, while protecting the environment. A needs-based approach allows development to become a more directed and purposeful process, one that is likely to result in qualitative improvements in people's lives.

Unfortunately, much development is primarily economic, and rather blind at that. It assumes that creating economic opportunities—such as more jobs—is all that is needed for a community to prosper. But heedless economic development strategies can backfire. When new growth, for instance, fails to pay its way, it becomes a strain on the local economy. Many communities become trapped on an economic-growth treadmill, constantly and blindly pursuing growth that raises the cost of infrastructure in excess of revenues. Economic development that destroys the amenities of a place—the clean air, uninterrupted view, and social relations—contributes to a spiraling decline. Rural communities are advised to look carefully at their social, economic, and environmental goals and to devise strategies to achieve them.

Principle 12: Greater Self-Reliance

A small contingency within the sustainable development movement promotes an idea that is as intriguing as it is controversial: the notion of increasing self-reliance to promote economic stability in communities, states, and nations (Power 1992). Supporters of the principle of self-reliance contend that economic development policies and practices that promote greater self-reliance in the production of goods and services promotes more socially and economically sustainable communities (Chiras 1992, 1995).

Contradicting the widely held belief in the preeminence of trade as a means of economic salvation, self-reliance depends in large part on tapping into local knowledge, skills, and resources to create more self-contained and potentially sustainable economies and societies. Despite the nearly unquestioned acceptance of the merits of a global marketplace, the future of the developed nations and their many cities and towns and rural communities may lie in quite the opposite direction: rather than globalization, more local or regional self-reliance—that is, nations, states, and communities providing a broader range of goods and services for local or regional consumption.

It is important to point out that greater self-reliance does not mean total self-reliance. Proponents of this view are not promoting isolationism. Nor are they advising communities to aspire to complete self-sufficiency. Rather, they argue that community economic stability, with its many social and environmental benefits, would very likely be furthered if local and regional economies were more self-sufficient than they are today. In short, this strategy promotes social and economic sustainability. If we think of the benefits a person gains by reducing home energy consumption by 50% through cost-effective measures such as insulation, we can imagine the social and economic benefits of community-wide efforts of similar nature that would accrue to an entire community.

At the very least, increasing self-reliance may mean becoming much more efficient in the use of energy and other resources. Optimizing resources can save residents considerable sums of money and can significantly strengthen the economic base by ensuring that dollars circulate more often within the community (Lovins 1990).

Increasing self-reliance may also require steps to reduce imports to a region, a strategy known as import substitution. Reducing the importation of food, building materials, and consumer goods by supplying needs locally can result in a huge economic boost to local and regional economies. Local/regional production for local/regional consumption puts people to work in their own communities, providing goods and services needed by neighbors. Like efficiency measures, import substitution enhances the circulation of money within the community or region. By becoming more self-reliant, local economies can buffer themselves against the vagaries of international politics and economics. Increasing self-reliance reduces a community's social and economic vulnerability.

OPERATING PRINCIPLES OF SUSTAINABLE DEVELOPMENT

The principles of sustainable development outlined in the previous section form the ideological foundation of sustainable development. Consequently, we refer to them as the directive principles of sustainable development. But ideology is not enough to create a sustainable future. We need action.

In the remaining pages of this chapter, we offer some suggestions for operationalizing many of these principles. To understand this method, one must first answer a larger question: Why are our communities unsustainable?

Human society is on an unsustainable course because it pollutes in excess of the planet's ability to absorb and detoxify waste and it exceeds the limits of resource supply. In short, we are unsustainable because we are exceeding the carrying capacity of the planet. The problem is one of design—the way our world is set up (Wann 1996). It is the result of unsustainable systems of transportation, energy supply, waste management, industry, and housing, among others, that were fashioned in an era of abundance and ignorance of limits and environmental impact (Chiras 1994b). Put another way, the reason for the present unsustainable state of affairs is that human systems—the physical infrastructure that makes up our communities, states, and nations—deplete resources and create levels of pollution that exceed local, regional, and even global capacity to absorb and render harmless that pollution. By depleting and polluting, ill-conceived human systems are eroding the planet's ability to support life.

The root causes of the present dilemma are that we are growing out of control; are inefficient and profligate in our demands for resources; recycle little of our waste; rely primarily on nonrenewable resources, especially fuels; tend to deplete natural systems and then move on without restoring them (Chiras 1992).

In cities and towns across the nation, signs of these actions are many. Intolerable levels of air pollution characterize many communities. Even if it does not present a local threat, pollution from rural communities contributes to global pollution levels that may be drastically changing the planet's climate, a global impact resulting in very severe local impacts. In Colorado, our home state, global warming could result in a drastic decrease in snowfall, a consequence that would be devastating to rural ski towns. In local communities, overconsumption results in severe resource depletion, such as deforestation, in other parts of the world. All around us are signs that humans are living beyond the carrying capacity of the environment. How can our communities be made to be sustainable?

The literature on this subject points to several key *operating* principles of sustainable development (Chiras 1992, 1993). Distilled from the advice of hundreds of books and articles, the message is clear. First, to be sustainable, humans must stabilize population. We cannot continue to grow indefinitely. Second, we must better manage *how* we grow. In other words, we cannot spread willy-nilly across the landscape, usurping prime agricultural land needed to feed people and provide other valuable services such as recreation. Third, we must use all resources, from fossil fuels to building materials to drinking water, with much greater efficiency. There are too many of us and too many demands on resources to be wasteful. Resource efficiency not only provides a great economic benefit to businesses and communities, it helps ensure adequate supplies for future generations. Fourth, we must turn to clean, renewable energy alternatives. The use of solar energy and wind may be the way of the future, for these provide the energy we need at a fraction of the current environmental impact. New technologies are producing energy from these sources at the same cost exacted by conventional and less sustainable sources, such as coal. Fifth, we must recycle everything we can. Waste is intolerable in a world of limited resources, and recycling can make good economic sense. It also helps to ensure a supply of resources for future generations. Not only must we recycle, we must manufacture a large portion of our goods with recycled materials. Sixth, we must restore natural systems. We cannot continue to abuse the land and move on as we have since the days of our pioneer ancestors. To support people and the other living things that share this planet with us, we must replant forests, rebuild wetlands, and restore grasslands and pastures. Seventh, we must manage resources sustainably. That is, we must do a much better job of managing forests, grasslands, and other natural systems. Short-term profit at the expense of long-term productivity only leads us farther from the path of sustainability.

Interestingly, these operating principles address many of the root causes of environmental unsustainability. By focusing on such principles, then, we operationalize an essential directive principle: the need to focus on root causes. The operational principles also help put into effect many directive principles. For example, using resources more efficiently is a clear recognition of our dependence on natural systems and the existence of limits (principles 1 and 2). It also helps us live within limits (principle 3). Wise use of resources reduces the impact of our actions and thus helps to ensure inter- and intragenerational equity (principles 5 and 6) and ecological justice (principle 7). It also promotes greater self-reliance (principle 12).

Recognizing the interplay of the directive and operating principles we have devised a simple, yet effective, method of applying the operating principles to basic human systems (Chiras 1993). This technique was conceived as a means of helping citizens, business owners, and government officials focus their efforts, given limited time and budgets. In our experience, the initiative required to launch such an effort and make it a success may depend on new leadership (principle 9) emerging from town council members, town planners, mayors, or citizens. To be successful, it also requires steps to encourage participation and cooperation (principle 8). As noted, this method focuses on root causes (principle 10) and ways to meet the needs of the community for a clean environment and a healthy economy (principle 11).

In summary, the operating principles and the process of sustainable community development help to operationalize the directive principles.

CREATING SUSTAINABLE COMMUNITIES: A SYSTEMS APPROACH

How do communities develop sustainably? What policies and actions are required? Which policies should they tackle first, given temporal and financial constraints?

Fortunately, there is a glut of information on sustainable development for communities to sift through. Dozens of books outline strategies. In our review of various programs and books on the subject, we have found that there is no comprehensive approach to sustainable community development. Moreover, most of what is available focuses on process. In our experience in the western United States, the main thrust of the sustainable community development is to develop action plans through a collaborative community-wide visioning process. This, of course, operationalizes principle 8, which calls for participation and cooperation. However, in our view, outcomes based on community visioning are often somewhat myopic and piecemeal. As a result, most cities, towns, states, and nations end up addressing symptoms of problems, rather than the underlying root causes. In short, they do not often do much to advance sustainability.

Attacking symptoms may lessen immediate pressures, giving an impression that something has been accomplished. However, stopgap solutions fail to create lasting solutions.

To help promote a more comprehensive and systematic rethinking of the nation's communities, we at the Sustainable Futures Society have created a strategy for rural communities that focuses on root causes of our communities' most pressing problems and inspires cooperation and participation.

Restructuring Human Systems for Sustainability

Our approach encourages communities to focus their efforts on ways to restructure systems, such as energy, waste management, transportation, and housing. Although the challenge of creating a sustainable society involves much more than simply restructuring systems, it is our contention that one of the most important tasks a community can undertake is a systematic rethinking and restructuring of its basic systems.

Figure 6.1 illustrates a community's physical infrastructure. For the sake of convenience, we have categorized human systems as either primary or secondary.. The primary systems are so named because they supply energy and materials required to support a community—its people, schools, hospitals, businesses, and government. The primary systems also

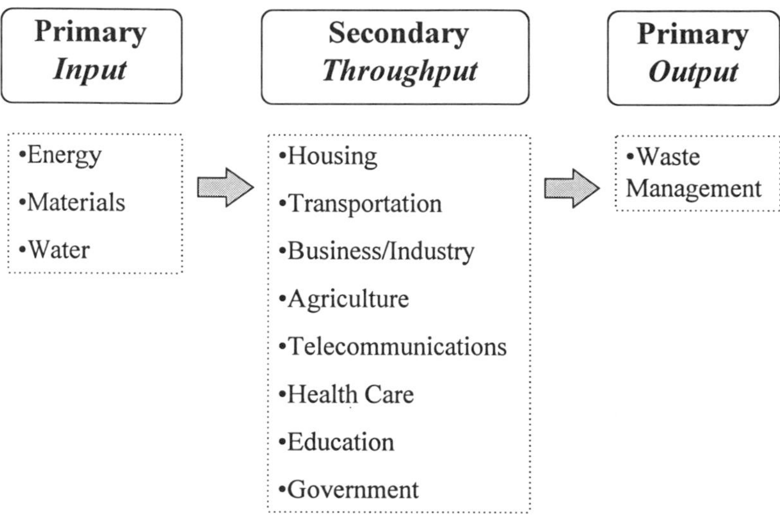

Figure 6.1 Human systems.

include a waste management system consisting of recycling and composting facilities, landfills, and incinerators.

Secondary systems are those supported by the primary systems. They are systems that resources flow to and from, including agriculture, industry, transportation, housing, and the like.

Supporting this network of interconnected systems of infrastructure is the natural environment, which provides the food we eat, the energy we consume, the air we breathe, the water we drink, the materials we need to support our economy, and much more. One of the main challenges of creating an environmentally sustainable future is to make human systems, our infrastructure, compatible with the biological infrastructure—fitting human civilization comfortably into the economy of nature.

The systems that make up our communities are highly interconnected and complex. For example, the energy system is a vital component of virtually all other systems, making it difficult to tease it out of the rest.

Another critical factor is that "ownership" and control of components of the various systems is often mixed. For example, a transportation system may have components that fall within the control of the public sector, that is, within the ambit of various government entities. In a rural community, highways may be owned and maintained by county, state, and federal entities, and some roads may be privately owned. Other components of a transportation system (e.g., the vehicles that travel on it) are mostly privately owned and operated.

An offshoot of multiple-level ownership is that there are many stakeholders. The cooperation and participation of these individuals and entities is essential for sustainable development. A further implication is that a community must recognize where its authority lies and tailor programs within its sometimes narrow realm of influence. Wherever possible, though, it can enlist the aid of other players (e.g., county or state government) to play a supporting role in—or, at the very least, to not impair—local initiatives for sustainable community development.

Communities consist of more than transportation systems, energy systems, and the like. There is also a social system—composed of various government entities, social organizations and clubs, not-for-profit organizations, schools, and the like—that is a vital part of each community. The social system is part of the glue that binds a community together, as is the economy (the flow of dollars and the economic relationships that develop in a community). It also provides an avenue for change.

Costanza and Daly (1992) identify three types of capital: natural capital (the environment), human-made capital (infrastructure and the like), and social capital (knowledge, skills, and social infrastructure). All three, they contend, are essential components of sustainable development and must be part of any sustainable development strategy. In this chapter, we deal primarily with the physical infrastructure, knowing full well that changes in the infrastructure will require social and political changes. Ultimately, our goal is to encourage changes in man-made capital to protect natural capital. The avenue for change lie in our social capital.

A Systems Approach to Sustainable Community Development

Understanding the principles of sustainable development—the ideological foundation and the key operating principles outlined in this chapter—is an essential first step in achieving sustainable communities. The early hours of our workshops concentrate on discussions of these aspects. But how can these principles be brought to bear on the task at hand?

Figure 6.2 is a matrix that we use to help workshop participants and communities organize the task of restructuring community infrastructure for sustainability. As shown, the vertical column of each matrix lists the systems (e.g., energy—for simplicity, only a few are included). The top row lists the main operating principles of sustainability (such as efficiency) that apply most directly to human systems. Related principles are combined to reduce the number of columns in the matrix.

Each box in the matrix represents the policies and practices that operationalize the principles of sustainability listed in the top row and described in the preceding sections. We invite practitioners to customize this chart, to include other essential operating principles they deem necessary. However,

Principle of Sustainable Development ⇨ Systems: ⇩	Conservation (efficiency and frugality)	Recycling and Composting	Renewable Resource Use (especially energy)	Habitat Protection, Restoration, and Sustainable Management	Growth Management (controlling growth patterns and stabilizing growth)
Transportation (surface, air, and water-based systems)	*Fill in each box with ideas for policies and actions that implement the principle in each system.*				
Housing and Other Buildings (i.e., construction systems)		*Note that some principles may not apply to certain system.*			
Agriculture, Food Processing, and Distribution (i.e., food production system)			*Some policies and actions may overlap–i.e., have effects on more than one system.*		
Business (commercial and industrial)				*Some systems should be considered together.*	

Figure 6.2 Sustainable development: A systems-based approach.

for this work to be most effective, we encourage communities to include operating principles rather than the less-tangible directive principles.

In our workshops with citizens, business owners, government officials, and representatives from not-for-profit organizations, we have found that this matrix stimulates an outpouring of ideas, often very creative and practical, on how to approach their communities' concerns and effect long-term vision.

We also encourage participants to suspend critical thinking when they first begin to work with the matrix. That is, the early ideas on policies and practices to operationalize the principles of sustainable development should be considered part of brainstorming and not subject to criticism or prioritization. After a full set of ideas has been generated, however, we encourage participants to choose the strategies that are most interesting, desirable, affordable, and cost-effective to the community.

Encouraging all stakeholders to join in brainstorming sessions and priority-setting efforts greatly increases community participation, cooperation, and ownership, all important components of sustainability. Because these subjects have received so much attention in books, articles, and practice of sustainable community development, we will not belabor the point.

After community members have had a chance to fill in the matrix, either collectively or in various subcommittees, and have selected the policies and actions they would like to tackle, the boxes in the matrix can be expanded to include a variety of additional considerations, such as those listed in Table 6.1. This work may include a simple step to determine whether policies and actions further the community's goals and visions, as articulated in meetings, as well as resources needed to effect the desired change (e.g., key players in the community). It might include special educational programs through schools, businesses, and government, thus engaging the social system. It might include the establishment of indicators needed to plot progress in achieving the community's goals. For example, steps to measure air quality or the number of water-efficient shower heads installed each year will help communities to monitor their progress toward sustainability and to adjust programs accordingly. The matrix might also include obstacles to the tasks prescribed by the group and ways to address them. A time line for completing various tasks is also helpful. As an example, consider the box in the matrix that includes policies and actions needed to apply the conservation principle (efficiency and frugality) to housing systems. One general goal of a community might be to cre-

Table 6.1 Suggested Additional topics for Community Consideration and Discussion (Figure 6.2)

1. Policies
2. Actions
3. Resources—information, people, and materials
4. Obstacles
5. Time lines
6. Work assignments
7. Capacity-building needs
8. Check on community visions and goals
9. Systems overlap
10. Check on directive principles

ate an affordable housing base within 50 years. Energy efficiency measures can help by drastically reducing operating costs. Making houses more energy efficient might entail steps to retrofit existing homes with additional insulation, weather stripping, and energy-efficient lighting. To make this happen, changes in building codes may be needed. Or communities might embark on educational programs to promote energy efficiency by home owners and builders alike. The time line would list all of the desired actions, when they will be completed, and who will complete them. The list of resources might include builders who are attuned to green building practices, local business people who supply building materials, and citizens and not-for-profit organizations with expertise in this area. Obstacles might include misconceptions about the cost of retrofitting or a lack of funds on the part of homeowners to finance efficiency measures.

The systems approach was devised to streamline sustainable community development. It offers communities a framework to address complicated issues in a systematic fashion. In essence, it makes a very difficult task much more manageable.

The reader will note, however, that systems overlap. That is, some systems are parts of others. Energy, for instance, is essential to transportation, industry, housing, and agriculture. Moreover, decisions about one system often affect other systems. Thus, certain systems are best addressed in conjunction with others. For example, creating a sustainable transportation system cannot be achieved without discussing housing (at least the placement of housing subdivisions) and business development (notably, the location of business development). To streamline the process, participants are encouraged to identify systems that require integrated policies. Participants in this process are also encouraged to identify policies and actions in one system that may adversely affect others.

Readers may have noted that our approach omits some familiar issues, for example, land use, open space preservation, and air quality. The reason for this omission is that air quality and loss of open space are symptoms of a deeper problem: our reliance on unsustainable systems. To create lasting solutions, air quality is best addressed by rethinking and redesigning transportation, housing, and commerce. Encouraging alternative forms of transportation, energy-efficient housing, energy efficiency in local businesses, the use of renewable energy resources, more compact development, and other sustainable measures can actually solve air quality problems, unlike stopgap measures such as the use of pollution control devices.

Habitat and open space preservation, also essential to sustainable community development, must be an integral element of policies designed to restructure basic systems such as housing, transportation, and business. In other words, policies for systemic reform should include appropriate measures to ensure adequate protection of open space and wildlife habitat.

Growth management strategies, for example, may be far more effective and affordable to a community than open space acquisition.

Making Sustainable Development Work in Crisis Politics

Unfortunately, much of what transpires in community politics and planning occurs in a state of moderate-to-extreme urgency. Political analysts describe the current form of government as "crisis politics." Henry Kissinger put it best when he noted in a television interview that in government "the urgent often displaces the important."

Unfortunately, urgency often leads to stopgap solutions with little lasting value. Solutions patched together in a frantic effort may ameliorate immediate problems, but the same fate awaits those who use duct tape to repair leaks in the hull of a sailboat. Sooner or later the "solution" is going to give way. More likely than not, the leaks will have grown worse in the intervening time. In cities and towns, little problems that are "Band-Aided" often become major problems, more costly to fix with each passing year.

Clearly, one of the greatest challenges facing communities today is the melding of the reality of political life—the crisis atmosphere—with a sustainable approach. Can communities fashion sustainable solutions in an atmosphere of urgency?

We think so. In fact, we think we have to. In our sustainable community development workshops, we attempt to show ways that planners and other government officials can deal with "emurgencies" in a sustainable way. Our approach is relatively simple and straightforward.

First, citizens, planners, and public officials (or some combination thereof) are encouraged to spend some time brainstorming solutions, using the matrix—that is, they are asked to examine each of the systems and propose policies and actions that could make their communities sustainable. Wherever possible, of course, we recommend that communities take steps to incorporate new ideas into existing policies and action plans. We also suggest that the ideas be typed, distributed, and retained for later use. When problems arise, this gold mine of ideas formulated in a time of relative political calm could prove to be a valuable asset.

When an issue becomes acute, the first thing one must do is to identify the system or systems involved, then refer to the matrix for sustainable solutions. For example, suppose that a community is facing water shortages. Obviously, water supply is a distinct system. But like energy, this system permeates several others, such as housing, agriculture, and industry. The typical solution to water shortages is to develop new supplies, to build new dams on streams or rivers. But by studying the matrix of ideas already devised for the water supply system, housing, and business sectors, one would very likely find a host of sustainable solutions, ideas that would help communities meet demand without requiring the construction of costly dams. The city of Ashland, Oregon, found itself in such a bind in the late 1980s (Reed 1995). Faced with an imminent shortfall in water supply, the city contacted consultants, who recommended the construction of an $11 million dam on the nearby Ashland Creek. The town, however, sought the advice of a second consultant, who showed that simple water conservation measures in the community could "produce" the same amount of water at one twelfth the cost (about $825,000). This strategy not only saved the 20,000 residents more than $10 million, it made it unnecessary for them to dam the creek. It provided social, economic, and environmental benefits.

There is nothing revolutionary about this approach. It simply combines advance planning, based on a set of sustainability principles, with the nearly unavoidable dictates of crisis politics. An understanding of sustainability and sustainable design principles is essential to this process. Ultimately, we believe that communities that take the time to understand sustainability and sustainable development, and engage in creative brainstorming that seeks truly sustainable solutions, are in a much better position to bring about lasting results.

Sustainability will not occur, however, until communities get serious about restructuring human systems. It will not occur until we recognize that the problems many communities face are really just symptoms of deeper problems: systems that were designed without regard to their long-term sustainability. Such direction may come from outside consultants or, better yet, from leaders within the community who can articulate a vision and help unite disparate factions toward a common purpose—an essential principle of sustainable development.

Sustainable community development requires careful rethinking of our cities and towns and the various unsustainable systems that are at the root of our problems. By doing some advance thinking—thinking of ways to restructure systems based on an understanding of the principles of sustainability—and tracing the symptoms to the underlying systems, we think that America's communities can successfully operate, using strategies for sustainability even within the typical atmosphere of crisis.

NOTES

1. The Sustainable Futures Society is a national not-for-profit organization headquartered in Evergreen, Colorado. Its mission is to promote a broader understanding of the principles, policies, and practices of sustainable development and to encourage their adoption. The Society offers workshops on sustainable community development and community indicators, provides technical assistance, and publishes books and articles on sustainability.

REFERENCES

Chiras, D. D. 1992. *Lessons from Nature: Learning to Live Sustainably on the Earth.* Washington, D.C.: Island Press.

———. 1993. "Toward a Sustainable Public Policy." *Environmental Carcinogenesis. & Ecotoxicology Reviews.* C11(1):73–114.

———. 1994a. *Sustainable Development in Colorado: A Background Report on Indicators, Trends, Definitions, and Recommendations.* Evergreen, Col.: Sustainable Futures Society.

———. 1994b. "From Symptoms to Systems: A Case Study for Systemic Reform." *Sustainable Futures.* 1(2):1–2, 7.

———. 1995. "Principles of Sustainable Development: A New Paradigm for the Twenty-First Century." *Environmental Carcinogenesis & Ecotoxicology Reviews.* C13(2):143–178.

Costanza, R. 1992. "The Ecological Economics of Sustainability: Investing in Natural Capital." In *Population, Technology, and Lifestyle: The Transition to Sustainability,* edited by R. E. Goodland, H. E. Daly, and S. El Serafy. Washington, D.C.: Island Press.

Costanza, R., and H. Daly. 1992 "Natural Capital and Sustainable Development." *Conservation Biology* 6:37–45.

Goodland, R. 1992. "The Case That the World Has Reached Limits." In: *Population, Technology, and Lifestyle: The Transition to Sustainability,* edited by, R. E. Goodland, H. E. Daly, and S. El Serafy. Washington, D.C.: Island Press.

Gardner, J. 1990. "Leadership and the Future." *The Futurist* 24(3):9–12.

Gilman, R. 1990. "Sustainability: The State of the Movement." *In Context* 25:10–12.

Lovins, H. S. 1990. "Abundant Opportunities: An Interview by Robert Gilman." *In Context* 25:20–24.

Meadows, D. H., D. L. Meadows, and J. Randers. 1992. *Beyond the Limits: Confronting Global Collapse, Envisioning a Sustainable Future.* Post Mills, Vt.: Chelsea Green.

Postel, S. 1990. "Toward a New 'Eco'-Nomics." *World-Watch* 3(5):20–28.

Power, T. M. 1992. *The Economic Pursuit of Quality.* New York: Sharpe and Armonk.

Reed, D. 1995. "Negadam Power." *In Context* 41:10.

Regenstein, L. G. 1991 *Replenish the Earth.* New York: Crossroads.

Reilly, W. K. 1992. "The Road from Rio." *EPA Journal* 18(4):11–13.

Renner, M. 1992. "Creating Sustainable Jobs in Industrial Countries." In *State of the World 1992.* New York: Norton.

Schmidheiny, S. 1992. *Changing Course: A Global Business Perspective on Development and the Environment.* Cambridge, Mass.: MIT Press.

Wann, D. 1996. *Deep Design: Pathways to a Livable Future.* Washington, D.C.: Island Press.

Weiss, E. B. 1990. "In Fairness to Future Generations." *Environment* 32(3):7–11, 30–31.

World Commission on Environment and Development. 1987. *Our Common Future.* New York: Oxford University Press.

PART TWO

Rural Diversity and Diversity of Approaches to Community Sustainablity

Seven

Moving from Principles to Policy: A Framework for Rural Sustainable Community Development in the United States

Owen J. Furuseth and Deborah S. K. Thomas

INTRODUCTION

The principles underlying sustainable development are practical notions that are not new to the latter portion of the twentieth century. As currently used, *sustainable development* refers to development that meets the needs of present generations without compromising the ability of future generations to meet their resource needs. Geographers and other social scientists have long noted the interdependence between human activity and natural systems and the tension created by human attempts to maximize resource outputs without exceeding the bounds imposed by environmental systems (Rees 1992).

Although the principles of sustainable development are neither new nor scientifically unmanageable, the application of the concept to rural community development and planning in the United States remains ephemeral and uneven. The issues and problems that accompany the translation of sustainable development from theory to practice are the topic of this discussion. In turn, these issues present a framework for the chapters, based on empirical research, presented in Part Two. The unifying theme of Part Two is diversity: diversity in rurality and diversity in community-based approaches to sustainable planning. Clearly evident in the following chapters is that the conceptualization and implementation of sustainable development is not a template that can be casually moved from place to place. The meaning of sustainable-based planning and development must be guided by locally framed, community-based needs and vision. The critical importance of the research findings contained in this section lies in the lessons and models they provide for other rural communities. The authors present examples that can assist in defining the questions, issues, and strategies that shape future sustainable development initiatives.

BACKGROUND

The growing interest in pursing a sustainable approach to rural development and community planning has occurred at the end of a century marked by two contradictory experiences. On one level, human activity and the pattern of resource use during the twentieth century have resulted in massive environmental degradation and the coincident disruption of natural systems. The consumptive life-style that has evolved in Western societies has increased per capita use of physical resources (e.g., energy, mineral resources, wood products, land) in the worlds' richest nations. Perhaps more important, the international marketing of Western goods and the consumptive life-style in less developed areas have fostered the diffusion of mass consumerism to new and larger markets. As we move into the next millennium, the promise for international capitalists is expanding markets in Asia, Africa, and Latin America.

Yet although environmental disruption has been accelerated, our ability to measure and understand the character of these changes has been greatly enhanced. Because of advances in scientific knowledge and technological improvements, contemporary policymakers have better information about environmental conditions today than at any other time in human history. Although the scientific knowledge base and predictive capabilities are not infallible and clearly not able to render environmental disruption harmless, policy makers do have a growing array of options for addressing these issues. It was, therefore, the convergence of increasing human-induced environmental disruption and enhanced detection and understanding of this phenomenon that led to socially and technically based inquiries to develop an alternative development path. The challenge was to balance short-term and long-term human needs with physical boundaries. The phrase *sustainable development* evolved from these processes in the 1960s (Williams 1994).

The widespread use of the term, as well as attempts to establish a working definition and guidelines for implementing the concept, largely evolved in the 1980s. The *World Conservation Strategy* (IUCN/UNEP/WWF 1980) is generally considered to be the first technical document to use the term *sustainable development*. It was followed in 1987 by the World Commission on Environment and Development (WCED) document, *Our Common Future* (WCED 1987). This report formally articulated sustainable development in the broader sense in which it is used today and popularized the concept (Blowers 1994). Most critically, the WCED report, commonly called the Bruntland Report, transformed sustainable development from an ecologically focused notion to a human-centered concept.

Although many of the requirements suggested by the *World Conservation Strategy* are included in the Bruntland Report, the focus in the latter document is on social issues. The natural environment remains an integral element in maintaining human quality of life, but social and economic opportunities cannot be overlooked. The Bruntland Report established the link between social, economic, and environmental issues. The environment was recognized as a fundamental element in any growth equation, rather than as an obstacle to growth.

Not long after the Bruntland Report was published, preparations began for the United Nations Conference on Environment and Development (UNCED), perhaps better known as the Earth Summit or the Rio Summit. With 178 governments represented, the Earth Summit was unprecedented in its international focus on the environment and development. Five agreements—the Framework Convention on Climate, the Convention on Biological Diversity, Agenda 21, the Rio Declaration, and Forest Principles—were signed at the Earth Summit.

Although all of these agreements deal with the environment, Agenda 21 addresses the concept of sustainable development most directly. Agenda 21 is an action plan addressing general issues of social and economic development, including resource management; the role of major citizen groups, such as nongovernmental organizations; and the means for implementing the ideas embodied within sustainable development. Although Agenda 21 is not a binding agreement, it opened a dialogue among the world's governing bodies and established a precedent for future development. In the fall of 1992, the United Nations established the Commission on Sustainable Development to oversee the implementation of Agenda 21, with the first review to be completed by 1997.

Taken together, the Bruntland Report and Agenda 21 introduced sustainable development as an alternative paradigm to status quo development and provided a better understanding of the concept. Subsequently, a number of national governments, including Canada, Costa Rica, the United Kingdom, Holland, New Zealand, and Zambia, have been engaged in ongoing efforts to articulate the implementation of national sustainable development plans (Berke and Kartez 1995). Within these frameworks, local and regional governments have sought to develop complementary programs (Furuseth and Cocklin, 1995a, 1995b).

In the United States, President Clinton has pledged support for the concept of sustainable development at the federal level (President's Council on Sustainable Development 1994). In 1993, Clinton established the President's Council on Sustainable Development with a charge to recommend a national action strategy for implementing sustainable development in the United States. Underpinning the work of the council was a commitment to moving American thinking "from conflict to collaboration and adopting stewardship and individual responsibility as tenets by which we live (President's Council on Sustainable Development 1996, 1). The council's report, published in February 1996, is a broad, comprehensive document. Contained within the report are 10 goals for the achievement of sustainability, with accompanying indicators for measuring progress. Further specificity is added by 38 policy recommendations and associated actions. The proposed recommendations cover topics ranging from ecosystem integrity to educational reforms. These proposals are directed at all levels of government, the private sector, and nongovernmental organizations.

Widespread implementation and support for the council's report are not likely. In particular, at the federal level it is improbable that sustainable development initiatives would pass Congress with a Republican majority. The Republican congressional leadership has pledged to allow the market and private institutions to regulate themselves. Federal leadership in the sustainable development area appears to be at an impasse for the near future. Indeed, it seems that in the United States sustainable development has to develop in a bottom-up fashion, with individual communities embracing sustainable options. This can be followed by cooperation and coordination between local governments with the diffusion of sustainable approaches to a larger scale.

LIMITED APPLICATION OF SUSTAINABLE DEVELOPMENT

Although the notion of sustainable development at the microscale has been widely discussed in North American planning literature, the application of the concept to specific community planning situations in the United States has been minimal. Only a limited number of cities and metropolitan counties have adopted sustainably based community development plans. Portland, Oregon, and Seattle, Washington, are often cited as leaders in this regard (Beatley 1995). In general, application of a sustainable planning approach is less than community-wide and involves one or more elements of the planning program. Sustainably based transportation, housing, or open space plans are common applications. Among the communities implementing a partial sustainable planning approach are San Jose and Sacramento, California; Austin, Texas; Boulder, Colorado; Brunswick, Maine; Chattanooga, Tennessee; Cleveland, Ohio; and Olympia, Washington (Corson 1995).

We believe that a critical reason for the lack of widespread application is definitional ambiguity. The broad scope of sustainable development—and the integration of social, economic, and environmental issues associated with the concept—impedes precise definition of the term *sustainable development*, and thus it remains ambiguous. Accordingly, implementation of the concept at any scale is rendered difficult. Ambiguity, in turn, makes selection and implementation of sustainable development policies and plans especially problematic in the United States. Because of Constitutional restrictions, particularly those relating to due process and the application of police powers, planning and community development in the United States require specific terminology and implementation procedures (Cullingworth 1993). Consequently, use of a concept such as sustainable development is difficult within the constraints of the existing planning process.

If sustainable development is to become an accepted approach to community planning in the United States, the definitional limitations will require attention. Perhaps the first step is to effect enhanced recognition of the concept, which will involve the elaboration of a more precise definition within the context of community development and planning, as well as increased empirical research. The latter would allow communities to better understand the impact of sustainable development on privately owned property. A related component is public education. As these changes take place, implementation of sustainable development at the local scale will, no doubt, become a more common practice. Although local communities cannot independently achieve total sustainable development, they can contribute to sustainability at larger geographic scales. Local communities pursuing sustainable options must remember that they are part of a larger whole (Campbell and Matusz 1994).

THE SUSTAINABLE COMMUNITY

Although sustainable development is clearly not defined with enough precision to provide a strict basis for a theory of actions, many communities are recognizing that the way in which resources are currently being allocated and the interaction between humans and the physical environment are not working (Beatley and Brower 1993). Within this framework, community leaders are seeking alternative ways of approaching development that will improve local quality of life. Not surprisingly, sustainable development has recently gained widespread attention as a paradigm for community growth and planning (Beatley 1995; Corson 1995). As drawn from the preceding discussion, certain key concepts to planning for a sustainable community become apparent. Table 7.1 summarizes a number of principles embodied by a sustainable community.

Use of the sustainable development model at the local level requires a systems approach for successful sustainable community development. Specialized areas within planning systems must be integrated as policies and programs are developed and implementation begins. For example, environmental planning should not be conducted separately from transportation planning or human resource programming. The boundary between environmental and economic or social issues and policies is artificial. The interrelationships between environmental, social, and economic impacts

Table 7.1 Key Concepts of a Sustainable Community

Integrative	• Adopts a systems approach • Addresses ecological, economic, and social perspectives together
Diverse	• Promotes ecological, economic, and social diversity • Provides ability to adapt • Provides flexibility
Interdependent	• Links with surrounding communities of similar scale • Ties to other scales
Equitable	• Advocates democratic approach to problems—community involvement and education • Furthers intragenerational and intergenerational equity
Environmentally Sensitive	• Recognizes dependence on health of natural environment for sustainability • Reduces impact on the environment
Visionary	• Champions an improved quality of life • Develops long term vision • Advances vision guide to growth • Politically resolves to pursue vision

on human life must be captured and embedded in the planning process. Admittedly, this is no easy task, considering the broad scope of such planning. This requirement should not, however, be interpreted as a call for the elimination of specialization within planning organizations. Rather, it simply recognizes that different areas of community development must complement one another and reinforce the ideas embodied in the vision of the community.

Williams's (1993) "'living systems" approach to sustainable community planning illustrates this concept of integration. The 12 systems or dimensions shown in Figure 7.1 interact and contribute to the overall sustainability of any landscape, community, or region. Specialized planning activities cannot ignore any of the 12 components, because any type of planning affects all of them. In other words, in conducting any specialized community planning, the interaction between the components of the system should be taken into consideration and the impacts on all 12 areas should be understood.

The idea of a living system also implies diversity within the system, a dynamic rather than static system, and interdependence between components. These are all key elements of a sustainable community. Diversity, as applied to a sustainable community, involves not only diversity in the natural environment, which is commonly identified with all environmentally

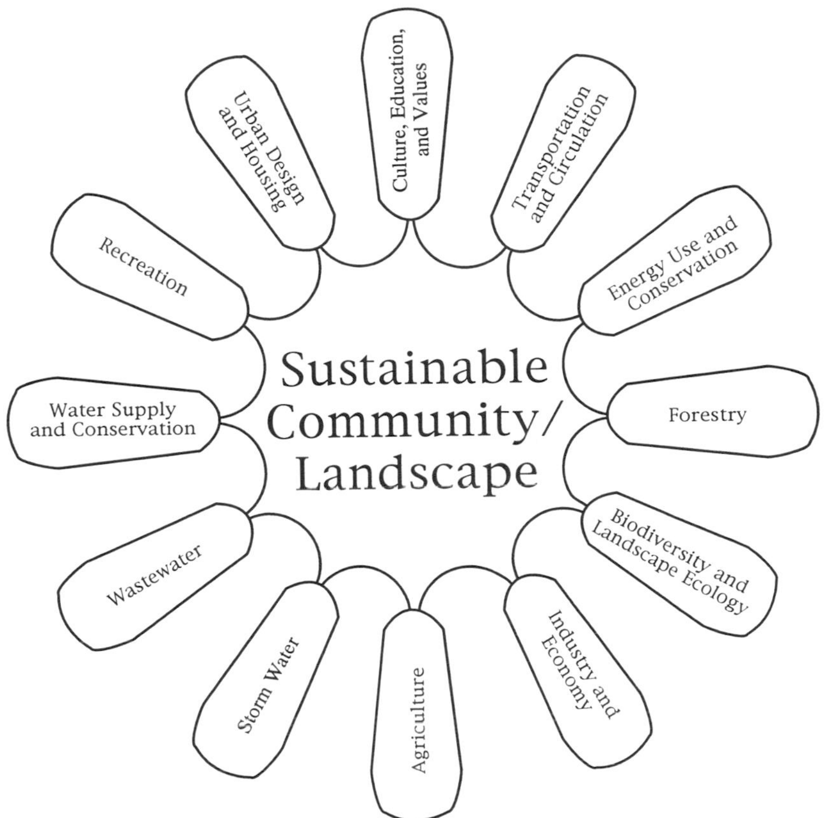

Figure 7.1 Living Systems of a sustainable community. *(Source:* Williams, 1993)

oriented plans, but also social and economic diversity. For a community to be truly sustainable, it must be composed of a wide variety of people from various socioeconomic backgrounds. Similarly, a community should have a broadly based economic system. Such a plan can ensure that if one segment of the economy becomes dysfunctional, the community as a whole will still be able to support itself through other sectors. This provides a broad employment base and a broad tax base, the benefits of which are far reaching.

Unfortunately, in the United States, the concepts of *sustainable planning* and *sustainable community* have generally ignored the requirement for social and economic sustainability. Too often, residential developers have latched on to the term *sustainable development* to help market environmentally friendly or neotraditionally designed developments (Duany and Plater-Zyberk 1991; Calthorpe 1993; Arendt 1994). Mixed-use developments and neotraditional urban design may be closer to the goals of sustainability than postwar American suburban subdivisions, but most still fail to address fundamental social and economic issues. Communities that segregate residents by income level or class or do not allow for a full range of economic activity are inherently unsustainable.

In natural environmental systems, a diversity of species provides the means by which the system as a whole can adapt to changes in the environment; this principle applies to human—economic and social—systems as well. Diversity provides the ability to adapt in any context and is no less vital to the survival of a viable community.

Furthermore, no community can exist in isolation of other towns, cities, regions, or even countries. Sustainable development should not be interpreted to imply a completely insular community. Instead, interconnectedness between other places, economically, socially, and environmentally, is vital in striving toward sustainability. As a result of an increasingly global economy and the global nature of many environmental problems, expanding cooperation between communities, regions, and countries seems to be the only alternative. Indeed, local communities cannot independently plan for completely sustainable development. They can, however, maximize their contribution to supralocal sustainability by reducing land and resource consumption, reducing spatial inequities, and enhancing livability.

The key to the success of sustainable development is an approach that integrates the concept at all levels, from the microscale to the global scale. Because of the democratic ideas that sustainable development embodies and the lack of state or federal laws fostering sustainable development, the local level seems to be the appropriate place for sustainable planning to be embraced in the United States. This assessment recognizes the potential political and institutional obstacles resulting from microlevel application of sustainable development. States and localities have the expertise, experience, and regulatory authority needed to design and direct sustainability (Saporito 1992). Most important, as local implementation of sustainable principles occur and evolve over time, so will the commitment of citizens and leaders to making dramatic changes in the current development status quo. These shifts cannot be imposed from higher levels of government but, by necessity, must rise from the neighborhood and community levels.

CREATING CONSENSUS FOR A COMMUNITY VISION

The relationship between citizen involvement, consensus building, and sustainable communities is critical. In examining this connection, Walz (1995, 37) notes that "the processes used to develop public policy will affect the sustainability of the result." Particularly in the United States, where there are few federal or state mandates, the empowerment of communities to create a better future is key to the use of the sustainability paradigm as a basis for community growth and development. If this is to occur, then citizen participation in the process and education of the community are vital to the success of shaping a sustainable community.

Involving the community in creating a strategic framework for future growth and change, commonly known as a community vision, is essential for sustainable development. Too often, current planning and community development practices fail to develop a more broadly based, long-term community vision. Comprehensive planning in the United States tends to be linear. Once a development plan is adopted, it is the community development guide for the period of time for which the plan was created. Planning goals must take a longer view—a vision not just for the next 10 years, but considering the needs for 50 or even 100 years. This is not to say that a community vision should be binding, that it should not be revisited and revised. Rather, the vision is an established set of broader, longer-term goals or purposes for the community. Essentially, the vision should act as a guide, reflecting how community residents would like their town to look and function over the long term. The vision should not be a prescription. As individual planning documents and initiatives emerge, a community would revisit its vision in order to ensure that it is working toward its dynamic goals.

Just as difficult as promoting public participation, developing a common vision for the future of a community can be precarious, presenting many challenges of its own. Defining the desirable characteristics or selecting the "good" attributes of a community in concrete terms can be a formidable task. What may be a positive characteristic to one person may, in fact, be unacceptable to another. For example, one segment of the citizenry may dislike the idea of Wal-Mart coming to the community because it disrupts the small town atmosphere and may negatively affect downtown merchants. Yet, others in the community may find it to be an attractive economic amenity, a place where they can shop for a wider variety of goods and services and pay lower prices. In short, individuals and groups have different needs and desires that affect the vision they may have for their community. Understanding individual perspectives is vital to the success of the community visioning process.

Balancing Individual Rights and Collective Needs

The difficulty associated with individuals' perceptions emphasizes the conflict between individual rights and needs and collective needs. In the United States, the rights of the individual are of utmost importance and often take precedence over the common good. In contrast, in other industrial societies, planning for the good of the community as a whole is of much greater priority than individual concerns. That is not to say that these nations are less democratic, only that the values concerning community good are given precedence. Embracing sustainable development necessitates creating a community vision composed of qualities that the community as a whole cherishes. Creating a vision and implementing that vision will require finding a balance between community rights and individual rights.

Public Education

Understanding society's and individuals' beliefs about the environment and environmental protection will be extremely important to the success of sustainable development (Milbrath 1989, 118). Clearly, if sustainable development is to operate as an alternative to the standard growth-based development model, then a transformation of values will have to occur (Ruckelshaus 1989). Public education and dialogue concerning sustainable development will be as vital to its success as public involvement. Community values can be transformed through education. Citizens can come to better understand the relationship between the good of the community and individual needs and, through this understanding, arrive at a balance ultimately embodied in the community vision. Most important, sustainable development will not succeed as an alternative growth model if the public does not support its principles.

Geographic Uniqueness

Different communities face varying challenges in the future. The community in a fast-growing rural resort area or urban fringe has a very different set of challenges than a community in an economically declining area. One community may be gaining population at an unprecedented rate, while another may be losing population and businesses. The geographic setting of a community will dictate the priorities and those things that are important to that community. Each community should create an appropriate vision for its local circumstances; clearly, it should not try to become something it is not (Hester 1995). The vision should be realistic. Accordingly, key to any strategy is an inventory of existing community resources that can be used as indicators for determining strengths and weaknesses.

Inventory of Resources and Indicators

The notion of sustainable community may still seem somewhat vague, as based on the criteria discussed thus far. The issues of the local quality of life and equity involved in creating a community vision reveal the difficulty of quantifying many of the principles on which sustainable development is based. For instance, how can the quality of life be measured? How can the needs of future generations be ascertained? These are just two of the questions that underlie the creation of an accounting system to be used for sustainable development. Certain indicators within a community, however, can provide a measure of the status of the community as a whole (Cutter 1985). These variables can be used to focus public attention on the issues or concerns that a community must address to attain sustainable development and on how to actually move forward toward a sustainable growth model.

The principle that the ultimate goal of sustainability is to create and maintain a certain quality of life now and for the future demonstrates how critical quality-of-life indicators are in applying sustainable development to a community. First, indicators within the community constitute a way to measure how well a community is meeting the needs and expectations of its residents. This represents a quantification of the local quality of life. Second, the element of time is exceedingly important, because sustainable development aims to leave a legacy for future generations. Consequently, measuring the quality of life and then monitoring it over time is the only way to determine the success of efforts to achieve sustainability. Even before indicators can be appropriately determined, however, a good inventory of existing resources must be available (Felleman 1994).

According to a report by Sustainable Seattle (1993), good quality-of-life indicators should exhibit several criteria. First, they should reflect the long-term economic, social, and environmental health of a community. *Long-term* means positive change over generations, not just for the current time period or short-term future. Indicators that can be used over a long period of time to assess the successes or deficiencies of a sustainable development plan are critical. Second, the chosen indicators should be understood and accepted by the community as valid signs of sustainability. These factors must reflect widely held public beliefs and opinions concerning appropriate development standards and criteria for the individual community. Third, indicators should have media appeal. In other words, the indicators should be usable by local media in reporting the successes or shortcomings of sustainable practices. The local media can play an important role in educating and informing the community about sustainable development policies and programs. If the media can not communicate such information, then public understanding and participation are weakened. Finally, indicators should be statistically measurable and, preferably, drawn from commonly collected data sets. This would allow comparison with other communities and foster diffusion of the sustainable approach and cooperation among communities that are oriented toward sustainability.

Dependence on appropriate technology is a key component in improving the sustainability of an economic system and maintaining or improving environmental quality (Loening 1990). Unfortunately, many people look to technology as a replacement for nonrenewable resources. Although society cannot and should not depend on technology as the only answer to environmental problems, certainly it will provide some alternatives (Thayer 1994; Fri 1991). For example, research in the area of solar power may provide alternatives to the use of gas and electricity for warming homes and fueling automobiles. However, advances in technology and temporal availability of technology cannot be predicted with accuracy. Thus, reliance on technological fixes to remedy existing deficiencies or eliminate barriers that limit development options is counterintuitive to the principles of sustainable development (Blowers 1994).

The current thinking among many sustainable development researchers is to look toward appropriate, existing technology for answers to environmental problems. Too many policymakers, however, leave undiscovered technology as the only possibility for future generations. Unfortunately, technology that appears to be a solution can often be the cause of environmental problems or economically infeasible (Thayer 1994). If sustainable development is to be achieved, research into new technologies should be supported, but should not be looked to as a sole solution to present-day environmental problems.

ENVOI

Sustainable development is a complex concept. Only the future will reveal how fully sustainable development will be embraced into the mainstream of American rural community development and planning. Certainly, implementing such a complex concept at any geographic scale will be a struggle requiring unusual commitment, as well as trial and error. Nonetheless, given the general failure of our traditional development model to solve planning and development problems in many rural areas and the potential positive outcomes associated with the sustainable development paradigm, the effort and risks associated with adoption of this new approach seem acceptable. Indeed, it must be remembered that 50 years ago the suburban subdivision development was seen as the American dream. If sustainable development is not to end with the same failings as the current planning framework, then much research and refinement of the concept must occur.

According to Walz (1995, 36), "sustainability is a concept of global significance, but effective action toward sustainability must occur locally." Particularly in the U.S. where national sustainable development legislation is unlikely, movements at the local level will be vital to the success of sustainable development. The creation of sustainable communities can contribute to the sustainability of larger geographic scales. It will be important to take what works from the current planning system and integrate it with other environmentally oriented disciplines in order to operationalize sustainable development.

The remaining five chapters in Part Two provide valuable insights for rural community leaders and planners engaged in the process of creating sustainable development strategies. The authors report on active community-based sustainable development initiatives in the United States and in the international arena. The approach and focus of the chapters reflect the diversity of problems facing rural America, as well as the complexity and variety of meanings assigned to sustainable development. What links these disparate chapters is their focus on mobilizing rural resources—that is, establishing processes for involving and activating rural citizens and leaders in programs and implementing strategies to affect sustainable social, economic, institutional, and environmental changes in rural communities.

Chapters 8 and 9 document the work toward implementing locally directed community development and planning in the Great Plains and the Midwest. Joseph Luther reports on strategies to sustain small towns in a rural region that has been slowly experiencing economic decline and loss of population. Luther offers positive development futures as alternatives to the so-called Buffalo Commons strategy proposed for the Great Plains. In a somewhat parallel fashion, Gary Green examines a community-based self-development approach as a strategy for rural sustainability. Using data from nearly 100 rural communities, Green explores the outcome of self-development programs and presents an assessment of the implications for rural sustainable development.

The important role that building social capital plays in maintaining and creating rural sustainability is the topic of Chapter 10. In their text, Lionel Beaulieu and Glenn Israel build on Coleman's model of family social capital to examine the role of educational and job-skill procurement among rural young people. Their research findings offer specific strategies for enhancing rural sustainability through strengthening human resources (community social capital).

The theme of Chapter 11 shifts to an examination of political relations, particularly the impact of technological and communication innovations on rural governance. In this chapter, James Seroka discusses what stronger intergovernmental coordination can mean for rural governing bodies and the support that cooperative programs can provide for sustainable rural development.

The final chapter in Part Two returns to the theme of involving and empowering citizens in community decision making. Here James Segedy examines processes for engaging citizens and allowing them to become informed planning participants. He discusses community-based workshop and charrette techniques and suggests how these tools can be used to educate and involve local communities in planning for sustainable development.

Sustainable rural development and planning calls for re-creating a sense of place and reestablishing people's connections with both the physical and human environments. The remaining five chapters in Part Two embrace this notion and present working evidence of locally oriented development strategies. Their product constitutes valuable guidance and lessons for other communities.

REFERENCES

Arendt, R. L. 1994. *Rural by Design*. Washington D.C.: American Planning Association Planners Press.

Beatley, T. 1995. "Planning and Sustainability: The Elements of a New (Improved) Paradigm." *Journal of Planning Literature* 9:383–395.

Beatley, T., and D. J. Brower. 1993. "Sustainability Comes to Main Street." *Planning* 59:16–19.

Berke, P. R., and J. Kartez. 1995. *Sustainable Development as a Guide to Community Land Use Policy*. Cambridge Mass.: Lincoln Institute of Land Policy.

Blowers, A. 1994. *Planning for a Sustainable Environment*. London: Earthscan Publications, Ltd.

Calthorpe, P. 1993. *The Next American Metropolis: Ecology, Community and the American Dream*. Princeton, N.J.: Princeton Architectural Press.

Campbell, A., and M. Matusz. 1994. "Sustainable Planning." *Arizona Planning* 2(July/August): 10–13.

Corson, W. A. 1995. *Defining Progress: An Inventory of Programs Using Goals and Indicators to Assess Quality of Life, Performance, and Sustainability at the Community and Regional Level*. Washington D.C.: Global Tomorrow Coalition.

Cullingworth, J. B. 1993. *The Political Culture of Planning: American Land Use Planning Comparative Perspective*. New York: Routledge.

Cutter, S. L. 1985. *Rating Places: A Geographer's View on Quality of Life*. Washington D.C.: Association of American Geographers.

Duany, A., and E. Plater-Zyberk. 1991. *Towns and Town Making Principles*. New York: Rizzoli.

Felleman, J. 1994. "Deep Information: The Emerging Role of State Land Information Systems in Environmental Sustainability." *Journal of the Urban and Regional Information Systems Association* 6(2):11–24.

Fri, R. W. 1991. "Sustainable Development: Principles into Practice. " *Resources* 102:1–3.

Furuseth, O. J., and C. Cocklin. 1995a. "An Institutional Framework for Sustainable Resource Management: The New Zealand Mode.," *Natural Resources Journal*. 35:243–273.

———. 1995b. "Regional Perspectives on Resource Management: Implementing Sustainable Management in New Zealand. *Journal of Environmental Planning and Management* 38:181–200.

Hester, R. T., Jr. 1995. "Life, Liberty and the Pursuit of Sustainable Happiness." *Places* 9(3):4–17.

IUCN/UNEP/WWF (International Union for Conservation of Nature and Natural Resources/United Nations Environment Programme/World Wildlife Fund). 1980. *World Conservation Strategy*. Gland, Switzerland: International Union for Conservation of Nature and Natural Resources.

Loening, U. 1990. "The Ecological Challenge to Growth." *Journal of SID* 3(4):48–53.

Milbrath, L. W. 1989. *Envisioning a Sustainable Society: Learning Our Way Out.* Albany: State University of New York Press.

President's Council on Sustainable Development. 1994. *A Vision for a Sustainable U.S. and Principles of Sustainable Development.* Washington, D.C.: President's Council on Sustainable Development.

———. 1996. *Sustainable America: A New Consensus for Prosperity, Opportunity, and a Healthy Environment for the Future.* Washington, D.C.: President's Council on Sustainable Development.

Rees, J. 1992. "Markets—The Panacea for Environmental Regulation." *Geoforum* 23:383–394.

Ruckelshaus, W. 1989. "Toward a Sustainable World." *Scientific American* 261(3):166–174.

Saporito, G. 1992. "Global Warming: Local Governments Take the Lead." *Public Management* 74(7):10–13.

Sustainable Seattle. 1993. *"The Sustainable Seattle 1993: Indicators of a Sustainable Community,"* Seattle, Wash.: A Report of Citizens on Long-Term Trends in Our Community.

Thayer, R. L., Jr. 1994. *Gray World Green Heart: Technology, Nature, and the Sustainable Landscape.* New York: John Wiley & Sons.

Walz, K. 1995. "Consensus Building for Sustainable Communities." *Carolina Planning* 20(1):36–43.

Williams, J. M. 1993. Sustainable Agriculture Facilitation Programme 1993. Seminar Presentation I, Ministry of Agriculture, New Zealand.

———. 1994. Sustainable Development Seminar, Department of Geography and Earth Sciences, University of North Carolina at Charlotte, November 21.

World Commission on Environment and Development. 1987. *Our Common Future.* New York: Oxford University Press.

Eight

Still Life on the Plains: Strategies for Sustainable Communities

Joseph Luther

INTRODUCTION

Change is occurring on the Great Plains, as well as most of rural America. But this change is not the apocalyptic vision of abandoned rural small towns and subsequent government conversion of significant parts of the Great Plains into the "Buffalo Commons" as prophesied by Deborah and Frank Popper (1987) or the governmental condemnation of rural small towns advocated in the community triage proposal of Tom Daniels and Mark Lapping (1987). It is rather an evolving adaptation of communities to their changing environments. There are many success stories of rural and small towns on the Plains that are aggressively seeking to survive and persist. Contrary to rumor, there is still life on the Plains.

Unusual businesses are appearing in unexpected places as these rural and small towns search for alternatives to the Buffalo Commons. Leaders are actively involved in developing the knowledge and skills needed by their communities to identify and recruit new economic activities. Rural villages, such as Lusk, Wyoming, are creating the infrastructure necessary to connect to the emerging information superhighway. These Great Plains communities are aggressively seeking information on how to survive and persist into the twenty-first century. Publications, such as those of Wall and Luther (1992a, 1992b), as well as concomitant training programs, are much in demand as a response to the sensationalized images evoked by the publicity-propagated message of the Buffalo Commons.

Central to the process of community survival is the concept of sustainability. The significance of sustainability began to be appreciated with the Brundtland Report (World Commission Environment and Development 1987), which defined sustainable development as development that meets the needs of the present without compromising the ability of future generations to meet their own needs. This definition became the basis for a general strategy that involves improving the quality of human life within the absolute capacity limits of the life-support functions provided by the environment. As rural and small towns evaluate and choose future community economic development strategies, intelligent choices must be made on the basis of long-term sustainability. There are many possible futures that a community may envision. Of these, the Buffalo Commons is but one alternative.

Relying almost exclusively on secondary demographic data and statistics, the Poppers hypothesized that the Great Plains will become depopulated, "the small towns in the surrounding countryside will empty, wither, and die. The rural plains will be virtually deserted. . . . Little stands in the way of this outcome" (Popper and Popper 1987, 12). The original Popper proposal caused a significant amount of controversy. Since 1987 these authors have written and spoken about the Buffalo Commons, and the sensationalized aspect of the concept continues to receive extensive coverage in the popular press.

In time, the Poppers' work regarding the Great Plains came to be reviewed by some of their academic colleagues. Geographers De Bres and Guizlo (1992) criticized the Poppers' work from a scholarly standpoint. They noted that "the Buffalo Commons proposal suffers from three major flaws: perceptual assumptions, methodological problems, and a failure to recognize and articulate the proposal's implications for Great Plains residents"(170). However, little, if any, of such scholarly criticism of the Buffalo Commons proposal made its way to the popular press.

There was, and continues to be, a fear that the extensive reporting of the Buffalo Commons theme will have a deleterious effect on the rural small towns of the Great Plains. According to De Bres and Guizlo (1992, 170), "At the outset of the original article, the Poppers create an image of a dying region by a vivid writing style and by exaggeration and misstatements found in nearly every paragraph on the first page." In effect, the widely reported dismal future portrayed in the Poppers' work, in concert with their colleagues' proposals for small town triage and the redlining of small rural villages, had the realistic potential of creating a self-fulfilling prophecy (Luther 1994; Fitchen 1991; Shepard 1994). For many residents of the Great Plains, the only news about their communities was that of probable extinction. Disheartened, some residents began to leave, thus strengthening such a likelihood.

This phenomenon may well be part of the larger national anxiety regarding the loss of the American frontier and the concurrent loss of a unique quality of the American character (see Wrobel 1993).

THE FUTURE HISTORY OF THE SMALL TOWN

The rural and small towns of the Great Plains were located primarily as a result of Euro-American settlement under conditions of a subsistence economy (Veregge 1995). The regional pattern of settlement derived from a nineteenth-century agrarian economic system in which the cost of transportation, in terms of time and distance, were primary determinants. The rationale for their location and the character of these small rural communities on the Plains has been described and explained in the economic geography theories and models of land economists such as von Thunen, Christaller, and Losch (Hoover 1971).

In this agrarian economic paradigm, the rationale for existence and the essential value of the rural small town on the Great Plains are directly related to the town's role as an agricultural service center. As the importance of the agricultural economy diminishes, the importance of the town may also diminish. In the end, when the grain elevators and the farm implement stores are gone, there may be no reason for the continued existence of the community. At least, this seems to be the thesis of the Buffalo Commons. There are other perceptions.

According to Popper and Popper (1987, 12), "over the next generation the Plains will, as a result of the largest, longest running agricultural and environmental miscalculation in American history, become almost totally depopulated." The U.S. Department of Agriculture (USDA) clarifies the fallacy in the Popper's prediction by noting that each year a new set of figures appear proclaiming the eminent demise of rural America owing to an unprecedented decline in population. However, according to the USDA, there is nearly a complete failure to qualify these figures by reporting that it is the "farm" segment of the rural population that is in decline. In fact, USDA (1989) points out that farm population in rural America reached its peak between 1910 and 1920.

This is not only a Great Plains phenomenon. By 1990 only about 9% of all rural people worked on farms and ranches. Fewer than 19% of rural Americans now derive their livelihoods from activities associated with agriculture. Only 500 of America's 2,400 rural counties are now considered "agriculture dependent" (Economic Policy Council 1990).

What these data tell us is that agricultural production continues, but the agrarian society is undergoing fundamental changes. The numbers of people and the numbers of towns required to support these agricultural pursuits continue to decline. As farm-to-market centers, the communities have reached the apparent end of their life cycle. The traditional economic activities in these small rural towns have been disappearing as a consequence of changes in agricultural practices and markets (Ekstrom and Leistritz 1988; USDA 1989).

On the Great Plains of North America, rural small towns are facing desperate choices. Most of these communities are less than 150 years old and originated in the agricultural settlement movement of the mid-to-late 1800s. Communities all across the Great Plains are in search of solutions to their plight. Such solutions, however, are often at variance with the traditional agrarian paradigm of the plains (Swanson 1990). Nonfarm enterprises are becoming essential to the local community's future (Economic Policy Council 1990). It follows that for these communities to persist, they must diversify their economy.

Contrary to the Buffalo Commons hypothesis, more and more studies reveal that there are possibilities for the economic survival of the small rural community.

The National Governors' Association Study

John and associates' (1988) study of rural development strategies for communities and states notes that it is very difficult to predict changes in the economic vitality of different and varying rural communities. They found that readily available secondary data, such as used by the Poppers, cannot be used effectively to target development funds to those areas where growth is most likely to occur, thus disarming the triage proposal of Daniels and Lapping (1987). The study clearly shows that even if a community appears to face constraints to its development, it may still be able to achieve significant economic growth.

The Center for the New West Study

In 1992, the Center for the New West conducted a comprehensive examination of the economic, social, political, and cultural dynamics of the Great Plains Region.[1] Funded in part by the State Rural Policy Program of the Aspen Institute and the Ford Foundation, as well as by the U.S. Department of Commerce, Economic Development Division, the Great Plains Project sought "to provide a window on the transformation of the region, including its assets and liabilities, the outlook of its civic leadership and some likely future scenarios" (Shepard et al. 1992, iii).

One of the more revealing aspects of this project was the methodical review of quantitative indicators. The research team examined a wide range of indicators used to describe the phenomena and processes of the Great Plains region. They found that many of these indicators measured distress, but few measured vitality; many measured pathologies, but few measured well-being.

The study concluded that "the models and indicators used in quantitative analysis of economic and social change require serious re-examination. To further reinforce the need to look beyond secondary census data, the research team noted, "The use of methodologies that are simply reproducible (and familiar) with little regard for validity is not enough when the lives of people and communities are at stake" (Shepard et al, 1992, 4).

The Center for the New West discovered that many casual observers of the Great Plains are misreading change as decline. Clearly, the Center challenges existing rural economic development paradigms and proffers new paradigms and new approaches to the way we measure and predict development potential.

New Localism is a concept innovated by the Center for the New West to describe a process whereby local communities are taking responsibility for their own economic well-being. To this end, the cities and towns of the Great Plains are breaking away from the traditional land use economic hierarchy patterns modeled by Von Thunen, Christaller, and Losch. The Center describes a future in which small rural towns are no longer lost in a hinterland dominated by central cities. The Great Plains communities of New Localism may build their own future from an extended web of networks organized around the activities of everyday life.

The research of the Center for the New West clearly describes and explains the relationship between community and place. As many writers and organizations are now beginning to pick up the theme of "sense of place," the Great Plains Project has explored and described the transformation that is changing how people relate to each other spatially and how we make and use our towns and cities.

The report's corollary is that, rather than being in decline, this is a region in transformation. Community leadership is the sine qua non in determining community survival, but there must also be a fundamental community determination to survive. There is a need to recognize change and adapt, but collaborative behavior is the *principia media*. Most critical, the report recommends that the state not rely on traditional indicators to describe and explain communities. Instead, policymakers should look at leadership, institutions, and activities of everyday life at the local level. The policymaker is also reminded that development is not necessarily a function of population growth; there are many ways a small town can grow without getting bigger. Quality-of-life factors are a major asset in a small town and, sometimes, the community's only comparative advantage. For these reasons, community and economic development are interdependent.

Change, Identity, and Survival in Rural America

Fitchen (1991) has produced a credible study of the situations troubling small rural communities. Focusing on the problems and opportunities facing these rural villages, Fitchen describes the slow death spiral in which many communities are now caught. More important, she makes valid suggestions for community survival through local self-help and self-determination.

With a strong message to policymakers, Fitchen unequivocally urges them not to discard small communities through triage nor to consign those who dwell in sparsely populated rural areas to abandonment. Either of these actions may be akin to prematurely issuing death certificates—and may even hasten the demise of these places.

She has clearly identified the effects of federal and state "overmandating and shortchanging" small towns as the policy problem of rural America. These separate federal and state decisions have had cumulative and damaging impacts on our rural communities. The adverse effects of such policies have been compounded by the limited federal recognition of rural community needs. Fitchen's assessment strengthens the growing chorus of voices seeking a more appropriate federal agency for rural community development. She makes a persuasive argument for the separation of farm policy from rural community policy.

Fitchen believes that the federal government has generally failed to perceive rural problems in their broader sense, beyond agriculture and beyond economic problems. The federal government, she asserts, has not sufficiently realized that rural development must involve more than economic development.

Basic Paradigm Shifts

The primary paradigmic problem is that of urban growth bias. Thompson (1978) describes this perception of the basic structure of civilization as a dialectical tension between the center and the periphery. Ever since the advent of human settlements, the city has been perceived as the center of civilization, the place of education, culture, government, finance, and beauty. The rural hinterland labors to support the urban place. It is as if there were a scale along the radius extending from the city center to the rural hinterland; this would be a measure of "hickness."

A related paradigmic problem is that of the apparently axiomatic future of communities. They seem to be destined either to grow up to be urban places or to fail. Think about how our towns are labeled as first-class, second-class, or third-class communities. If a small town grows up, it may become a first-class community, but until that time it is destined to be second or third class. As planners, we have been programmed to see the world as an urban place. The indicators of progress by which we measure the quality of a community are most often urban criteria, such as population growth, traffic density, shopping demand, and so forth. For many small towns, there may be a compulsion to have at least a traffic light, parking meters, or the Golden Arches as clear evidence of urban sophistication.

New paradigms are needed whereby small towns in rural regions can explore and evaluate alternative futures—futures that are achievable and sustainable. Urban centers and farming are not the only futures for the Great Plains. Luebke (1990) noted that, in recent years, technology has dramatically expanded the possibilities of what can be done to transcend environmental limitations. The consequence is that the boundaries of appropriate human behavior in the Great Plains are not well understood. He hypothesizes that one might even argue that the history of the Great Plains has been a search for those limits. Luebke believes that the region is thus enmeshed in economic, political, and social uncertainty. For Luebke, this is the heart of the problem of the Great Plains.

The Worth of a Place

Duncan (1993) provides a rare opportunity to hear the West defined by its own people in their own voices. Long after the agrarian economic rationale for their existence has disappeared, the communities of the Great Plains seek to survive and persist for cultural reasons. This is a different way of looking at the worth of a place—a different paradigm. Certainly the Lakota Sioux have a different way of perceiving their Great Plains. They have a strong mythology that deals with a landscape more spiritual than physical. The spiritual landscape found within the heart of the learning community (Luther 1994) determines how they perceive and value their place, their town. We cannot say a community will die simply because it no longer has economic value.

The relationship between the settlement of the community and the ecology of the site is an enduring bond that has strong cultural connotations. The identity of a place derives from the historical integration of human activity with its natural environment. This kind of place identity is the basis for the cultural identity of the community. Although the communities of the Great Plains have evolved under conditions of a subsistence economy, the cultural landscape is constantly being changed through a process of sustained reciprocal modification (Veregge 1995).

Cultural survival may mean the evolution and development of alternative economic bases within the small town. It may mean the need to import a radically new economic activity that holds the promise of employment and needed tax revenues—but at what price?

Offered the prospect of jobs—in areas of high unemployment and falling tax revenues in the face of obsolete and crumbling infrastructure and inadequate services—community leaders are often motivated by short-term benefits. These leaders have few tools with which to examine the true costs of new development and the concurrent changes in their towns and regions. Apparently, long-term costs are not so easily envisioned and may be discounted by the present generation.

PLANNING THE SUSTAINABLE COMMUNITY

The future of the small town of the Great Plains may well be bound up in the concept of sustainable development. Sustainable development is a new concept of economic growth: a process of change in which all policies are economically, socially, and ecologically sustainable. It requires more equitable distribution and equal opportunities. Environmental concerns must become an integral part of decision making at all levels (Panos Institute 1987). In this approach, economic growth may be seen as sociocultural, economic, and environmental transformation (Rees 1995).

In its barest essence, a sustainable society is "one that satisfies its needs without jeopardizing the prospects of future generations. Inherent in this definition is the responsibility of each generation to ensure that the next one inherits an undiminished natural and economic endowment. This concept of intergenerational equity, profoundly moral in character, is violated in numerous ways by our current society" (Brown et al. 1990).

The primary goal of any economic or environmental policy should be sustainable development. Environmental design must take its place alongside cost, safety, and health as a guiding criterion for development (National Commission on the Environment 1993).

Small towns of the rural plains must balance social, economic, and environmental values in a planning and development process allowing for its long-term survival and persistence. Strategies developed at national, state, and local levels should enable and encourage this concept of sustainable community development. The New Localism concept urges local areas to take responsibility for their own economic well-being. The sociocultural, economic, and environmental transformation of sustainable development will require long-term views of possible futures which may bend the trend of historical patterns of growth and development.

The Buffalo Commons proposal, with its urban growth bias, fails to develop a measure of social and ecological sustainability of the Great Plains region (De Bres and Guizlo 1992). Sustainability is a critical choice for a Great Plains community that seeks to survive and persist beyond the current generation of residents. The dilemma unfolds before the leaders of the rural small town. Do they have an obligation to the next generation? Is there a sense of intergenerational equity?

Even more elusive is the impact of choice on one's neighboring communities. Although the concept of multicommunity collaboration is only now beginning to flower on the plains, the reality is still one of dog-eat-dog competition for whatever scraps of economic development may be available in the short term. The success of one community venture comes often at the expense of other communities. There is little legislative or cultural compulsion to cause a community to examine the long-term, regional effect of development.

However, multicommunity collaboration is seen as one of the few viable means of survival and persistence for rural small towns. Baker (1993) has conducted significant research on the role of multicommunity collaboration in restructuring rural communities. He notes that "the concept of multicommunity collaboration assumes that individual small communities may be able to sustain themselves more effectively and often maintain or improve their own identity and viability by collaborating with neighboring communities." Only through multicommunity collaboration can essential services be continued to regions of economic transformation and declining population. "By several communities working together, leadership skills, tax revenues, political influence, and other factors may be pooled to undertake relatively larger initiatives, both economic and social in nature. Also, development may become more sustainable" (12).

A Holistic Environmental Systems Approach

Traditional urban and regional planning appears to hold little promise for small towns of the Great Plains. Given the choice of hiring an economic development specialist who is clearly pro-growth or a city planner who appears to be anti-growth, many communities on the Plains are opting for the development specialist. Even the Poppers reveal this urban growth bias, as they indicate that a growing urban settlement pattern is to be preferred and promoted over nonurban settlement (De Bres and Guizlo 1992, 171–172). As a consequence of the demand for economic development, less and less comprehensive planning is practiced and the short term becomes the focus of the community's future.

McHarg (1969, 25) aptly encapsulates this short-term view: "We have but one explicit model of the world and that is built upon economics. The present face of the land of the free is its clearest testimony, even as the gross national product is proof of its success. Money is our measure, convenience is its cohort, the short term is its span, and devil may take the hindmost is the morality."

To be sustainable, the focus of the community planning and development program must be on the long term. The planning must be comprehensive and must consider the full range of environmental costs and benefits as advocated by the National Environmental Policy Act of 1970, which was created to ensure that unquantifiable environmental amenities and values are given appropriate consideration in decision making, along with economic and technical considerations. Today we are at a critical threshold where we must ensure that social, cultural, and environmental values, as well as economic development considerations, are the basis for shaping the character of the future of our communities.

Indicators of Sustainability

Just as the National Environmental Policy Act requires disclosure and evaluation of possible environmental impacts of a proposed action, communities can employ a similar checklist approach to disclose and evaluate the probable effect of change on the sustainability of a community. Environmental checklists may serve as a model for developing such an approach. With a wider range of indicators or measures of sustainability, the checklist not only provides a means for disclosure and evaluation, but also serves as an educational device for the community. Use of such an approach can be a catalyst and can stimulate thinking and learning about the costs and benefits, the sustainability, of proposed actions (Luther and Borner 1995).

Many communities are hard at work today developing indicators of sustainability. The basic questions are, How do we know when we are practicing sustainable development, and how do we know when we are doing it well? Hart (1995) reviewed the indicators of sustainability being used by many communities and evaluated the appropriateness and effectiveness of such indicators. This study provides an excellent list of communities using indicators of sustainability, as well as a qualified list of indicators to be used.

The Community Sustainability Resource Institute (CSRI) is working to organize the Community Sustainability Network. This is a collaborative project, designed to create a national clearinghouse for information about community sustainability issues, projects, and resources and to provide a forum for the discussion of this topic. CSRI is launching a quarterly journal and is in the process of organizing its third annual conference. The CSRI newsletter lists more than 60 groups from around the country—with names, addresses, contact people, and descriptions of the work they are doing (A. Jones, e-mail to author, July 1, 1994).

Sustainable Seattle (1993) produced a report that contains 20 fully documented indicators of sustainability for the Seattle area and an additional 20 indicators that are under development. More than 150 volunteers have been involved in developing indicators over several years. The Seattle indicators include the major categories of environment: population and resources, economy, and culture and society. These are not the detailed categorical impact topics found on an environmental checklist, but broader indicators of quality of life. The Seattle indicators include attributes such as total population with annual growth rates, gallons of water consumed per capita, tons of solid waste generated and recycled per capita per year, percentage of employment concentrated in the top 10 employers, hours of paid employment at the average wage required to support basic needs, percentage of children living in poverty, and usage rates for libraries and community centers, among others.

The Sustainable Community Roundtable is a grass-roots effort that grew out of a one-day seminar held by the City of Olympia, Washington. Roundtable organizers attempted to include a great diversity of interests in these discussions, including minority groups, business leaders, educators, and government officials. Their report (Craig and Baker 1994) first gives some very broad community visions that came out of the workshops. They then identify at least one primary indicator that is already being measured in the Olympia area and that can continue to be measured in the coming years to determine whether the community is moving toward sustainability. Primary indicators now include, among others, water consumption per capita, electricity consumption per capita from nonrenewable sources, sales of locally produced food, and percentage of commuters driving alone.

The effective use of indicators in the performance evaluation of proposed plans and actions can be facilitated by the use of a indicator question format (Luther and Borner 1995). As shown in Figure 8.1, indicators may be arrayed into a checklist format. Decision makers may ask these questions in a review process. The sustainability checklist then becomes a disclosure document.

The Minnesota Environmental Quality Board has a project called the Minnesota Sustainable Development Initiative. This Initiative was launched in January 1993 and has produced—with the help of seven diverse citizen teams—a year-end report that charts the course for sustainable development in Minnesota (Nordstrom 1994). The Minnesota Initiative is now entering its second phase.

The Northwest Policy Center (NPC) at the University of Washington is currently developing a workbook to help Northwest Communities apply the concept of sustainability locally. Johnson and Baven (1994, 1) of NPC note that,

> At its heart, sustainability challenges communities in two ways. One is to think about the long-term consequences of today's actions and decisions: Do they enhance or detract from the community's ability to prosper into the future? What will be their effects on later generations, both locally and around the world? A second challenge is to think broadly, across issues, disciplines, and boundaries. Sustainability suggests that creating economic vitality, maintaining a healthy environment, meeting human needs, and building healthy communities are closely related.

Guidance on identifying elements of a sustainable community is given by Roseland (1992), who illustrates how communities can apply the concepts of sustainability in governmental functions. Community sustainability can also be enhanced by the incorporation of policy statements into a town's comprehensive plan or economic development plans. Blowers (1993) offers a wide range of policy statements for sustainable development. These recommendations include encouraging development to provide maximum accessibility to jobs and leisure opportunities while also creating attractive living and working conditions; discouraging redevelopment when rehabilitation would require less use of scarce natural resources; encouraging a greater diversity of new development; achieving energy efficiency through better design and greater self-containment for energy and food supplies; and encouraging an increasing level of economic self-suffi-

Environmental Factors

1. **Earth:** Will the proposed action change areas of unique geologic or physical features?
2. **Air:** Will the proposed action change the amount of air pollutants from passenger vechiles?
3. **Water:** Will the proposed action change surface water quality?
4. **Flora:** Will the proposed action change biodiversity in the region, including birds, land animals, reptiles, fish, shellfish, bethnic organisms, insects and microfauna?
5. **Fauna:** Will the proposed action change biodiversity in the region, including trees, shrubs, grass, crops, microflora and aquatic plants?
6. **Land Use:** Will the proposed action offer land use plans and development schemes that encourage mixed use development?
7. **Transportation:** Will the proposed action change the number of commuters living more than 30 minutes from work?
8. **Natural Resources:** Will the proposed action change areas and quanity of non-renewable natural resources?
9. **Risks to Human Safety:** Will the proposed action change the risk of release of hazardous substances (including, but not limited to oil, pesticides, chemicals and radiation) in the event of an accident or upset condition?

Social Factors

10. **Quality of Life:** Will the proposed action change the percentage of people who feel safe in the community?
11. **Population**: Will the proposed action change the existing of population density?
12. **Housing:** Will the proposed action change the percentage of buildings designed for long life, adaptability, and low resource consumption?
13. **Human Health And Well-Being:** Will the proposed action change obstacles to health care?
14. **Education:** Will the proposed action alter per capita state and local K-12 expenditures?
15. **Archeological and Historical**: Will the proposed action change the number of protected sites and structures?
16. **Recreation**: Will the proposed action change the number of recreation options?
17. **Culture:** Will the proposed action change the number of public art sites in the community?
18. **Public Safety:** Will the proposed action change the number of reported crimes per 100,000 population?
19. **Public Welfare:** Will the proposed action change the number of residents living below the poverty line?

Economic Factors

20. **Employment:** Will the proposed action change the number of people employed in locally owned businesses?
21. **Economic Development:** Will the proposed action change the number of products and services local businesses buy from other local businesses?
22. **Public Services:** Will the proposed action change the leadership rating for local government?
23. **Energy:** Will the proposed action change the percentage of existing buildings meeting full-energy efficiency standards?
24. **Utilities:** Will the proposed action change the percentage and volume of waste recycled to beneficial uses?

Figure 8.1 Examples of some indicators of sustainability in a checklist format.

ciency in terms of the capacity of the region to provide a greater variety of jobs and to supply daily goods and services from local sources.

A Participatory and Anticipatory Community Learning Process

Traditional community planning and development is firmly founded on the principle of continuity. That is, trend analysis is the basis for projection and forecast. These images of the future assume that trend is destiny and that the future, as given, exists within a fairly narrow range of scientific probabilities. In this traditional approach, the planning task is to devise a means of reacting to a given future. The anticipatory planning process, in contrast to the traditional approach, seeks first to create a vision of desired future and then proactively devise and evaluate a preferred future for the community. The community then tests alternative futures to determine the costs and benefits of each option. Evaluation requires knowledge regarding the existing environmental system within which the community exists (Luther 1994). This sustainable community planning process is shown in Figure 8.2.

A community can learn about its total environmental system and make intelligent choices regarding its future. But such action, based on holistic knowledge, must be a process of community learning (Luther 1994). This community learning process is based on the theories and procedures of community organizing and community development. Members of the community actively participate in the exploration and description of the existing environmental system. From such activities derives not only holistic knowledge about the community, but also community self-actualization. As this information is shared in community forums, new perceptions of reality, and thus new values, are formed. In this manner the economic, social, and environmental values requisite for sustainability are made available to the community (Luther 1990a).

The role of the professional or the academic in this community development process is primarily that of educator rather than consultant. The objective is that technical assistance be rendered in a participatory educational manner rather than by the traditional consultant delivery. The role of the consultant is to develop the capacity and ability of the community to plan for itself over the long term. If consultants do this well, they will never have to return (Luther and Luther 1981).

More than Just Planning

Planning as we know it may not be relevant to rural small towns. To be meaningful, planning must change its character, and planners must develop knowledge and skill in new roles. If they are to be credible, small town planners must become effective in economic development. If they are to be able to work with small towns, planners must become effective in

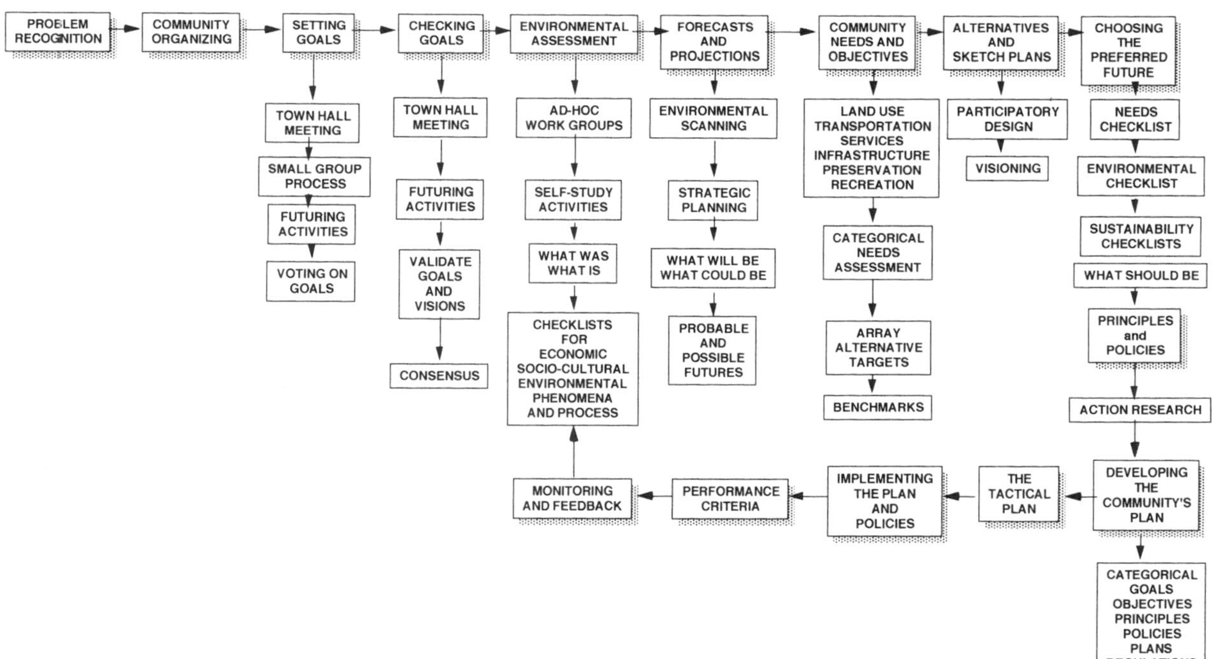

Figure 8.2 The sustainable community planning process.

community organizing and community development. They must also become futurists with the ability to seek out and visualize new possibilities; they must learn the attributes of new roles in organizational development and facilitation.

Lippitt (1973) identified a series of roles for clients and consultants that have been adapted (Luther and Luther 1978) to the needs of the small town planner. Among these new roles are advocate, expert, trainer, alternative identifier, collaborator, process helper, catalyst, and resource linker. Needless to say, development of the new knowledge and the skill needed to undertake these new roles requires fundamental changes in the offerings and accreditation of educational programs in planning. Such changes also require new frames of reference.

Possible Futures

The principal vision of the future derived from planning practice is the scientifically valid population projection. Utilizing statistical and/or mathematical models, the planner extrapolates trends of historical population data. From these estimates of future population are derived corresponding levels of residential, commercial, manufacturing, and other land use demands, with attendant demands on transportation and other services.

The problem with this paradigm is its sole reliance on statistical probability to forecast a future state. With some appreciation of quantum mechanics (Zukav 1979), planners must recognize that there exists, at any moment, a seemingly infinite range of possible futures available to the small rural community. With this approach, the role of the planner is radically altered. The probable future derived from trend extrapolation becomes only one of many possibilities to be considered. Such imagination is a key component of the sustainable community.

The role of the planner is to be proactive rather than reactive. Rather than asking the community to conform to a statistical probability, the planner works with the community to visualize, evaluate, and choose among alternative futures. The plan becomes the strategy or road map to this future state.

The sustainable community planning process, as shown in Figure 8.2, actively involves the community in a number of self-assessment activities, which take place in town hall meetings, in field studies, in community assessment surveys, and in small group work. Overall, these activities and tasks are undertaken by members of the community through local self-help projects rather than technical consultants. Studies focus on the three primary areas of transformation: economic, sociocultural, and environmental. By working together and sharing information among themselves, community members engage in a process of social learning and community self-actualization.

Economic Survey and Analysis

Economic values must be analyzed in terms of the community's existing situation. This survey and analysis will describe the existing economic phenomena and processes and provide information about unsatisfied potential. The process will also provide marketing information that can be used to recruit new economic activities to meet the unsatisfied potentials.

An economic survey and analysis can also allow the community to conduct contingency analyses to determine the answers to what-if questions. Having set up the description of existing economic situations in a spreadsheet program, it is very easy to use a computer to discover and describe what would happen if the population grew or declined at varying rates. Community analysts can also determine what would happen if their town consumed all the regional market in any given category.

The capability to model regional economic impacts of community market decisions is critical in determining the sustainability of the regional community. This is also important to multicommunity collaboration. The capability to visualize potential economic change enables the learning community to anticipate the consequences of its actions. Moral and ethical issues of multicommunity collaboration may arise when a learning community must decide whether to take business away from a competing community, thus adversely affecting that community's sustainability.

Issues regarding the evaluation of these economic activities, including cost-benefit analysis, true-cost analysis, sustained returns, intergenerational economic welfare, and green economics, are emerging in a number of publications, such as Tisdale (1993).

Sociocultural Survey and Analysis

A sociocultural survey and analysis seeks to determine a community's attitudes or values regarding its existing and potential situation. Goals statements derived from town hall meetings, in combination with community attitudinal surveys, can answer a number of questions. The community analyst can discover not only what types of change are desirable or undesirable, but also where within the community such change is acceptable. Maps and sketches can aid in enabling the community to accept or reject various types of development or changes in various geographic locations. These sociocultural values can then be arrayed as a series of questions to be used in a sustainability checklist.

Environmental Survey and Analysis

An environmental survey and analysis generally proceeds on the basis of an environmental checklist, such as those employed in an environmental impact assessment. As these data are collected, it is important to give them social value. One effective means of giving social value to intangible and unquantifiable environmental phenomena and processes is to use a graphic device for storing, manipulating, and disseminating these data. In fact, the use of such graphics can allow the community analyst to convert data into meaningful information.

The graphic device is known as McHargian analysis, which employs a mapping technique to record the character and distribution of environmental phenomena and processes (McHarg 1969). These data are displayed on standardized base maps, allowing stacking of the maps into composites to yield information regarding synergistic groups of phenomena and processes. The strength of this approach is that these phenomena and processes cannot be ignored: they are rendered explicitly in color. There is also the power of locality relevance. The learning community is able to determine how proposed changes will affect environmental attributes. Similarly, the community may learn how the existing environmental attributes will affect a proposed action. The invisible is rendered visible.

The McHargian approach is useful in providing a basic understanding of the historical and existing environmental situation of a community. Environmental processes and phenomena are mapped and their social value established by the community. As a learning process, the McHargian approach enables a community to learn about and visualize previously intangible environmental values that may be affected by change. The maps and information provide a foundation for the evaluation of all future proposals. This approach is strengthened by the use of a sustainability checklist—a major asset to the learning community. As these analyses are conducted by ad hoc work groups within the community, a great deal of knowledge is generated and shared with credibility. The McHargian approach relies on existing scientific data that are timely and accurate.

These three categories of community survey and analysis provide the basis of understanding what was and what is in the community's environmental domain. From this base, projections and forecasts may be devised to learn about what could be and, more important, what should be.

Participatory Design

Participatory design empowers the learning community to create visual images of its environmental domain. The drawing pencil is removed from the hand of the professional architect, landscape architect, or planner and placed firmly in the hands of members of the community design work group (Luther 1990b, 33–56).

The learning community uses ad hoc work groups to create a series of sketch plans of possible future conditions, illustrating land use, transportation, and services that will meet projected needs. These illustrations take the form of colored maps and three-dimensional sketches of critical elements of the landscape and townscape. At this point, the community is involved in participatory design.

The learning community can and should rediscover the traditional townscape elements that represent its heritage. These historical elements form an important basis of a unique sense of place in time and space. Visual techniques, as well as images, for this activity may be found in Arendt's *Rural by Design: Maintaining Small Town Character* (1994).

The three-dimensional sketches, like the environmental analysis, provide the foundation from which to evaluate change. For each of the alternative sketch plans, the human-scale, eye-level renderings of street scenes, landscapes, and building facades are modified to show, realistically, the consequences of such changes.

A good handbook for the participatory design approach is Hester's *Community Design Primer* (1990). This book not only provides knowledge about community design as a participatory activity, it also teaches skills in drawing and problem solving by design.

The power of the three-dimensional illustration easily exceeds that of the traditional two-dimensional map of the planner. The learning community can visualize what it would be like to walk and drive through this townscape, to work in this future town, to play in this future landscape.

Moreover, the community design work group can enhance power to this vision of the future by rendering the images in a four-dimensional aspect (i.e., change is shown over time, rather than in the typical one-shot image of the completed state). Such time-series illustrations enable the learning community to understand that change is incremental and happens at a certain pace rather than all at once. Thus, the shock of change may be lessened.

The Preferred Future

The learning community must now make an informed and intelligent choice among the alternative sketch plans—the alternative images of the future of their town. How do they proceed?

The sustainability checklist is employed as a learning and decision-making device, as shown in Figure 8.2. Each alternative is evaluated according to the checklist. The costs and benefits of each sketch plan are described and explained. Examples of the indicators and proposal review questions that can be used in such a checklist are shown in Figure 8.1 (Luther and Borner 1995).

At another community-wide town hall meeting, the alternative sketch plans and the analysis of their sustainability are displayed to allow examination by the learning community. Each alternative sketch plan is presented in turn and the findings of the sustainability analysis are discussed.

Members of the learning community may select one alternative, may reject all alternatives, or may synthesize several alternative sketch plans into a new vision. By the end of the meeting, the learning community has selected an image of its preferred future; it has achieved a single, shared vision.

However, although such images of a future for the small town may be possible, they may not be preferable. The community, in its learning process, must evaluate all alternatives in terms of social, economic, and environmental costs and benefits, ever seeking that goal of "satisfying its needs without jeopardizing the prospects of future generations."

THE ENTREPRENEURIAL COMMUNITY

The image of a preferable future for the small town of the Great Plains must include a viable economy. Today, given the fundamental restructuring of the agricultural economy, the economic viability of these rural communities depends significantly on the entrepreneurial abilities of its community leaders (Luther 1990a).

An entrepreneurial community is one whose leadership demonstrates a highly positive attitude and has the ability to motivate the community to identify and constructively exploit new opportunities for its survival. This finding is borne out by the field research and professional practice of the Heartland Center for Leadership Development, which conducted a number of case studies on rural and small towns of the Great Plains (Luther and Wall 1986; Wall and Luther 1989).

Burgess (1988) notes that rural America can make it in the new economy if it can learn to be more productive and adaptive to shifting markets and changing technologies. Burgess believes that the effect of successful community economic development activity is new hope, increasing wealth, and expanding choices for people, communities, and enterprises. The many assets of rural America, according to Burgess, include a skilled and adaptable work force, a strong work ethic, lower operating costs than found in many larger urban areas, clean air and open spaces, and good market systems. As the Center for the New West likes to point out, these are the qualities of life that many companies are looking for today as means of attracting and retaining highly skilled workers (Burgess 1988).

For the small rural community, the exploration of alternative futures is really a search for solutions. This search seeks to identify not just probable futures, but possible futures that may include components radically different from the historical economic base of the community. John and associates (1988) make the point that in many rural areas there is great interest in finding industries that are compatible with rural conditions. These include industries that rely on telecommunications and can locate wherever such facilities are available. Examples include food-processing and other value-added industries, import replacement, home-based manufacturing, alternative crops such as vegetables and flowers, and low-pesticide crops.

Strategies and Policies

Choy and Rounds (1992) examined strategies for rural community economic development on the coterminous Great Plains of North America. They provide an excellent overview of a number of successful strategies for small towns on the Great Plains, as well as some policy guidance. Policymakers should note that communities perceived support from both regional and federal governments as being only about half of what rural communities needed or wanted.

Echoing the growing sentiment that agricultural restructuring has drastically and rapidly changed the relationship between farming and rural communities, they emphasize that agricultural sustainability and development are quite distinct from rural sustainability and rural community development. They assert that such a dichotomy requires change in many of the conventional ways of viewing rural areas, and legislation and regulations must allow for a wider variety of possibilities from the rural regions.

These authors conclude that rural revitalization is dependent on addressing the issues of community apathy, appropriate financial assistance, and expanding options for the development of rural areas. They also find that traditional agriculture must see itself, and must be seen by others, as only one element of what must be included in a more diversified region. According to Choy and Rounds, such paradigmic shifts are required if rural communities on the Northern Plains are to be sustained.

Strategies and policies to support the economic basis of the sustainable rural community may be developed at the national, state, and local levels. Only a few critical possibilities are noted here.[2] At the national level, perhaps the strategy most essential to acheiving sustainable rural communities is to act on the recognition that farm policy is no longer an effective rural development policy. The USDA notes that until a decade or two ago rural policy and farm policy could often be considered synonymous. However, such a congruency is no longer defensible. Today we find that other economic influences, besides those related to farming, now exert more important effects on the rural economy (USDA 1987).

On June 10, 1994, Secretary of Agriculture Mike Espy kicked off a team approach to the development of the 1995 Farm Bill. Within this team, subgroups have been formed; one, chaired by Undersecretary Bob Nash, focuses on rural development and protection (RDP). Nash created teams to work on eight issues, one of which is sustainable rural development. Secretary Espy drafted a report that guides the overall RDP effort and advocates a new vision of the federal role in rural development. This new vision is to provide assistance, based on inclusive development initiatives, to help rural communities become more competitive in a world marketplace through creating sustainable economic opportunities for all residents. The report also states priorities that address the increased viability of rural communities in declining, sparsely settled regions such as the Great Plains (G. A. Bernat, e-mail to author, June 28, 1994).

There is a clear need for national policy to recognize that farm-based communities constitute distressed regions shared by states trying to shed their agricultural image and diversify their economies (Strange et al. 1990). There are reasonable arguments regarding the needs of the distressed farm town versus the needs of the inner city neighborhood. If farm-based communities represent disadvantaged populations, so-called triage policies would manifest overt discrimination.

Hence, public officials and, particularly, professional planners are compelled by their codes of ethics to expand choice and opportunity for all persons, to recognize a special responsibility to plan for the needs of disadvantaged people, and to urge changing policies, institutions, and decisions that restrict choices and opportunities.

New policies must include programs of the cooperative extension service, which is desperately needed to implement nonfarm rural development assistance programs in community and regional planning, as well as community economic development. Federal support for research and extension activities at land grant universities must be increased significantly; over the last decade such rural development funding has declined. The federal administration has shifted its focus away from rural areas, and fewer dollars have been made available for rural development (Bradshaw and Blakely 1987).

Because of the shift from federal to state responsibility, the state now has the significant role in designing and administering specific policies supportive of sustainable rural communities. Unfortunately, in some cases this has led to the politicization of rural development and economic development policies and programs. Such turf wars among state agencies, educational programs, and institutions have resulted in situations in which no rural community can be effectively served on a sustained basis.

Scarce resources, including specialists and funds, imply the need to coordinate technical assistance to help small towns develop the capacity to become self-sustaining. One means of effectively coordinating the delivery of services to these rural regions is the linkage model (Moe 1975).

The Linkage Strategy

Cooperative behavior appears to be a key to the concept of sustainable communities. States should promote collaboration among adjacent rural communities, helping each to identify a specialized role as a "neighborhood" in the surrounding region (USDA 1987). The linkage strategy is based on Havelock's (1969) concept of planning for innovation through dissemination and utilization of knowledge. For the achievement of sustainable rural communities, states should work to relate existing and available knowledge and skill in a federation that can be directly applied over a sustained period in direct association with community leaders.

By itself, the linkage strategy represents a cooperative-collaborative model of institutional behavior in which the design and invention of new linking mechanisms and new organization arrangements can make it possible for agencies or organizations with overlapping goals or functions to work together in the achievement of objectives (Lassey 1979).

Such a linkage model, applied within the context of the foregoing sustainable community concept, represents the normative-reeducative strategy of rural development, which emphasizes the deliberate creation of new methods and rules while helping people to learn the required changes in values, behavior, and roles through direct experience (Luther and Luther 1978).

By utilizing universities, state agencies, and professional organizations in a linkage strategy, states may provide reliable, individually tailored expertise and information, on a coordinated and sustained basis, to rural communities to help them prepare and implement their own plans and development strategies. This linkage strategy seems to fulfill the promise of the New Localism and appears to be politically appropriate to the emerging trend of increasing state responsibility in such development programs.

Employment of the linkage strategy will require some significant changes in the traditional evaluation and rewards systems at universities and some agencies. Such changes have long been advocated as a means of providing more effective linkages of knowledge and action (King 1980).

Even more effective use of the linkage strategy may be achieved if states encourage rural communities and counties to pool their resources and form linkages for regional planning and development. Regional planning should be reconsidered for these rural areas.

Sources for Solutions

There are many resources available for assistance to small rural communities on the Great Plains: business service programs, community assistance programs, economic development programs, and information resources. A recent count showed 91 organizational resources in Colorado, 105 in Kansas, 75 in Nebraska, and 55 in Wyoming. The following paragraphs discuss a few of the more significant resources available to rural small towns on the Great Plains.

The Heartland Center for Leadership Development of Lincoln, Nebraska (an independent nonprofit organization) provides training for leaders in communities, businesses, and organizations to help them deal confidently with change. In addition to publishing practical guides and booklets, the center offers individualized workshops and hands-on programs to meet the specific needs of communities and organizations.

The Rocky Mountain Institute is an independent, not-for-profit research and educational organization that presents a wide range of economic development information through its Economic Renewal Program. The program provides information, practical tools, and workshops to help small communities tackle the challenge of revitalizing their local economy. The institute assists communities in developing their own economic renewal, publishes how-to workbooks and case books on economic renewal, and offers a four-part economic development training program for elected officials and staff.

The Partnership for Rural Nebraska is a cooperative commitment by the State of Nebraska, University of Nebraska, the U.S. Department of Agriculture, and other stakeholders, to address rural opportunities and challenges identified by rural Nebraskans. Together, these institutions have the potential to leverage additional resources to benefit rural Nebraska, and to help make even more effective use of existing resources. These resources include the Nebraska Rural Development Commission, the Nebraska Community Foundation, the Nebraska Development Network, the State Department of Economic Development, and the University's Center for Rural Community Revitalization and Development, as well as USDA's three mission areas of Natural Resources, Rural Development, and Research and Extension.

The mission of the Huck Boyd National Institute for Rural Development is to provide an integrated and coordinated approach to addressing current and future needs of rural Kansas by helping residents help themselves. Most of the institute's work deals with encouraging interagency coordination, informing citizens of rural development resources, and studying issues and approaches vital to the success of rural communities.

The Kansas Center for Community Economic Development is designed to bring university expertise to rural-based communities throughout Kansas. The center works within the framework of state policy and other businesses and service improvement efforts to help stimulate development. It accomplishes this through enhancement of strategic planning, evaluation of existing strategies, and development of information outreach and database support. The center is a joint effort of the University of Kansas Institute for Public Policy and Business Research and Kansas State University's Center for Rural Initiatives. The center provides several services to enhance economic development, which include strategic planning assistance, consultation in community development projects, and information dissemination.

The Nebraska Community Improvement Program (NCIP), operated through the State Department of Economic Development, provides technical assistance with community development goal-setting and resource programs. Through cosponsorship with the University of Nebraska, the NCIP also offers design and planning assistance to communities through the College of Architecture and seeks funds for community design and improvement programs.

The Nebraska Development Network is a partnership of public and private agencies, businesses, and organizations involved in Nebraska's communities. The network seeks to create and support community capacity to use public and private resources to create healthy businesses and a healthy community environment. Among the elements of the Nebraska Development Network are an online computer bulletin board service (Nebraska Online), access to electronic databases, electronic mail, bulletin board referral, electronic news services, and a directory for Nebraska development services.

In addition to the aforementioned resources available to rural small towns of the Great Plains, there is a multiplicity of resources, which include business advancement centers, small business development centers, enterprise zones, cooperative educational services, university cooperative extension, state agencies, state municipal leagues, state community block grant programs, councils of government programs, regional planning commissions or agencies, resource conservation and development organizations, and other independent not-for-profit organizations. The problem lies in connecting knowledge with action—linking the right resources with the right clientele and asking the right questions.

ASKING THE RIGHT QUESTION

There have been many sensationalized press accounts of the demise of small towns on the Great Plains. The decline of such communities seems to derive from misguided strategies and self-fulfilling prophecies that ask only, Who goes? The key to sustainable rural communities, however, is found in the following question: In an age of increasingly scarce resources available for small rural towns,what can we do? The capacity of communities to do it themselves for themselves is critical to their survival and persistence. In an activity that may be described as community self-actualization, the rural small town can explore, discover, describe, and explain the elements, attributes, and synergy of its total operating environment. The learning community not only conducts this survey and analysis, but learns about itself in the process.

Resources for both knowledge and skill, which offer planning and development strategies for the survival and persistence of these small towns, are available within each state. Collaborative efforts through linkage strategies can derive effective and efficient use of these resources.

Nationally, planning and development professionals must recognize the distinct character and needs of the rural community. Development policies should include a clear obligation to sustain small communities. If our national policies continue to comparatively disadvantage small communities, it is unrealistic to expect them to thrive. The Center for Rural Affairs notes these current development policies for small towns are likely to fail. Such failure can bolster the forecasts of doom that seem to pervade the current discussions about these small rural communities (Strange et al. 1990).

For planners and community economic developers, the responsibility to nurture and foster sustainable communities is axiomatic. It is our clear, compelling, and collective responsibility to recognize the fundamental rights of the rural small town resident to a sufficient and sustainable environment of quality that will permit a life of dignity and well-being in the community, now and in the future. There is still life on the Plains.

NOTES

1. The Center for the New West conducted telephone surveys of small communities that are actually surviving and succeeding. The research team worked through state agencies in 12 states to identify small communities that appeared to be successfully fighting decline. Telephone interviews were conducted with towns in seven states. Community Profiles were developed for 12 Great Plains towns. Opinion surveys were conducted with economic development practitioners. Some 30 experts from nine states participated in Roundtables on the Future of the Great Plains. More than 100 individuals in 10 states participated in a Survey of the Future of the Great Plains. The yearlong literature review included academic and professional journals, books, government and private reports, newspapers, and magazines. In-depth case studies, based on extensive field research, were prepared. The case study methodology was developed by Wall and Luther of the Heartland Center for Leadership Development based in Lincoln, Nebraska. Two communities were examined: Superior, Nebraska, and Brush, Colorado. Superior, Nebraska, is located in a county identified as distressed by the Poppers' Buffalo Commons study (Shepard et al. 1992).
2. For an excellent review of existing and needed strategies, see Strange and associates (1990), which includes a review of state economic development policies for small agricultural communities The reader might also review John and associates (1988).

REFERENCES

Arendt, R. 1994. *Rural by Design: Maintaining Small Town Character.* Chicago: American Planning Association.

Baker, H. 1993. *Restructuring Rural Communities, Part 1: With Special Emphasis on Multicommunity Collaboration.* Saskatoon: University Extension Press.

Blowers, A., ed. 1993. *Planning for a Sustainable Environment: A Report by the Town and Country Planning Association.* London: Earthscan Publications Ltd.

Bradshaw, T., and E. Blakely. 1987. "Unanticipated Consequences of Government Programs on Rural Economic Development." In *Rural Economic Development in the 1980s: Preparing for the Future.* Washington D.C.: USDA Economic Research Service.

Brown, L., C. Flavin, and S. Postel. 1990. "Picturing a Sustainable Society". In *State of the World*. New York: W.W. Norton.

Burgess, P. 1988. "Bootstraps and Grassroots: The Role of Private Institutions in Rural Development." In *Rural Development Policies for the 1990s*. Symposium proceedings. Congressional Research Service and the Joint Economic Committee of the U.S. Congress, Washington D.C. Published as *Center Reports*. Denver: Center for the New West.

Choy, K. A., and R. C. Rounds. 1992. *Community Development Strategies on the Northern Plains*. Brandon, Mani.: Rural Development Institute, Brandon University.

Craig, D. P., and B. K. Baker. 1994. *State of the Community: A Sustainable Community Roundtable Report on Progress Toward a Sustainable Society in the South Puget Sound Region*. Olympia, Wash.: Sustainable Community Roundtable.

Daniels, T. L., and M. B. Lapping. 1987. "Small Town Triage: A Rural Settlement Policy for the America Midwest." *Journal of Rural Studies* 3:273–280.

De Bres, K., and M. Guizlo. 1992. "A Daring Proposal for Dealing with an Inevitable Disaster? A Review of the Buffalo Commons Proposal." *Great Plains Research* 2:165–178.

Duncan, D. 1993. *Miles from Nowhere: Tales from America's Contemporary Frontier*. New York: Viking Penguin.

Economic Policy Council. 1990. *Rural Economic Development for the 90s: A Presidential Initiative. Working Group on Rural Development*. Washington D.C.: Office of the President of the United States.

Ekstrom, B. L., and F. L. Leistritz. 1988. *Rural Community Decline and Revitalization: An Annotated Bibliography*. New York: Garland.

Fitchen, J. M. 1991. *Endangered Spaces, Enduring Places: Change, Identity and Survival in Rural America*. Boulder, Col.: Westview Press.

Hart, M. 1995. *Criteria and Ranking Scheme for Indicators of Sustainability*. North Andover, Mass.: QLF/Atlantic Center for the Environment.

Havelock, R. 1969. *Planning for Innovation Through Dissemination and Utilization of Knowledge*. ISR. Ann Arbor: University of Michigan.

Hester, R. T., Jr. 1990. *Community Design Primer*. Mendocino, Cal.: Ridge Times Press.

Hoover, E. M. 1971. *An Introduction to Regional Economics*. New York: Alfred A. Knopf.

John, D., S. Batie, and K. Norris. 1988. *A Brighter Future for Rural America? Strategies for Communities and States. Center for Policy Research*. Washington D.C.: National Governors' Association.

Johnson, K., and R. Bauen. 1994. "Sustainability: Guiding Community Action." *Community and the Environment*, a supplement to *The Changing Northwest* 6(2).

King, G. W. 1980. *Rural Development and Higher Education: The Linking of Community and Method*. Battle Creek, Mich.: W. K. Kellogg Foundation.

Lassey, W. 1979. *Partnership for Rural Improvement: An Emerging Rural Planning and Development Model*. Pullman: Washington State University.

Lippitt, G. 1973. *Visualizing Change*. Fairfax, Va.: NTL Learning Resources Corporation, Inc.

Luebke, F. 1990. "Back to the Future of the Great Plains." *Montana: The Magazine of the West* 40 (Autumn).

Luther, J., and W. Borner. 1995. "Planning the Sustainable Community: Performance Indicators and Standards." Paper presented at conference, Four Corners Region and Utah State Chapter of the American Planning Association and the Western Planner organization.

_______ . "The Learning Community: Survival and Sustainability on the Great Plains." In *Issues Affecting Rural Communities*. Townsville, Australia: James Cook University of North Queensland.

Luther, J. 1990a. "Participatory Design: Vision and Choice in Small Town Planning." In *Entrepreneurial and Sustainable Rural Communities*, edited by F. W. Dykeman. Sackville, N.B.: Rural and Small Town Research and Studies Program, Mount Allison University.

_______ . 1990b. "The Future of the Great Plains: A Sustainable Vision." *The Western Planner Journal* 11:3–4.

_______ . 1986. "Visions from the Heartland: A Program of Alternative Images for Rural America." Paper presented at national conference of the American Planning Association.

_______ . 1981. "Transactive Planning as a Principia Media in Rural Planning Education." Paper presented at national conference of the American Planning Association.

Luther, J., and V. Luther. 1978. "The Partnership for Rural Improvement: Linking Knowledge and Action." Paper presented at national conference of the American Institute of Planners.

Luther, V., and M. Wall. 1986. *The Entrepreneurial Community: A Strategic Planning Approach to Community Survival*. Lincoln, Neb.: Heartland Center for Leadership Development.

McHarg, I. L. 1969. *Design with Nature*. Garden City N.Y.: Doubleday/The Natural History Press.

Moe, E. 1975. "Strategies in Rural Development." Working paper. Washington, D.C.: USDA.

National Commission on the Environment. 1993. *Choosing a Sustainable Future*. Washington, D.C.: Island Press.

Panos Institute. 1987. *Towards Sustainable Development*. London: Panos Publications.

Popper, D. E., and F. J. Popper. 1994. "Great Plains: Checkered Past, Hopeful Future."*Forum for Applied Research and Public Policy* 9(4):89–100.

_______ . 1987. "The Great Plains: From Dust to Dust: A Daring Proposal for Dealing with an Inevitable Disaster." *Planning 53(12):12–18.*

Rees, W. 1995. "Achieving Sustainability: Reform or Transformation?" *Journal of Planning Literature* 9:343–361.

Roseland, M. 1992. *Toward Sustainable Communities: A Resource Book for Municipal and Local Governments*. Ottawa, Ont.: National Round Table on the Environment and the Economy.

Shepard, J. C. 1994. "Grassroots Response from the Great Plains." *Forum for Applied Research and Public Policy* 9(4): 101–105.

Shepard, J. C., C. B. Murphy, L. D. Higgs, and P. M. Burgess. 1992. *The Great Plains in Transition: Overview of Change in America's New Economy*. Denver, Col.: Center for the New West.

Strange, M., P. Funk, G. Hanson, and D. Mack. 1990 *Half a Glass of Water: State Economic Development Policies and the Small Agricultural Communities of the Middle Border.* Walt Hill, Neb.: Center for Rural Affairs.

Sustainable Seattle. 1993. *Indicators of Sustainable Community.* Seattle, Wash.: Metrocenter YMCA.

Swanson, L. F. 1990. "Rethinking Assumptions About Farm and Community." In *American Rural Communities,* edited by A. E. Luloff and L. E. Swanson. Boulder Col.: Westview Press.

Thompson, W. 1978. *Darkness and Scattered Light: Speculations on the Future.* New York: Anchor Press/Doubleday.

Tisdale, C. 1993. *Environmental Economics: Policies for Environmental Management and Sustainable Development.* New Horizons in Environmental Economics Series; Aldershot, UK: Ashgate.

U. S. Department of Agriculture. 1987. *Rural Economic Development in the 1980s: A Summary.* Agriculture Information Bulletin No. 533. Washington D.C.: Agriculture and Rural Economy Division, Economic Research Service.

———. 1989. *Signs of Progress: A Report on Rural America's Revitalization Efforts.* Washington D.C.

Veregge, N. 1995. "Sense of Place in the Prairie Environment: Settlement and Ecology in Rural Geary County, Kansas." *Great Plains Quarterly* 15:117–132.

Wall, M., and V. Luther. 1989. "Rural Communities Cultivate Entrepreneurial Businesses." In *Visions from the Heartland.* Lincoln, Neb.: Heartland Center for Leadership Development.

———. 1992a. *Clues to Rural Community Survival.* Lincoln, Neb.: Heartland Center for Leadership Development.

———. 1992b. *Six Myths About the Future of Small Towns.* Lincoln, Neb.: Heartland Center for Leadership Development.

World Commission on Environment and Development. 1987. *Our Common Future.* New York: Oxford University Press.

Wrobel, D. M. 1993. *The End of American Exceptionalism: Frontier Anxiety from the Old West to the New Deal.* Lawrence: University Press of Kansas.

Zukav, G. 1979. *The Dancing Wu Li Masters: An Overview of the New Physics.* New York: Bantam.

Nine

Self-Development as a Strategy for Rural Sustainability

Gary P. Green

INTRODUCTION

Research on sustainability has focused primarily on the ecological stance against economic growth and development (Mitlin 1992; Rees 1995). At its core, sustainability refers to the extent to which development is either self-undermining or self-renewing. This literature has focused on whether, and how, economic development can produce a sustainable environment. Sustainability is especially important for rural areas that historically have been dependent on the extraction of natural resources (i.e., forestry, agriculture, fishing, mining) as their economic base. It also is relevant, however, for rural communities that are no longer economically dependent on the extraction of natural resources as a source of jobs and income. Increasingly, rural areas are valued not as a source of natural resources for production, but for their aesthetic worth. Thus, preservation of the aesthetic quality of rural areas continues to have an important economic dimension as well.

It is possible to take a broader view of rural sustainability. To create a sustainable economy, communities must also develop long-term strategies for the viable use of financial, human, and social resources, as well as environmental resources. Economic restructuring has produced significant population and employment changes in rural communities. Rural America, as compared with urban areas, faces several disadvantages as the U.S. economy continues to shift rapidly away from the production of goods toward delivery of services and as markets become increasingly global.

Development of a postindustrial economy disproportionately benefits urban areas that have large concentrations of people and businesses. Although advances in communication and transportation have improved access to many rural communities, the primary beneficiaries are those communities on the urban fringe. The more isolated communities continue to

Material in this chapter is based on work supported by the College of Agricultural and Life Sciences at the University of Wisconsin-Madison (Hatch Grant #3677).

struggle to survive. These communities are searching for development strategies that will generate a sustainable economy.

Although most rural communities have responded to globalization and postindustrialism by intensifying their industrial recruitment efforts, a few localities are employing community-based strategies (referred to here as self-development strategies) to generate jobs and income. These strategies stand in stark contrast to traditional economic development efforts, such as industrial recruitment, to attract employment and population to the community. Self-development strategies minimize dependence on external organizations and actors by promoting local ownership and control of resources (especially land, labor, and capital). These efforts are creating demand among extralocal actors, which produces local benefits and returns surplus created back to the community. They emphasize a decentralized and more egalitarian social order based on ecological principles.

This chapter examines self-development efforts toward rural sustainability. Self-development strategies offer a holistic approach to development that generates local benefits, balancing various objectives such as job creation, citizen participation, and environmental quality. Based on data collected from approximately 100 rural communities involved in self-development efforts, their contribution to rural sustainability is assessed.

COMMUNITY SUSTAINABILITY IN THE RURAL CONTEXT

A common theme in the literature on American communities is loss of autonomy and absorption into the larger culture and economy as a consequence of urbanization, industrialization, and bureaucratization (Stein 1960; Vidich and Bensman 1958; Warren 1978). As Bender (1978) shows, however, ties between communities and the larger social and economic system were established early in American history, but small towns probably have become much more dependent in recent decades. Several types of dependency may affect local development prospects. First, some rural communities are constrained by their economy's dependence on natural resources and commodities late in their product or profit cycle. Second, most rural communities are dependent on a few employers in their economic base. Finally, most rural communities are dependent on external sources of capital, which have become increasingly mobile and abstract. Self-development efforts typically address one or more of these dimensions of community dependency. Summers and associates (1988, x) suggest that the key to economic vitality is the capacity to ensure a flow of jobs and income over time. Self-development may offer a more reasonable opportunity for accomplishing these objectives than do other economic development strategies in which rural communities might engage.

Many of the problems facing rural communities in the 1990s are rooted in their dependency on industries that are vulnerable to wide fluctuations in prices, especially those based on natural resources (Freudenburg 1992) or manufacturing industries late in the product or profit cycle (Markusen 1985). Historically, these economies have been dominated by agriculture, forestry, mining, or fishing, but a growing number of rural communities are becoming dependent on manufacturing for employment. Industries that are likely to move to rural areas generally provide low-skilled jobs and low pay. Dependence on these two types of commodities (natural resources and mass-manufactured goods) tends to create periods of boom and bust for rural communities.

Rural communities also face a second type of dependency that constrains their economic development efforts. Because their economies are not diversified, many rural communities are dependent on the fate of a few employers, and a sudden shift in a single market may create high levels of unemployment and poverty.

There is a third, and probably more debilitating, form of dependency that constrains development efforts, that based on the relationship of communities to external sources of capital. Globalization and the accelerating mobility of capital have generated new sources of vulnerability in rural areas. What is distinct about the current period of restructuring is that communities are becoming similar to commodities. Firms now play one community off another in an attempt to obtain the most attractive deal. Communities are evaluated on the basis of how much they can contribute to corporate profits (exchange value) rather than how good they are as places to live (use value). Gunn and Gunn (1991, 2) characterize the contradiction between capital and community in this way:

> Capital wants profit; communities want development. Communities want well-paying jobs for their residents; investors are driven to pay the lowest possible wages relative to capital costs at given levels of productivity. Capital seeks an environment free of costly regulation; communities require a life-sustaining ecology. Communities are defined by place and stability; capital is concerned with location primarily as a factor in transportation and transaction costs.

For some analysts, the rational choice for rural communities is to enter the competition for capital. Because some communities are providing incentives (e.g., tax breaks) and subsidies (e.g., loans), others must enter the "arms race" for jobs. Communities that are not growing are unable to provide public services as cheaply as growing communities. With relatively high taxes or low levels of public services, communities lose population, which weakens their viability even further.

However, an increasing number of rural communities are adopting innovative approaches to economic development that involve a wide spectrum of groups and organizations in the locality (see Flora et al. 1991). These grass roots approaches to local economic development address more directly the problem of sustainability. By taking control of local resources, communities are attempting to reduce their dependency on the resources and decisions of external actors and organizations. Some communities seek to minimize their dependency on external markets, others attempt to become less dependent on a single industry, and a limited number of communities strive to restructure their relationship with capital through "decommodifying" local land, labor, and capital.

Rather than withdrawing from the larger economy or society, or advocating change from above, self-development communities seek to manage their relationship to the larger economy or society. One approach to developing a sustainable community is to decommodify the major factors of production—land, labor, and capital. Polanyi (1957) argued that attempts to fully commodify land, labor, and capital in market societies have been unsuccessful. Society has not relied (and cannot rely) entirely on markets to allocate these scare resources. According to Polanyi, many societal needs would go unmet if markets were totally unregulated. Thus, self-development can be conceptualized as an effort by communities to manage these market relations in a way that better satisfies local (societal) needs. A basic assumption of this approach is that markets must be embedded in social relations in order to generate benefits that are sustainable (Granovetter 1985).

There is a conceptual link, then, between community sustainability and self-development. By placing ownership and control at the local level, communities have greater potential to develop strategies acknowledging the relationships between development, the environment, and the social needs of their citizens. Self-development establishes a logic that broadens the profit-maximization goals of the market. Profits are still important, but economic decisions are embedded in social and environmental objectives as well. To put it a bit more concretely, the goals of the development strategy focus not only on increasing profits, but also on producing jobs at a livable wage and on practices that do not degrade the environment. A key element of self-development is the allocation of development decisions to the local level, where relationships between development, the environment, and social needs are most visible.

RELATIONSHIPS OF COMMUNITIES TO MARKETS AND THE STATE

Most approaches to economic development of rural communities emphasize either market or bureaucratic (state) solutions to their problems. Market approaches typically focus on the costs of space (distance) and the need for communities to reduce the costs of production or to increase productivity as a means of attracting basic employment. According to this view, communities are inevitably engaged in a competition for capital and population, which leads them to pursue development policies. Bureaucratic approaches to community economic development stress the need for state intervention to overcome these problems. State subsidies are necessary because market mechanisms will not solve the problems of rural communities and intervention will allow the market to operate more equitably.

Advocates of self-development argue, however, that pure market or bureaucratic approaches will fail to produce a sustainable community. Markets and the state are essential elements of self-development, but change must come from within communities. The following paragraphs briefly describe each of these approaches to community economic development.

Market-Oriented Approaches

Rural communities have been rather slow to enter the competition for jobs. Increasingly, however, local leaders have begun to provide a wide range of incentives and subsidies to attract industry to their localities. The most frequently used tools for promoting growth are tax breaks, providing buildings and land to firms, and discounting loans through industrial development bonds. Traditional economic development strategies attempt to make the community a more attractive commodity in the marketplace.

Industrial recruitment has come under fire in recent years. Research on the effectiveness of these financial incentives and tax policies is inconclusive (Wilson 1989). Much of it indicates that such incentives have little or no effect on location decisions, but recent research by Bartik (1991) suggests that the incentives may have an effect on job growth. For many rural communities the probability of success is very low, and they may end up paying more in incentives than they receive in benefits. As competition for jobs has increased among localities, the number of incentives offered has grown dramatically.

One of the weaknesses of traditional economic development strategies—what Eisinger (1988) refers to as supply-side approaches—is that capital has become increasingly mobile and likely to shift locations. This means that even rural communities successful in attracting manufacturing plants to their localities may lose jobs after a short period. Exacerbating this tendency are corporate mergers that also contribute to rapid movement of capital. The result is that it is becoming increasingly difficult for communities to develop a viable economy through traditional economic development strategies.

Bureaucratic Approaches

Local government involvement in economic development is a relatively new phenomenon. Historically, economic development has been the responsibility of federal and state governments. State government involvement in economic development can be traced back to the 1930s, when the state of Mississippi began offering industrial revenue bonds to attract firms (Cobb 1982). Beginning in the 1950s, the federal government contributed to regional development by building the interstate highway system and by investing in other types of physical infrastructure. During the 1960s the federal government extended its involvement in economic development by increasing federal aid to state and local governments. From 1950 to 1970, federal aid to state and local governments grew from $2.3 billion to $24.4 billion.

During the early 1960s, the Area Redevelopment Administration was responsible for much of the policy relating to rural areas in the United States. One of the key elements of this program was the requirement that citizens and local government officials draft an economic development plan prior to receiving any federal grants; yet participation by citizens was largely nominal. Purely bureaucratic approaches to community development, however, fail to produce rural sustainability. These programs frequently do not identify local needs or incorporate local citizens into the decision-making process. Although most bureaucratic solutions today require some form of participation, most of it can be characterized as "pseudoparticipation" (Pateman 1970). This process requires citizens to provide input, but it does not provide a real basis for citizens to influence decisions affecting their community.

Government has become involved in rural development for several reasons. The public sector may intervene in markets because of market failures or for the public welfare. Market failures can be the result of monopolistic control, imperfect information, or other external factors. Government may also intervene to ensure a basic level of living for its citizens. In many instances, the purpose of government involvement in rural development has been to provide basic needs, such as electricity, which would be difficult to provide through market solutions.

State or bureaucratic solutions, however, present several problems. Obviously, there is growing concern about the costs of government programs. In addition, there is concern that many government programs do not produce a viable economy in the long run because they do not empower local citizens (McKnight 1995). Moreover, government programs are seldom based on local needs or strengths.

Self-Development Strategies

Self-development strategies attempt to overcome some of the weaknesses of relying primarily on market or bureaucratic solutions to community development. Self-development efforts do not deny the importance of market or state activities; rather, they attempt to embed them in the context of community strengths. A basic premise of this approach is that development strategies must be community based, emanating from within the community to provide sustainable solutions.

A central characteristic of self-development is that communities rely primarily on local resources to stimulate demand (Bruyn and Meehan 1987). Self-development strategies seek to minimize dependence on external organizations and actors by promoting local ownership and control of resources (i.e., land, labor, and capital). Strategies to promote growth from within the community via locality-based development efforts simultaneously contract and expand community ties with the larger society. Contraction means that the community attempts to become more self-reliant (but not self-sufficient). By taking control of their local resources, communities may be able to reduce their vulnerability to sudden shifts in the economy or to decisions by nonlocal organizations and institutions. Expansion means that communities extend their economic activities to national and international markets. Put more simply, community-based strategies are generating demand among extralocal actors that produces local benefits and returns the created surplus to the community (Gunn and Gunn 1991).

Community-based development strategies do not limit themselves to employing only local resources or withdrawing from nonlocal markets. Instead, they generally use a mixture of local and nonlocal resources to create community owned and controlled institutions and organizations. Both markets and the state are important components of this approach. By organizing local resources and institutions, they provide a means of building the social organization necessary to leverage outside resources.

SELF-DEVELOPMENT IN RURAL COMMUNITIES

To examine the prospects of self-development strategies in rural communities, it is useful to consider data collected from a study of approximately 103 self-development projects in rural America. A survey was sent to a key informant in each of these projects to identify how self-development projects were initiated, financed, and organized; the obstacles they faced; and the benefits and costs to each project. In-depth studies of 8 of the 103 projects yielded information on the processes of self-development in rural areas (Flora et al. 1991). We also combined the data with census data to assess the characteristics of communities adopting these strategies and how their social and economic characteristics influenced their success (Green et al. 1993).

The data suggest that most rural communities turn to self-development in response to an economic crisis. This finding implies that it may be difficult to move communities toward sustainability without an external impetus. Furthermore, we found that communities with ties to other communities, organizations, and institutions are more likely to adopt self-development strategies. Such ties provide important sources of information and access to resources that are essential to the community. Self-development communities also are characterized by a lack of strong factions that can make it difficult to obtain community support.

Several factors were identified as influencing the success of self-development efforts. The ability to leverage local resources to gain access to external sources of credit is crucial. We found that small communities were more successful in implementing self-development strategies than were large communities. It may be that it is more difficult to gain broad community support for such an effort in large communities.

The following paragraphs briefly describe three projects that focus on controlling local resources—land, labor, and capital. It should be pointed out, however, that many projects involve more than one activity. For example, Ganados del Valle, a Hispanic collective enterprise in New Mexico, includes a sheep-grazing cooperative, a weaving cooperative, an enterprise that sells organic lamb to area restaurants, and a general store.

Labor

In response to increasing numbers of plant closings, a limited number of communities have invested in community- or worker-owned firms (Stern et al. 1979). The literature suggests that community- and worker-owned firms have several advantages when compared with conventionally owned firms, including higher profit rates and greater productivity growth rates (Conte and Tannenbaum 1980). They also generate more jobs than conventionally owned firms (Rosen and Klein 1981). Community- and worker-owned firms are particularly attractive under two conditions: (1) when an owner is ready to retire and has no one to pass on the business to, which thus reduces the risk of start-up, and (2) when a company is purchased by workers or the community as a strategy to avert a plant closing. Such a buy out may be feasible even when the parent corporation has decided to shift operations to another site that is more profitable. The establishment may be making a profit, but the firm can earn a higher return on its investment by restructuring the operation. One of the major obstacles to employee/community ownership is access to capital. In many situations, workers' wages are not high enough to allow them to purchase the firm directly. Other options are trust-based ownership and employee stock ownership plans (ESOP).

An example of a community-owned business is Community Store, Inc. in Frederick, South Dakota. In 1984 the owner of the Frederick grocery store could not sell the business and was ready to close the doors. The community, not wanting to lose its only grocery store, held public meetings and began fund-raising for the business. A for-profit corporation was formed with assistance from an attorney in Aberdeen, and contributions were converted into $1 shares in the corporation. Approximately 200 people (out of a total population of 320) contributed $36,000 to purchase the building and inventory in 1985. By 1988 the corporation made a profit of more than $9,000.

There are numerous examples in rural areas of businesses that are closing, and workers and residents are considering buying out the owners. However, they face many obstacles in these situations. Most worker- or community-owned firms lack access to capital, commercial lending institutions are generally reluctant to make loans to them, and workers' wages are often too low to provide much financing. A further obstacle in these self-development projects is that workers and residents may not have the managerial expertise to operate a business. Managerial skills were sorely missing in many of the worker- and community-owned businesses examined in this study. Finally, most of the worker- and community-owned businesses confronted a contradiction in having both economic and social objectives.

We have very little evidence at this time to determine what types of businesses (i.e., resource extraction, manufacturing, services, commercial, farming, etc.) represent the greatest opportunities for worker ownership. Given the capital constraints, however, industries with low capital requirements may provide the greatest opportunity, but these operations also provide lower profits and earnings. Yet a more important consideration may be the type of market in which the business is located. Businesses in niche markets may provide more opportunity inasmuch as they may not be competing in global markets.

Capital

Most commercial financial institutions shift capital to the most profitable location or economic sector, regardless of where the investors or savers are located. Rural communities generally do not lack capital; they are usually net exporters of capital even when there is a demand locally for capital. An alternative way of operating financial institutions is to generate profits while meeting community needs. This approach is frequently used by community-oriented finance institutions, which take into consideration the social, as well as the economic, benefits of investment.

There are several innovative types of community-oriented finance institutions in rural areas, including community development loan funds (CDLF), community development credit unions (CDCU), microenterprise loan funds (MLF), royalty investment funds, and revolving loan funds (RLF). Each financial institution model has its own set of advantages and disadvantages (Parzen and Kieschnick 1992). CDLFs are not-for-profit corporations that take loans from individuals and institutions and allocate this capital to community development projects within their geographic areas

(Swack 1987, 82). A CDCU is a cooperative, not-for-profit corporation, created by and for people, that reinvests members' savings back into their community for development purposes. There are approximately 400 CDCUs in the United States. An RLF is a pool of money used for the purpose of making loans and loan guarantees. As the loans are repaid by borrowers, the money is returned to the fund to make additional loans. Thus, the fund becomes an ongoing or "revolving" financial tool. The Economic Development Administration capitalizes many community RLFs, but a variety of other sources also capitalize such funds. A central characteristic of these community-oriented finance institutions is that they are governed by a wide range of community representatives (Swack 1987). In most cases, RLFs provide debt capital, but not equity capital. Community-oriented institutions direct their lending to enterprises and activities that have difficulty accessing traditional capital sources. The borrowers are usually locally owned businesses that will provide community benefits.

A successful and innovative community-oriented financial institution is the Association for Regional Agriculture Building the Local Economy (ARABLE) in Oregon. ARABLE is a not-for-profit community investment association designed to provide investment capital from member lenders to member borrowers for socially and environmentally sound enterprises within the local agricultural economy. The organization, founded by organic farmer activists in the Eugene area, is one of the emerging set of new institutions supporting organic agriculture in the state of Oregon.

Our research on community-oriented financial institutions suggests that they can be an effective mechanism for directing capital to locally owned businesses. Yet, currently, they are seriously limited in the number and size of loans they make. Although they can help support some small businesses, community-oriented financial institutions usually do not have the capital to invest in larger firms. In most cases these institutions have difficulty in becoming self-supporting and must rely on foundations and other sources of capital. Self-sufficiency is a serious problem for microenterprise loan funds. Thus, community-oriented financal institutions have to rely on external institutions until they can become viable in the future.

Land

Community land trusts have been developed in many urban and rural communities to deal with problems of land speculation. If land is allowed to operate purely as a commodity, its exchange value takes precedence over its use value. In many rural areas, the high cost of land is a major obstacle to beginning farmers and poor people seeking to buy homes. Land trusts remove land from local markets and place it in the hands of the community.

An example of a land trust in a rural community is H.O.M.E., Inc. in Maine. In 1978, H.O.M.E. started the Covenant Community Land Trust and has helped to build 14 family farmhouses. Wood for the houses is harvested and processed in H.O.M.E.'s own sawmill and shingle mill, and cabinets are made in the woodworking shop. The houses come with 10 acres, which allows low-income people to supplement their earnings by gardening, gathering firewood, and raising animals. Although the Hancock County shore land is under intense development pressure, with the inland area mostly owned by paper companies, the land owned by the community land trust is kept out of the speculative market. As a result, low-income people have greater access to affordable housing.

Land trusts obviously face a number of obstacles, particularly in regard to equity capital. Several land trusts have been financed through community reinvestment agreements, but this source is not likely to be available in the future. Although more information is now becoming available, there is still a need for some form of technical assistance in establishing and organizing a land trust.

WHAT WORKS? WHAT DOESN'T?

Although a growing number of communities are adopting self-development strategies, many of these projects fail. What factors influence success? First, the self-development project must be appropriate for the locale; it should build on community strengths and the skills of the work force. Some communities identify projects that have proven successful elsewhere and assume they will work for them. Second, successful self-development projects rely on local funds to leverage outside sources of capital. Almost all of the self-development projects examined in the study relied on external sources of capital, and funding was a strong predictor of the number of jobs or the amount of income generated in the community. Third, rural communities that have organized self-development projects frequently lack technical information. In many cases the information required was in the area of production techniques. For example, a group of herb growers in North Carolina sought information on growing and marketing herbs but could not find any that was appropriate. In other instances, legal or organizational information was required. Self-development communities frequently need assistance with incorporation (legal advice) or with organizational management (e.g., how to form a not-for-profit or how to organize a cooperative). Fourth, self-development activities appear to work best in small towns, where it is easier to gain broad community support for the project. However, small towns also have small cliques or networks, which may limit interaction between various groups.

One of the interesting findings from the survey of the 103 self-development cases is that about half were located in the Midwest. This finding can be explained in several ways. Obviously, this region experienced a considerable amount of economic stress in the 1980s with the farm crisis and the restructuring of basic industries. Another interpretation, however, is that there is something about Midwestern communities that facilitates self-development. A relatively strong sense of community may contribute to community-based development. The lack of sizable ethnic groups in most Midwestern communities, however, may be a factor as well. We also found that a disproportionate number of the projects were related to agricultural production, which may also explain the high incidence of self-development in the Midwest.

There are several possible ways to increase the likelihood of success for self-development efforts. Supporting organizations, either regional or national in scope, are playing an increasingly important role in assisting self-development efforts. For example, there is a lending institution in North Carolina that provides funding to worker-owned businesses. The Institute for Community Economics provides technical assistance to communities establishing land trusts. The development of additional organizations that support grass roots programs is essential. These support organizations might also establish networks among self-development projects, which would facilitate the exchange of information and ideas between communities.

State and federal regulations frequently constrain self-development efforts. For example, a legal fee is required for issuing stock. Very small projects may not have the funds necessary to meet such requirements. In another instance, there was one community in which a small locally owned meat processing plant could not operate because of federal statutory regulations that were established for the large packing houses.

Aside from providing technical assistance, development organizations can play an important role in facilitating self-development in rural areas. One of the most important needs is support for community development practitioners in rural areas who can help citizens formulate and implement strategies. Research suggests that a critical component of such community strategies is follow-up, which is often best ensured by the involvement of local practitioners.

CONCLUSIONS AND DISCUSSION

Several questions are raised by this analysis of self-development strategies in rural communities. Probably the most obvious is whether self-development can generate enough jobs and income to replace jobs and income being lost to economic restructuring. In the short run, I would have to say no. Several structural forces, such as global markets and agglomeration effects, continue to place rural communities at a competitive disadvantage in relation to urban areas. Small towns do not have the skilled labor force or market size to compete with most urban areas. Nor can small towns compete in terms of labor costs with Mexican or Caribbean communities. Thus, rural areas are being squeezed by the growth of the postindustrial economy and the shift in capital to lower-cost areas.

If self-development strategies are evaluated solely in terms of the number of jobs created and/or saved or the amount of income generated in the economy, the conclusion must be that these efforts have produced modest results for a relatively small number of communities (Green et al. 1994). Self-development projects, however, produce several benefits beyond simply jobs and income to the community. First, self-development projects provide communities with greater control over their local economy by shifting ownership of land, labor, and capital. In the short run, increasing local control may not lead to many new jobs, but in the long run it may save jobs. Decommodification of key resources (land, labor, and capital) provides communities with the basis for allocating resources to meet community needs (use value) rather than to satisfy market demand (exchange value). The emphasis is on creating local wealth (Nozick 1992).

Second, self-development often contributes to increased levels of participation among citizens in local economic development decisions. Self-development extends democratic principles from the political arena to the economic arena. Democratization is an important component of sustainability, and rural sustainability is based largely on local knowledge and input from citizens (Kloppenburg 1991).

Finally, I believe self-development may produce a more favorable benefit-cost ratio than do traditional economic development activities. The literature on rural industrialization suggests that local residents frequently do not benefit from new industry, because inmigrants take the available jobs when a branch plant locates in the community (Summers et al. 1976). In addition, with global competition for industrialization communities must offer increased levels of subsidies or incentives, thus reducing the net benefit to localities. Self-development generally does not involve such costs to the community.

The focus of this chapter has been primarily on the economic and social dimensions of rural sustainability. I have argued that self-development strategies offer communities an alternative path of development that appropriately responds to the restructuring process occurring in rural areas. A characteristic of these self-development approaches is that they are holistic, recognizing the strong linkage between economic development, environmental quality, health care, housing, and other areas of the community. This feature, I believe, is critical to any approach toward rural sustainability.

Self-development goes beyond state-versus-market debates and places greater emphasis on civil society. By promoting self-regulation, these community-based solutions have a better chance of creating sustainable communities in the future.

REFERENCES

Bartik, T. J. 1991. *Who Benefits from State and Local Economic Development Policies?* Kalamazoo, Mich.: W. E. Upjohn Institute.

Bender, T. 1978. *Community and Social Change in America.* Baltimore: Johns Hopkins University Press.

Bruyn, S. T., and J. Meehan, eds. 1987. *Beyond the Market and the State: New Directions in Community Development.* Philadelphia: Temple University Press.

Cobb, J. C. 1982. *The Selling of the South: The Southern Crusade for Industrial Development, 1936–1980.* Baton Rouge: Louisiana State University Press.

Conte, M., and A. Tannenbaum. 1980. *Employee Ownership.* Ann Arbor: University of Michigan Survey Research Center.

Eisinger, P. K. 1988. *The Rise of the Entrepreneurial State: State and Local Economic Development Policy in the United States.* Madison: University of Wisconsin Press.

Flora, J. L., G. P. Green, E. A. Gale, F. E. Schmidt, and C. B. Flora. 1992. "Self-Development: A Viable Rural Development Option?" *Policy Studies Journal* 20:276–88.

Flora, J. L., J. Chriss, E. Gale, G. P. Green, F. E. Schmidt, and C. Flora. 1991. *From the Grassroots: Profiles of 103 Rural Self-Development Projects.* Economic Research Service, Staff Report No. 9123. Washington, D.C.: U.S. Department of Agriculture.

Freudenburg, W. R. 1992. "Addictive Economies: Extractive Industries and Vulnerable Localities in a Changing World Economy." *Rural Sociology* 57:305–32.

Granovetter, M. 1985. "Economic Action and Social Structure: The Problem of Embeddedness." *American Journal of Sociology* 91:481–510.

Green, G. P., J. L. Flora, C. B. Flora, and F. E. Schmidt. 1993. *From the Grassroots: Results of a National Study of Rural Self-Development.* Agriculture and Rural Economy Division, Economic Research Service, Staff Report No. AGES9325. Washington, D.C.: U.S. Department of Agriculture.

Green, G. P., J. L. Flora, C. B. Flora, and F. E. Schmidt. 1994. "Community-Based Economic Development Projects Are Small But Valuable." *Rural Development Perspectives* 8:8–15.

Gunn, C., and H. D. Gunn. 1991. *Reclaiming Capital: Democratic Initiatives and Community Development.* Ithaca, N.Y.: Cornell University Press.

Kloppenburg, J., Jr. 1991. "Social Theory and the De/construction of Agricultural Science: Local Knowledge for an Alternative Agriculture." *Rural Sociology* 56:519–548.

Markusen, A. R. 1985. *Profit Cycles, Oligopoly, and Regional Development.* Cambridge, Mass.: MIT Press.

McKnight, J. 1995. *The Careless Society: Community and Its Counterfeits.* New York: Basic Books.

Mitlin, D. 1992. "Sustainable Development: A Guide to the Literature." *Environment and Urbanization* 4:111–123.

Nozick, M. 1992. *No Place Like Home: Building Sustainable Communities.* Ottawa: Canadian Council of Social Developments.

Parzen, J. A., and M. H. Kieschnick. 1992. *Credit Where It's Due: Development Banking for Communities.* Philadelphia: Temple University Press.

Pateman C. 1970. *Participation and Democratic Theory.* Cambridge: Cambridge University Press.

Polanyi, K. 1957. *The Great Transformation: The Political and Economic Origins of Our Time*. Boston: Beacon Press.

Rees, W. 1995. "Achieving Sustainability: Reform or Transformation?" *Journal of Planning Literature* 9:343–361.

Rosen, C., and K. Klein. 1981. "Job Creating Performance of Employee Owned Companies." *Monthly Labor Review* 106(August):15–19.

Stein, M. R. 1960. *The Eclipse of the Community: An Interpretation of American Studies*. Princeton: Princeton University Press.

Stern, R. K., H. Wood, and T. H. Hammer. 1979. *Employee Ownership in Plant Shutdowns*. Kalamazoo, Mich.: W. E. Upjohn Institute for Employment Research.

Summers, G. F., L. E. Bloomquist, T. A. Hirschl, and R. E. Shaffer. 1988. *Community Economic Vitality: Major Trends and Selected Issues*. Ames, Iowa: North Central Regional Center for Rural Development.

Summers, G. F., S. D. Evans, F. Clemente, E. M. Beck, and J. Minkoff. 1976. *Industrial Invasion of Non-Metropolitan America*. New York: Praeger.

Swack, M. 1987. "Community Finance Institutions." In *Beyond the Market and the State*, edited by S. Bruyn and J. Meehan. Philadelphia: Temple University Press, 79–96.

Vidich, A. J., and J. Bensman. 1958. *Small Town in Mass Society: Class, Power and Religion in a Rural Community*. Princeton: Princeton University Press.

Warren, R. L. 1978. *The Community in America*. Chicago: Rand McNally and Company.

Wilson, R. 1989. *State Business Incentives and Economic Growth: Are They Effective? A Review of the Literature*. Washington, D.C.: Council of State Governments.

Ten

Strengthening Social Capital: The Challenge for Rural Community Sustainability

Lionel J. Beaulieu and Glenn D. Isreal

INTRODUCTION

Since the release of the Bruntland Commission report in 1987, increased attention has been given to the important subject of sustainability. The Commission, established by the General Assembly of the United Nations under the aegis of the World Commission on Environment and Development, was charged with the responsibility of articulating the actions needed to guide the economic development activities of countries, while simultaneously working to preserve each country's environmental integrity (Kane 1992). Among the important products of the Commission's report was the formulation of a definition of sustainable development: "development that meets the needs of the present without compromising the ability of future generations to meet their own needs" (World Commission on Environment and Development 1987, 8).

The Commission's report has been instrumental in awakening individuals, businesses, and governments to the importance of considering sustainability in their decision-making processes. Within the agricultural community, for example, production of food and fiber products in an environmentally sound manner has been pursued by an expanding pool of farmers (Aplet et al. 1993; Keeney 1989, 102). Similarly, planners have become increasingly involved in the design of communities that reflect a sensitivity to the localities' ecological limits (Beatley 1995a). Even in rural communities, economic development opportunities have been undertaken with an eye on protecting and restoring the areas' critical ecological resources (Stauber 1994).

The research described in this chapter is being funded as part of a National Research Initiative grant by the CSREES/USDA titled *Social Capital Attributes of Families, Schools and Communities* and is part of Florida Agricultural Experiment Station Project FLA-4-H-03436.

Though wise management of environmental resources is a necessary ingredient in sustainable development activities, sustainable development encompasses more than just the environmental integrity of a community. According to the Organisation for Economic Co-operation and Development (OECD), sustainable development incorporates two additional interdependent systems: the economic system and the social system (OECD 1995). The underlying theme of the OECD's statement is that movement to a sustainable society and community must attend to the economic and social systems, as well as to the ecological resources system. The economic dimension entails community efforts to establish an economic base that impacts minimally on the environment, and even works toward its restoration (Beatley 1995b). The social component embraces activities that enhance the capacity of local citizens to guide the emergence of a sustainable community by improving their understanding of policy activities, empowering people to be major actors in formulating policies, and engaging all segments of the community in creating a shared vision of their future (OECD 1995; Kline 1994).

Creating a human resource pool that has the capacity to effectively guide sustainable development initiatives constitutes one of most substantive challenges facing many communities. A fundamental ingredient for facilitating the emergence of such a pool is education. In many ways, education is the key to sustainable development (Youth Source Book on Sustainable Development 1995). For example, active participation in the political process is significantly linked to education. Better-educated individuals are more likely to vote and are better prepared to take an active role in the political, economic, and social lives of their communities (U.S. Department of Education 1994, 93). According to Kane (1992, 21), education supports sustainable development in that it helps individuals understand the linkage between their own life-styles and the condition of the environment. Indeed, investments in education are essential if people are to be seen as managers and protectors of the environment (OECD 1995).

Unfortunately, as one surveys the community landscape in America, it becomes all too apparent that not all localities are equally endowed with the quality human resources needed to undergird sustainable activities. For example, the proportion of adults 25 years of age and older with a baccalaureate degree or more is 10 percentage points lower in nonmetro than in metro areas (Killian and Beaulieu 1995). Furthermore, since the early 1980s, poverty rates among nonmetro residents have consistently surpassed the levels found in metro areas, often matching the rates found in central cities of metro areas (O'Hare 1988; Porter 1989). In essence, nonmetropolitan America continues to be the home of the most chronically poor individuals in the country and has the largest share of our nation's unemployed and underemployed workers (Beaulieu and Mulkey 1995;

Wimberley 1993). How can sustainable development be pursued in rural environments in which the human capital shortfalls are so profound?

It is no accident that the constraints on human resources present in many rural communities often lead them to pursue economic development strategies that could endanger the areas' long-term sustainability. It is not uncommon, for example, for rural areas to be the major targets of efforts to site landfills, hazardous waste incinerators, and other waste-deposit or storage facilities (Couch and Kroll-Smith 1994). The degree to which communities are receptive to these activities is shaped, in large part, by the state of their economies and by their normative educational aspirations (Humphrey et al. 1993). As Bourke (1994, 494) notes, "Health and environmental risks are more likely to be accepted when economic need is or is perceived to be very high. . . . Poor rural communities may view their only option for survival as accepting the risks associated with hazardous industries."

Given the decisive role of human capital in promoting the wise stewardship of our nation's resources, the question is, How can human capital enhancements be realized, particularly in rural areas? Human capital theorists would argue that the most direct way is investment in human capital, particularly through ensuring a formal education (Becker 1962; Schultz 1961). Many critics would argue, however, that such investment decisions are frequently the product of a set of preexisting productivity-enhancement attributes that place some people in a more favorable position than others to follow through on such investments (Marshall and Briggs 1989). For example, family background has much to say about why some people are more likely than others to invest in schooling (McCrackin 1984). Furthermore, the communities or neighborhoods in which individuals reside may prove important, given that they often represent the social context in which residents' educational aspirations are fostered (Semyonov 1981).

This chapter seeks to explore the set of factors beyond the individual that may prove significant in facilitating his or her human capital investments. Its specific focus, however, is on social capital, on the set of supportive interpersonal interactions within the family and community that improve chances that a child will complete his or her high school education. The extent to which social capital and high school completion may vary across rural-urban locales is also considered in this analysis. The decision to give specific focus to the academic success of youth is motivated by a recent comment offered by Richard W. Riley, the U.S. secretary of education: "Young people make up 20 percent of the population, but 100 percent of the future" (President's Council on Sustainable Development 1996, 70). If we succeed in improving the human resource capacities of our youth, we are, in essence, creating the foundation for a more informed and capable citizenry that will participate actively in the realization of healthy, sustainable communities in the years ahead.

HIGH SCHOOL DROPOUTS, SOCIAL CAPITAL, AND SUSTAINABILITY

"An educated public is one of America's most powerful resources to meet the challenges created by increasing environmental, economic, and social demands." This penetrating comment is contained in a 1996 report titled *Sustainable America: A New Consensus,* an important product of the President's Council on Sustainable Development. Although the document delineates the vital importance of education, the reality is that far too many young people today are failing to capture the most fundamental educational product necessary to be productive, contributing members to the sustainability agenda, namely, a high school education. Education correlates with citizenship—the higher one's educational attainment, the greater one's chances of being interested and actively involved in various political matters (Herrnstein and Murray 1994). A high school education serves as a critical beginning point for facilitating active engagement in civic activities.[1]

Figure 10.1 presents data profiling trends in status dropout rates over the course of the 1975–1993 time period. It shows the status dropout rates for young adults in the 16 to 24 age cohort living in a central city, suburban community, and nonmetro (rural) areas. Status dropout rates repre-

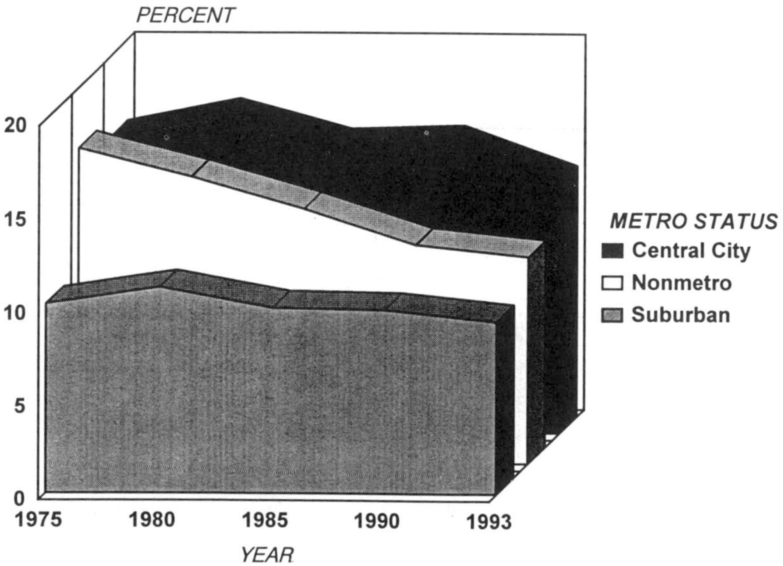

Figure 10.1 Status dropout rates by metropolitan status, 1975–1993. (*Source:* Dropout rates in the United States. U.S. Department of Education, National Center for Education Statistics)

sent the proportion of people in the 16 to 24 age cohort who have not completed a high school education and are not currently enrolled in school (as of October of that given year). These rates are important in that they reflect the extent of the dropout problem in the population (Kaufman et al. 1992, v).

Since 1975 dropout rates among residents of suburban areas have remained fairly stable, fluctuating only modestly over a period of 18 years. Throughout the 1975–1993 period, suburban areas have enjoyed the lowest percentage of persons classified as status dropouts. For nonmetro residents, the status dropout rates have been on a steady decline since 1975, hovering at 11.1% as of 1993. The most severe status dropout rates have continued in central cities, where 13.4% of the populace 16 to 24 years of age were classified as dropouts. However, since 1990 the status dropout rates in metro areas have shown impressive improvements, declining by more than two percentage points during the three-year period. Although the situation is definitely improving generally, the dropout problem continues to be more pervasive in the central city and rural areas of the country and to touch a much smaller segment of the 16 to 24 age cohort residing in suburbia.

As suggested earlier, one of the important ingredients in bringing about improvement in the educational credentials of youth is the presence of social capital. The process aspect of social capital is represented by the set of relationships within families and communities that serve to facilitate or constrain individual behavior in a manner consistent with the interests of both individuals and the social structure (Smith 1993, 55). Thus, a key feature of social capital is that it is invested in relationships that emerge through interpersonal interaction. The structure of relationships within the family and the community helps to determine the frequency, duration, and opportunities for such interpersonal interaction. Thus, both structural and process elements are an integral part of social capital, and they work in a complementary fashion to condition the environment for educational achievement (Smith et al. 1995).

Within the family context, a variety of structural characteristics can influence the likely emergence of social capital. These elements include the presence of one or both parents in the home and the number of siblings. Such components dictate the opportunity for interpersonal interactions between parents and children and give shape to the frequency and duration of such interactions (Smith et al. 1995). The process elements of family social capital represent those types of interaction intended to nurture or constrain children's behaviors. Nurturing activities are represented by parents assisting their child with homework, discussing important school activities with him or her, and encouraging their child to aspire to attend college. On the other hand, limiting the child's television viewing, having adult supervision for the child when she or he returns from school, and monitoring the child's homework, are key constraining activities intended to keep inappropriate behaviors in check.

As in the context of family, community social capital can be influenced by the presence and strength of structural and process elements within the locality. Among the structural attributes are size of place and population diversity. These structural features help give shape to the opportunities for interaction at the local level. The process components of community social capital are represented in the social norms, networks, and interactions between adult community members and children that support, enforce, or facilitate educational attainment. Community social capital can be demonstrated by the level of interest and caring that adult members of the community have in regard to the welfare of another person's child, and by the efforts of individuals and organizations to engage children in community programs and activities that make effective use of their time and energy (Beaulieu and Mulkey 1995; Coleman and Hoffer 1987; Smith et al. 1995). What is most unique about community social capital is that the participating adults' investments of time and resources are not intended to bring direct benefits to them. Rather, they are designed to contribute to the public good—in this case, educational progress among community youth.

SOCIAL CAPITAL AND PLACE OF RESIDENCE

It is a well-established fact that intact families have a positive influence on the academic success of their children (Bales 1979; Peterson et al. 1966). At the same time, the number of siblings present in the home can have negative consequences for a child, given that the frequency and duration of interactions parents are able to offer a child are likely to be limited as the number of siblings increases (Blake 1981; Zajonc 1976). The one structural component of the family that has had the most profound impact on the educational progress of children is parents' socioeconomic status (SES). SES serves to condition the environment of support for aspirations and achievement, in that children of higher SES are more often socialized to place a high value on educational achievement (Smith 1993; Wagenaar 1987).

There is some evidence to suggest, however, that family social capital may not be present in equal strength across different residential places. When these important family structural features are examined across space, one finds that rural families are far more likely than urban families to have traditional family arrangements, those in which both mother and father are present. At the same time, rural families tend to be larger because of their higher rates of fertility (Fuguitt et al. 1989). Family income and parental education are typically higher in urban areas, whereas a disproportionate share of U.S. families with limited education or whose incomes fall below the poverty line are located in rural locales (Hobbs 1991; Lichter and Eggebeen 1992; Lichter et al. 1993a; O'Hare 1988). Taken together, these family structural traits give shape to the quality and quantity of interaction possible between children and their parents and to the academic success and educational aspirations of children (Haller and Portes 1973; Kandel and Lesser 1969; Lichter et al. 1993b; Smith et al. 1995).

What of the presence and strength of community social capital? It is known that community forces can give shape to norms concerning educational aspirations and achievement, and many of these are likely to vary by size of place. As a general rule, educational aspirations tend to be higher among urban than rural students (Sewell 1964; Cobb et al. 1989). Rural students are more inclined to feel satisfied with less education and to have lower educational expectations than urban or suburban students (Hansen and MacIntyre 1989). However, with the increasing mobility of families with higher socioeconomic status into America's suburban communities over the course of the last two decades, the relationship between educational success and size of community of residence has taken on a curvilinear aspect (Cobb et al. 1989). As before, the lowest levels of aspiration tend to be found among rural students. However, urban students are next in line, followed by suburban students. Suburban students now register the highest levels of educational aspiration and achievement among all U.S. students (Smith et al. 1995). The lower educational success and aspirations of rural students may be rooted in the fact that rural residents are likely to place less value on education than their urban or suburban counterparts (Jensen and McLaughlin 1995). The reasons for this tendency, however, are far from simple.

Resistance to educational investment by rural residents is linked, in part, to the lower socioeconomic status of rural communities vis-à-vis suburban and urban communities. On average, rural communities are likely to consist of adults with lower educational attainment, lower income levels, and lower job-related skills (Hobbs 1995; Jensen and McLaughlin 1995). These community-level attributes can be pivotal in creating a milieu in which educational success is neither supported nor encouraged.

Yet other factors may contribute as well. According to Hobbs (1995), many rural communities are unable to fully capture the benefits of their investments in the educational progress of local children because many leave the community upon completion of their high school education. This creates a disincentive for rural communities, given that urban and suburban areas emerge as the major beneficiaries of their investment activities (Lichter et al. 1993a). Furthermore, rural areas tend to attract the lower-skilled, low-paying, routine production jobs, whereas more highly skilled managerial and technical positions tend to concentrate in more urbanized communities (Hobbs 1995). These local labor-market profiles are critical, because they tend to serve as a catalyst for human capital investments, especially in education. That is, the availability of quality jobs paying good wages is likely to generate increased interest on the part of individuals to pursue educational enhancements (Stallmann et al. 1995). If decent jobs are not available locally, the inducement to invest in one's human capital is likely to be weakened.

And what of opportunities for social interaction? Are they likely to differ across different-sized communities? Granovetter's (1973) notion of strong and weak ties is particularly informative for this discussion. Strong ties refer to intimate and ongoing interactions among family members and close friends. Weak ties, on the other hand, refer to more transitory and less intimate interactions among individuals. Wilkinson (1991) asserts that in both urban and rural communities there are a host of strong ties among residents. Unlike urban areas, however, rural areas are seriously deficient in respect to weak ties. This shortage is a consequence of the sheer physical distance between people in these settlements. Spatial distance inhibits the growth of such infrequent, more remote interactions. Yet both Granovetter and Wilkinson have asserted that strong and weak ties are both essential to the maintenance of community stability and residents' social well-being. Thus, it is possible that social interaction, which is one of the essential elements needed for social capital to emerge, could be depressed in rural settings as an outcome of the limited availability of weak ties.

In sum, there appears to be sufficient evidence that family and community characteristics vary across space and that these differences can effectively shape a child's chances of completing a high school education. This chapter outlines a study intended to further the understanding of the role of social capital in facilitating educational investments. In particular, its purpose is twofold: (1) to determine whether supportive interpersonal interactions within the family and the community facilitate a child's successful completion of a high school education and (2) to discern whether the connection between interpersonal interaction and high school completion varies across rural-urban places. If the evidence suggests that social capital is of vital importance to the completion of a high school education among students in various locations, then a compelling argument can be made that human capital advancements—deemed essential to fostering the emergence of sustainable communities—can be effectively realized if the strengthening of social capital, in the context of both the family and community, is recognized as one of the key and necessary strategies.

Methodology

The analysis is based on data collected as part of the National Educational Longitudinal Study (NELS) conducted by the National Opinion Research Center on behalf of the National Center for Education Statistics, U.S. Department of Education. The initial survey, conducted in 1988, involved a stratified national probability sample of more than 1,052 schools. A sample of eighth-graders was selected from each of these schools and surveyed, yielding a total of 24,599 useable responses. Each student completed a set of questionnaires that was designed to elicit information on individual/family background characteristics, school experiences, extracurricular activities, attitudes about family and school, and future plans. Linked to the student surveys were nearly 22,700 parent surveys, seeking information on family characteristics, parents' views of their child's school experiences, family life, and expectations for their child. Follow-up surveys conducted during 1990 and 1992 targeted a subsample of 1988 eighth-graders. A total of 16,489 (95% of the eligible sample) completed base year (1988), first follow-up (1990), and second follow-up surveys (1992). The intent of these follow-up surveys was to continue documentation of the secondary-school experiences of the students. Persons who remained in school completed a near duplicate version of the base year survey instrument. Students who dropped out of school received a specialized survey questionnaire focusing on the school environment, family life, aspirations, and work experiences since leaving school.

For purposes of this study, focus is limited to students enrolled in public-supported schools, because of our desire to assess variations that might exist in tax-supported schools located in different spatial areas. Because public schools are funded to no small degree by local citizens, the values and attitudes of families and communities can significantly influence the character of these schools and can orient children as to their future positions in society (Flora et al. 1992).

The number of students for whom complete information is available on all variables in our study is 8,772. Weights are included to correct for oversampling of policy-relevant strata, such as schools with disproportionate numbers of Asians and Hispanics (Owings et al. 1994).

Measurement of Variables[2]

Several measures employed in this study are intended to emulate the variables used by Coleman and associates in their examinations of social capital using the High School & Beyond data set (see Coleman 1988a, 1988b; Coleman and Hoffer 1987). Our study extends their earlier works by including additional variables from the NELS data files that were viewed as conceptually sound measures of family and community social capital. We recognize that social capital is an abstract concept and that the NELS data set was not specifically designed to measure this construct. Although we are confident that the measures of social capital used here are the best available in NELS, they are imperfect indicators and the results should be viewed as exploratory rather than definitive.

The series of *individual and family background* variables are used to assess what Coleman (1988a) labels the "traditional disadvantages" of background. Family income and parents' education reflect resources held by parents that can influence a child's academic aspirations and success. Racial minority status is designed to serve as a proxy for low levels of human capital. Of course, the literature offers ample proof of the importance of these variables in assessing academic performance. School dropout patterns have been found consistently to be inversely related to family socioeconomic status (Wehlage and Rutter 1986; Weidman and Friedmann 1984). Furthermore, blacks have a greater propensity to leave school early than white students (Ekstrom et al. 1986; Natriello et al. 1986).

The items associated with *family social capital* determine the opportunity for and process of interaction. The set of family structural factors measuring the opportunity for interactions include the number of parents in the household and the number of siblings. A third structural variable, included as a proxy for disadvantages that might exist in the family, is number of siblings who have dropped out of high school. Family process indicators are designed to measure education-relevant family interaction. These process factors include nurturing activities (parents expect child to attend college, parents help child with homework, child discusses school matters with parents, child talks to parents about planning high school program) as well as monitoring efforts (parents check on homework, how much parents limit TV viewing, amount of time the child spends at home alone after school with no adult present).

Among the structural attributes of *community social capital* included in our study are a regional identifier (i.e., South vs. non-South), percentage of the public high school student body that is of minority status, and the geographic homogeneity of the school's student population (i.e., whether all the students are drawn from the same geographic area or live in a variety of localities). All three factors are considered important, in that they give shape to the opportunity for social interactions and help influence the emergence of social norms and aspirations regarding educational performance. Because the South generally suffers from the highest rates of poverty, unemployment and underemployment, and their human capital stock (i.e., educational attainment) tends to be the lowest of any region in the country, it could be argued that communities located in this region are less likely to put a premium on human capital investments—in this case, completion of a high school education. The percentage of a school's student population that is of minority status is intended to serve as a proxy for the minority profile of the local community. Lyson (1995, 177) offers convincing evidence that racial and ethnic minorities suffer from inequality in regard to opportunities for education and employment. He suggests that lower levels of educational attainment among minorities, as compared with whites, could be the product of the lower community socioeconomic milieu in which many minorities are embedded. The geographic homogeneity variable is a measure of the opportunity that exists for interaction between students and adult members of a community. If students who attend a school physically live in the same geographic area as the school, then opportunities for interactions between students and adult members of the community are likely to be enhanced. This is far less likely to occur if students live some distance away from school.

Social integration refers to the social ties within and between groups that contribute to a person's attachment to these groups and to his or her desire to conform to the groups' norms and expectations; a student's lack of integration into the community can increase his or her chances of dropping out of school (Weidman and Friedmann 1984). Three measures designed to represent the notion of social integration are included in our analysis (number of times child has changed schools since first grade, student's participation in religious activities, and number of community organizations the student has been involved in). Children who move frequently from one school to another are often unable to develop a sense of integration into the social structure of a community and, as a result, are hampered in their abilities to establish long-term relationships with individuals in a community (Smith et al. 1995). Similarly, a student's involvement in a local religious organization facilitates social interactions between the child and other youth or adult members of that group. Likewise, the more groups (e.g., Scouts, Boys and Girls Clubs, sports programs) to which a student belongs, the greater the likelihood that the child will establish interactive ties with other youth and adults. Collectively, these resources provide the student with a valuable adult and peer support system beyond the family that she or he can tap when needed.

A fourth process component of community social capital included in this study is one that examines the degree to which parents know the parents of their child's closest friends. It is intended to reflect the strength of weak ties existing among these adults. Its importance lies in the fact that these links provide a means whereby parents can monitor their children through social norms that are mutually understood and enforced (Coleman 1988a; Lee 1993). In many respects, it represents a resource for the parents and child that can be activated to facilitate and guide a child's activities.

Analysis

Given that the dependent variable is dichotomous—each student either stayed in school or dropped out between 1988 and 1992—multiple logistic regression procedures were used. Multiple logistic regression allows for estimating the probability of a certain event's occurring, in this case dropping out of school. Further, a multiple logistic regression model was fitted for each type of residence—rural, suburban, and urban—to facilitate comparison.[3]

Interpretation of the results were clarified by estimating probabilities based on selected levels of independent variables in the fitted model. These probabilities are derived from the logits of dropping out, given one or more characteristics (e.g., student participates actively in church organizations), while controlling for the effects of other factors. The variables being controlled were set at their modal value in calculating the logits, with the exception of family income, which was set at its median value.

COMPARING THE EFFECTS OF FAMILY AND COMMUNITY SOCIAL CAPITAL ON RURAL, SUBURBAN, AND URBAN HIGH SCHOOL DROPOUT RATES

Table 10.1 presents the relative importance that the independent variables of major interest in this study have on chances of a person dropping out of high school. The results suggest that many of these variables appear to have statistically significant effect on an individual's dropout status. Although status attainment research (Blau and Duncan 1967) has found individual and family background characteristics to be important influences on high school dropout status, our findings prove to be much less definitive on this point. A child whose mother or father has attended college has a greater probability of completing high school. The effect of the father's education is about equal to the mother's for rural students, but the father's education has a much stronger effect than the mother's for suburban students. Contrary to expectation, being African American in rural and suburban areas shows a strong positive relationship with the likelihood of completing school once all other variables in the model are controlled for. Likewise, family income has little net effect on dropping out, except for suburban students, as detailed below.

Results presented under the Family Social Capital heading in Table 10.1 allow us to assess whether family social capital exerts a significant influence on the likelihood of a child's completing high school. The structural attribute of social capital found to be most important is whether the child has one or more siblings who dropped out of high school, particularly in rural and suburban areas. This may indicate a weak family environment in which support for academic progress is not nurtured. The number of siblings also proves to be a significant factor across all three residential settings. This would suggest that parental opportunities to provide their child with quality, uninterrupted time is difficult in the presence of many children in the home, a finding consistent with the literature on this subject (see Blake 1981; Zajonc 1976). The presence of both parents in the family improves the chances that a suburban or urban student will complete high school, but is found to be insignificant for students in rural areas.

Examination of the seven family social capital process attributes suggests that these family features are of significant importance in shaping the chances that a child will complete high school. Although the magnitude of the effects are found to vary among rural, suburban, and urban areas, it is clear that students are much more likely to drop out of high school if their parents do not expect them to attend college, if parents are unlikely to limit children's TV time, and if children rarely or never discuss school matters with their parents. A less vital, but significant, family social capital factor uncovered for rural students is the amount of time that a child spends alone after school with no adult supervision. The longer the period of time spent unsupervised, the greater the probability that the child will drop out of high school. While achieving statistical significance in at least one of the residential categories, three items are found to have only minor effects on dropout behavior—parents check on the child's homework, the child talks to parents about planning his or her high school program, and parents help with homework. Two of these latter activities (i.e., parents checking homework often and parents helping with homework often) actually boost a child's chances of dropping out of high school in suburban areas. Although, on the surface, these outcomes appear to run counter to expectations, it could be argued that parents who often assist with a child's homework may be doing so because of the lack of discipline or self-motivation a child may be demonstrating in regard to his or her schoolwork. Children who are self-starters and highly motivated may need little, if any, assistance from their parents in the completion of their homework assignments.

Results presented under the heading of Community Social Capital in Table 10.1 provide the basis for assessing whether community social capital wields an influence on the dropout rate of public high school students beyond that exerted by family social capital. The results reported in Table 10.1 suggest that the relationships between dropping out of high school and community social capital attributes are indeed significant. Two community structural features, namely, living in the South and the minority percentage of the student population, increase one's chances of dropping out of high school, the former for rural students and the latter for rural and urban students. For rural students, being part of a student body that is geographically drawn from the surrounding community reduces the likelihood of dropping out. Collectively, these three structural dimensions of community social capital have the greatest effects on the dropout behavior of rural students and the least on those residing in suburban areas.

The process characteristics associated with community social capital play an important part in shaping a student's chances of completing a high school education. Children subjected to frequent moves from one school to another since entering the first grade are much more likely to be high school dropouts, regardless of the spatial location of these schools. Several moves may inhibit the child's and parent's opportunities to develop an attachment with people and social organizations outside the family. As Coleman (1988a, S113) notes, for parents and children in mobile families, the social relations that constitute social capital are severed at each move.

Table 10.1 Logistic Regression Results for Effects of Family and Community Social Capital on the Probability of Dropping Out of Public High School Betwen the 8th and 12th Grades

	Rural		*Suburban*		*Urban*	
Variable	*Coef.*	*SE*	*Coef.*	*SE*	*Coef.*	*SE*
Intercept	-1.919	0.477	-0.810	0.467	-1.141	0.623
Individual and Family Background						
Black	-0.800***	0.226	-1.408***	0.328	-0.219	0.183
Family income	-0.011	0.006	0.001	0.004	-0.016	0.008
Father attended college	-0.374*	0.159	-0.561***	0.144	-0.049	0.199
Mother attended college	-0.334*	0.155	-0.196	0.137	-0.382	0.201
Family Social Capital						
Structural Attributes						
Both parent(s) in household	0.072	0.145	-0.461***	0.140	-0.490**	0.168
Number of siblings	0.094*	0.040	0.110**	0.041	0.184***	0.052
Sibling(s) dropped out of high school	1.006***	0.146	0.629***	0.163	0.224	0.212
Process Attributes						
Parent(s) expect college	-0.421***	0.063	-0.065	0.068	-0.432***	0.085
How often parent(s) help with homework	0.043	0.057	0.148*	0.060	-0.028	0.077
Discuss school matters with parent(s)	-0.089	0.054	-0.188***	0.054	-0.210**	0.074
Discuss high school plans with parent(s)	-0.089	0.047	0.054	0.051	-0.169*	0.068

Discuss high school plans with parent(s)	-0.089	0.047	0.054	0.051	-0.169*	0.068
Parent(s) check on homework	0.056	0.056	-0.138*	0.060	-0.059	0.081
How often parent(s) limit TV time	0.331***	0.060	0.206***	0.057	0.111	0.075
Time alone after school with no adult present	0.123**	0.045	-0.064	0.046	0.068	0.057
Community Social Capital						
Structural Attributes						
South	0.445***	0.118	-0.158	0.129	0.061	0.155
Percent minority	0.014***	0.002	0.003	0.002	0.015***	0.003
All students from area	-0.447*	0.204	-0.187	0.197	0.303	0.214
Process Attributes						
Number of moves since 1st grade	0.263***	0.037	0.326***	0.037	0.201***	0.053
Know parent(s) of child's friends	-0.053	0.056	0.058	0.063	0.051	0.084
Child involved in religious group	-0.287*	0.116	-0.090	0.118	-0.083	0.154
Number of other groups involved in	-0.072	0.045	-0.120**	0.046	0.045	0.059
Family income X know parents of child's friends	0.00008	0.002	-0.008***	0.002	0.00027	0.003
N	3,356		3,769		1,647	

*p<0.05. **p<0.01. ***p<0,001

Without a doubt, it takes a while for the uprooted individuals to establish ties in their new place of residence (Putnam 1995).

Involvement in religious activities tends to enhance a child's chances of completing high school, most notably in rural areas. This finding supports Coleman's (1988a, 1988b) argument that churches provide opportunities for interaction and support, activities that provide youth with ways to feel attached to the adult community beyond the family. Moreover, a child's involvement in other community youth groups and programs further augments his or her chances of completing high school, although this proves significant only for those in suburban areas. As a general statement, there is little evidence to suggest that parents who know the parents of their child's best friends are less likely to have children who drop out of high school. However, weak ties of this type appear to be pivotal for those in suburban areas when considered in combination with family income. That is, the benefits associated with knowing the parents of a child's best friends are likely to flow to suburban students of higher-income families.

The predicted percentages of students dropping out of high school for selected levels of independent variables, adjusted for all covariates in the full model, are presented in Table 10.2, according to residential location. The first item, for example, examines the percentage point difference in the predicted dropout rate (between grades 8 and 12) of rural, suburban, and urban high school students living in both two-parent and single-parent households (while controlling for all other variables in this model). The increased chance of a student from a single-parent household, versus one from a two-parent household, of dropping out of high school appears greatest among suburban students (a 4.6 percentage point difference), although urban students have the highest relative difference (60% higher for students in single-parent families). Two additional structural components of family social capital considered in Table 10.2 are number of siblings and having a sibling who has dropped out of high school. Having five or more siblings, versus having none, dramatically influences the dropout behavior of suburban students. But having one or more siblings who are high school dropouts increases the likelihood, for both rural and suburban students, that they will leave high school without a diploma.

Social capital attributes proving critical to the probability of rural students' dropping out of high school, in addition to having a sibling drop out, include parental aspirations for college, how often parents limit TV time, regional location, minority composition of the student body, and number of school changes since first grade. Rural students whose parents expect them to attend college are about half as likely to drop out of school as those whose parents have no such expectations. Rural students who had their TV time limited often are about two-thirds less likely to drop out as those who never had their TV limited. Students from southern communities and those within a student population that is half minority are 50% and 80%, respectively, more likely to drop out than students from non-South or low-minority communities. Rural students who have shifted schools four or more times since first grade are also predicted to have dropout rates three times higher than those who have never changed schools.

Among suburban students, the social capital elements that are found to have the most dramatic impact on dropout rates, in addition to the family structure variables mentioned earlier, are the frequency of the child's discussion of school matters with parents, how often parents limit TV time, level of mobility in terms of changing schools since grade one, and parents knowing the parents of the child's close friends. Within urban settings, few social capital attributes are found to have substantive effects on dropout behaviors. However, factors having modest impacts on the predicted dropout rates include number of siblings, parent's college expectations for their child, the extent to which the child has discussed school matters or high school plans with parents, the percentage of minorities in the school, and the number of times the child has changed schools since first grade.

The results reported in Tables 10.1 and 10.2 provide important guidance in portraying the set of factors that collectively influence the family and community social capital environments of students. Table 10.3 presents composite measures of *family social capital* and *community social capital*. For students whose families have a high level of social capital (determined on the basis of ten variables noted in the table footnote) and who live in a community where the presence of social capital is high (developed via the use of seven variables), the dropout rate is predicted to be 1.7% in suburban areas and even lower in urban and rural settings. If family social capital is high, but community social capital is low, the percentage of students who drop out of school at some point between grades 8 and 12 increases dramatically among rural and suburban students (24.7% and 23.5%, respectively), but only slightly for urban students (3.1%). If community social capital is high, but family social capital is low, the predicted dropout rate accelerates to 37.7% for suburban students, to 48.3% among rural students, and to 58.8% among those in urban locales. In cases where social capital is virtually absent in both family and community, the chances of dropping out of school surpasses 90% for both rural and suburban students and nears 78% for urban students. These results make a compelling case for the importance of high social capital being available to the student in the context of both family and community.

Another way to demonstrate this point is to begin with students in families having high levels of social capital. Going from a community with low social capital to one with high social capital reduces the chances of dropping out of high school substantially for rural and suburban students in families with high social capital. For example, for rural students with high levels of family social capital, the probability of dropping out of school

Table 10.2 Predicted High School Dropout Rates for Students Who Differ in Family and Community Social Capital, Controlling for Individual and Family Background Attributes

	Predicted Dropout Rate (%)		
Variables	*Rural*	*Suburban*	*Urban*
Presence of Parents			
One parent	4.8	13.9	4.8
Both parents	4.5	9.3	3.0
Number of Siblings			
None	4.4	8.4	2.5
5 or more	6.8	13.7	6.1
Sibling(s) Dropped Out of High School			
None	4.8	9.3	3.0
One or more	12.1	16.1	3.7
Parent(s) Expect College			
Parent(s) expect college	4.8	9.3	3.0
Parent(s) donít expect college	10.4	10.4	6.8
How Often parent(s) Help with Homework			
Parent(s) help nearly every day	5.0	10.6	2.9
Parent(s) seldom or never help with homework	4.4	7.1	3.2
Discuss school matters with parent(s)			
3 + times since school year started	4.8	9.3	3.0
Not discussed at all	6.7	17.8	6.7
Discuss High School Plans with Parent(s)			
Do talk about high school plans	4.0	10.2	2.2
Donít talk about high school plans	5.7	8.4	4.2
Parent(s) Check on Homework			
Often	4.8	9.3	3.0
Never	5.6	6.3	2.5
How Often Parent(s) Limit TV Time			
Often limit TV time	1.8	5.2	2.2
Never limit TV time	4.8	9.3	3.0

Table 10.2 (*Continued*)

Variables	*Predicted Dropout Rate (%)*		
	Rural	*Suburban*	*Urban*
Time Spent Alone After School with no Adult Present			
No time alone	4.2	9.8	2.8
3 + hours alone	6.7	7.8	3.7
Region			
Southern U.S.	7.3	8.0	3.2
Outside the Southern U.S.	4.8	9.3	3.0
Percent Minority			
3.0%	4.8	9.3	3.0
50.5%	8.8	10.6	5.8
All Students from Area			
Yes	4.8	9.3	3.0
No	7.3	11.0	2.2
Number of Moves Since 1st Grade			
None	4.8	9.3	3.0
4 + times	12.6	27.3	6.5
Know Parent(s) of Child's Friends			
Know none of them	5.5	15.0	2.5
Know five of them	4.3	6.6	3.4
Child Involved in Religious Group			
Participated	4.8	9.3	3.0
Did not participate	6.3	10.1	3.3
Number of Other Groups Involved in			
None	5.5	11.5	2.8
Three or more	4.5	8.3	3.1

*Includes the effects of the interaction with family income

Table 10.3 Predicted Dropout Rates Between Grades 8 and 12 for Students Whose Families and Communities Differ in Social Capital, Controlling for Individual and Family Background Characteristics

	Location of School		
Social Capital Attributes	*Rural*	*Suburban*	*Urban*
Low family/low community	96.4	91.3	77.9
Low family/high community	48.3	37.7	58.8
High family/low community	24.7	23.5	3.1
High family/high community	1.1	1.7	1.3

High family social capital is defined as: (1) Both parents/guardians are present in the household; (2) no siblings; (3) no family sibling has dropped out of high school; (4) parents expect the child to attend college; (5) child discusses school matters with the parent(s); (6) student talks to parent(s) about planning his or her high school program; (7) parent(s) often limit time allowed to watch TV; and (8) child spends no time alone after school without an adult present; (9) parent(s) seldom/often check child's homework; and (10) parent(s) seldom/never help child with homework.

Low Family Social Capital is defined as: (1) one parent present; (2) five or more siblings; (3) one or more siblings has dropped out of high school; (4) parent has no expectation for college; (5) child does not discuss school matters with parent; (6) student does not talk to parent about planning high school program; (7) parent never limits time allowed to watch TV; and (8) child spends 3+ hours alone after school with no adult present; (9) parent(s) often check child's homework; and (10) parent(s) often help childd with homework.

High community social capital is defined as: (1) child has never changed schools since entering first grade; (2) parent(s) know the parent(s) of child's closest friends; (3) child is involved in religious group; (4) child is member of three or more community groups/clubs; (5) 3% minority enrollment in child's school; (6) child's school is not in the southern region of the United States; and (7) all students in the geographic region attend the same school.

Low community social capital is defined as: (1) child has changed schools 4+ times since first grade; (2) parent(s) do not know any of their child's closest friends; (3) child is not involved in a religious group; (4) child is not a member of any community groups/clubs; (5) 50.5% minority enrollment in child's school; (6) child's school is in the southern region of the United States; and (7) not all students in the geographic region do not attend the same school.

is predicted to decline by 23.6 percentage points when they are located in communities with high social capital.

For urban students in environments of high community social capital, shifting from low to high family social capital environments markedly decreases the dropout rates (from 58.8% to 1.3%). Less sizable, but nevertheless significant, reductions are predicted for rural and suburban students (from 48.3 to 1.1% and from 37.7 to 1.7%, respectively). What of those students who reside in families and communities having low levels of social capital? In such situations, if family social capital remains low but community social capital shifts from low to high, the likelihood of rural students dropping out of high school is estimated to decline by 48.1 percentage points. For suburban students, the dip is even more sizable—53.6 percentage points—but the drop is less dramatic for urban students (19.1 percentage points).

In sum, Table 10.3 demonstrates the important, positive, and synergistic effect that takes place when high social capital is available to students in the context of both the family and community. Students in families and communities that are both high in social capital are virtually assured of successfully completing a high school education.

CONCLUSIONS

An ambitious blueprint, designed to set America on the path to a sustainable future, was recently advanced by the President's Council on Sustainable Development. This plan lays out 10 important goals that, if realized, would ensure economic prosperity, environmental protection, and social equity for the people and communities of the United States. Four of these goals are particularly apropos to the theme of this chapter (President's Council on Sustainable Development 1996, 12):

- *Economic Prosperity*: To create meaningful jobs, reduce poverty, and provide the opportunity for a high quality of life for all in an increasingly competitive world.
- *Sustainable Communities*: To encourage people to work together to create healthy communities where natural and historical resources are preserved, jobs are available, sprawl is contained, neighborhoods are secure, education is lifelong, transportation and health care are accessible, and all citizens have opportunities to improve the quality of their lives.
- *Civic Engagement*: To create full opportunity for citizens, businesses, and communities to participate in and influence the natural resource, environmental, and economic decisions that affect them.
- *Education*: To ensure that all Americans have equal access to education and lifelong learning opportunities that will prepare them for meaning-

ful work, a high quality of life, and an understanding of the concepts involved in sustainable development.

These four goals are, in no small degree, interconnected. It is our view, however, that the most fundamental of these is education. Education serves as the principal conduit for progress toward the achievement of the three other goals. However, raising educational levels depends, in no small way, on the social capital base of families and communities, as demonstrated in this chapter. Our analysis reveals that social capital can influence high school completion in a number of specific ways:

- Parents' aspirations significantly increase the probability that their children will complete high school. A child who perceives that both parents want him or her to attend college is far more likely to complete high school (especially a rural or suburban student).
- Parental monitoring, namely, limiting TV time and having adult supervision available when the child arrives home after school, improves high school completion rates of rural students.
- The negative consequences of having a sibling who has dropped out of school are enormous for both rural and suburban students, but less so for urban students.
- Children in rural communities located in the South, or that comprise a large proportion of minorities, are less likely to complete their high school education. In essence, these variables serve as proxies for the structural disadvantages associated with rural places, such as high levels of poverty, unemployment and underemployment, limited availability of good jobs or people with decent educational levels. These structural deficiencies severely limit a community's capacity to establish and enforce norms and values that place a premium on the educational success of local youth.
- Rural youth who are disconnected from the local community because they have moved frequently from one school to another, have had no involvement in any kind of local organization (particularly a religious group), or who happen to live in an area that is geographically distant from their high school, are at greater risk of dropping out of high school than are students who are far more integrated into the community.

It is worth noting that several family and community social capital measures, when considered in isolation, have only modest effects on high school dropout rates. However, when considered collectively (as shown in Table 10.3), the pool of family and community social capital attributes has a profound impact on high school completion. Irrespective of the spatial location under consideration, family social capital significantly alters the chances of a child's completing high school. Similarly, the elements that make up community social capital prove vital to students' chances for high school graduation, particularly for those from rural and suburban locales.

Given the central role of family and community social capital in promoting educational success and the critical importance of education in undergirding community sustainability, what can be done to build social capital in families and communities? No doubt, carefully focused efforts will be needed at the local, state, and federal levels. Although schools are primarily viewed as a local activity, state and national policies/programs are needed to set the stage for local efforts that seek to build social capital. For example, rural areas of the country have a decided disadvantage in regard to human capital resources. The pool of highly educated persons continues to be proportionally smaller in nonmetro than in metro areas of the United States (Killian and Beaulieu 1995). Furthermore, many areas of rural America continue to suffer from high rates of unemployment, underemployment. and poverty. Too often, rural economies are rooted in low-skill, poor-paying jobs, many of which have deleterious effects on the environment.

The *social costs of space* compounds the human capital shortfall of rural America. Wilkinson (1991, 114) writes that "dispersion of population and small population size limit the local resource base, and the rural setting tends to lack a complete array of services and facilities. It also tends to lack the complete networks of social interaction and relationships a community would provide to meet the social needs of people." These structural deficits create a drag on the development of social capital in rural areas.

From the federal and state arenas, policies that are designed to strengthen the human capital endowments of individuals must continue to receive attention and support. However, these types of investment strategies must be coupled with resources designed to build the capacity of families and communities. Human resource policies that focus strictly on individuals, without attending to the broader set of family and community environments in which these persons function, are likely to be ineffective. That is, federal and state policies and programs must approach human capital development from a more holistic perspective and give due consideration to strategies that can help families and communities to be stronger, more positive partners in shaping the educational aspirations and achievements of children. This approach should include efforts to broaden the economic base of rural communities so that good jobs can be created, jobs that can help to stem the outward migration of the best and brightest and to preserve the environmental richness of rural areas. We believe that the increasing interest in federal block grants may provide opportunities and resources to facilitate human and social capital development activities of this type in rural areas.

There is little doubt, given the fiscal constraints that are likely to persist at federal and state levels, that the hard work needed to strengthen social capital will have to come largely from within, from people and organizations situated in America's rural communities. We believe that there are local strategies that can facilitate the development and nourishment of social capital. A fruitful beginning point is with the local schools. It is critical that school administrators and teachers recognize that progress in raising educational attainment of students can be realized only when enhancement of family and community social capital is considered a key element in the equation. This implies that schools must embrace as part of their mission—and be active agents in—promoting family and community social capital (Coleman 1991). There are many steps schools can take toward this goal; among the possibilities are the following: (1) helping parents to understand the vital role of parental aspirations in instilling in their children a desire to succeed academically; (2) improving parents' skills, such as by educating them on the importance of nurturing and monitoring activities within the home; (3) connecting with the resources outside the school walls (i.e., businesses, local government agencies, community organizations) and actively engaging them in the education of local youth; (4) giving special attention to students with siblings who have dropped out of school, inasmuch as these individuals are at much higher risk of doing so as well; (5) offering special activities to new students, who may feel disconnected from the school, other students, and/or the community, so that their sense of belonging can be accelerated; and (6) facilitating the linking of parents with one another so that growth in what Coleman has labeled as "intergenerational closure," or Granovetter has referred to as "weak ties," can be cultivated. Coleman (1991) has even suggested that schools can enhance intergenerational closure by encouraging active parent-teacher associations and expanding parent-teacher communication using e-mail or other emerging technologies.

Within the broader community, schools can help develop social capital by organizing or making their facilities available for events and programs that involve students, parents, and other local adults. These activities "prime the pump" by creating or reinvigorating relationships that are available for other nonschool purposes (Merton 1968). Yet the burden of creating a milieu where community social capital can thrive cannot be borne solely by the school. Other local resources—religious institutions, civic/social organizations, government, and other public or private entities—must give leadership to these initiatives or, at a minimum, lend an active hand to the schools. The study in this chapter has shown how rural youth who have no adult supervision after school, or who are unattached to any type of community organizations, have a higher probability of dropping out of school. The community must provide youth with productive avenues to direct their after-school time. This is particularly problematic in rural communities. Too often, after-school programs are not available to local youth, or, in the event that they are available, they tend to be low on the list of funding priorities and, as a result, are cut when resources are limited. Unfortunately, many local leaders and citizens fail to appreciate the much broader role of these programs in the lives of children. Beyond providing children with organized activities for their free time, they allow valuable interactions to take place with adult members of the community, linkages that can foster values, norms, and expectations that promote aspirations and achievement. Many times, the availability of these sets of resources beyond the family can help place a child on a proper track of development.

Although many rural communities continue to experience constraints that make the evolution of social capital difficult, we concur with Kretzmann and McKnight (1993) that even the most disadvantaged communities have valuable organizational and institutional assets that can be identified and mobilized to address important local concerns. Activating and funneling these resources to create a strong social capital climate lets children know that local people truly care about their well-being and value their full participation in the life of the community. What better foundation can be laid for creating a sustainable future for that community?

NOTES

1. A concrete example can be offered in regard to voting behavior. High school dropouts in the 25 to 44 age cohort were 46% less likely than high school graduates to have voted in the 1992 presidential election. Furthermore, since 1988, voting rates in the U.S. presidential elections have improved for all educational attainment groups in the 25- to 44-year-old population, except for high school dropouts (U.S. Department of Education 1994).
2. A detailed description of the coding scheme employed for the variables in this study is presented in Table A.1 in the appendix to this chapter.
3. Multiple logistic regression calculates parameter estimates that are similar in interpretation to those generated in multiple linear regression. The questions posed in this chapter are addressed by estimating probabilities, based on selected levels of the independent variables in the fitted model. These probabilities are derived from the logits (log odds) of dropping out, given one or more characteristics, while controlling for the effects of other factors. A detailed description of the coding scheme employed for the variables in this study is presented in Table A.2 the appendix to this chapter.
4. A number of additional variables were included in the initial model, based on their expected importance in predicting high school dropout status. However, none achieved statistical significance. These variables included gender, Hispanic ethnicity, total school enrollment, and percentage of students receiving free or reduced-cost lunch at school.

Table A.1 Variables Used in the Analysis and Their Measurement (Variable names on the NELS *Electronic Codebook* are shown in parentheses)

Variables	*Coding Scheme*
Dropout status (F2EVDOST)	1 = Dropout (either no return or in alternative program); 0 = Did not drop out or returned to public school by second follow-up.
Black (RACE)	1 = Black, non-Hispanic; 0 = non-Black.
Family income (BYFAMINC)	Coded as the midpoint of one of 15 categories, ranging from 1 = none to 15 = $200,000 +.
Father attended college (BYS34A)	1 = Father attended college; 0 = father did not attend college or subject didn't know.
Mother attended college (BYS34B)	1 = Mother attended college; 0 = mother did not attend college or subject didn't know.
Both parents in household (BYFCOMP)	1 = Both parents/guardians in household; 0 = one parent/guardian in household.
Number of siblings (BYS32)	Range from 0= None; to 5 = 5 or more siblings.
Number of siblings who have dropped out of high school (BYP6)	1 = 1 or more; 0 = none.
Parent(s) expect college (BYS48A & BYS48B)	2= Mother and father expect 8th-grader to attend college; 1 = mother or father expects child to attend college; 0 = neither expects child to attend college or subject didn't know.
How often parent(s) help with homework (BYP69)	1 = Seldom or never; 2 = once or twice a month; 3 = once or twice a week; 4 = almost every day.
Discuss school matters with parent(s) (BYS36B & BYS36C)	Sum of scores of the student's response to two questions: (a) child's discussion of school activities with parents; and (b) child's talking to parents about things studied in class. Each question was coded as: 1 = not at all; 2 = once or twice; and 3 = 3 or more times since the beginning of the school year.

Discuss high school plan with parent(s) (BYS50 A & BYS50B)	Sum of the student's response to two questions: How often they discuss their high school program with (a)mother and with (b) father. Each question was coded as: 2=often; 1=sometimes; 0=never.
How often parent(s) check on homework (BYS38A)	1 = Often; 2= sometimes; 3 = rarely; 4 = never.
How often parent(s) limit TV time (BYS38C)	1 = Often; 2= sometimes; 3 = rarely; 4 = never.
Time alone after school with no adult present (BYS41)	0 = None; 1 = less than 1 hour; 2 = 1–2 hours; 3 = 2–3 hours; 4 = more than 3 hours.
Reside in south (REGION)	1 = Respondent's school is located in the southern region of the United States; 0 = respondent's school is located outside the South.
Percentage of minorities in the student's eighth grade class (G8MINOR)	Coded as the midpoint of 8 categories, ranging from 0 = 0.0% to 7 = {91% to 100%}.
All students from area (BYSC24A)	1= yes; 0 = no.
Number of moves since 1st grade (except for changes associated with promotion) (BYP40)	0 = None; 1 = Once; 2 = twice; 3 = three times; 4 = 4 or more times.
Know parent(s) of child's friends (BYP62B1 thru BYP62B5)	Range of scores from 5 = know parents of child's five closest friends; to 0 = do not know any parents of child's closest friends.
Child involved in religious group (BYP63E)	1 = Respondent participated in a religious group; 0 = did not participate.
Number of other groups involved in (BYP63A, B, C, D, F, G, H, I)	Range from 0 = none to 8 = Eight.
School (location)	1 = URBAN: Urban area or central city; 2 = SUBURBAN: area surrounding a central city within a county constituting an MSA; 3 = RURAL: outside MSA.

Table A.2 Means for Variables by Place of Residence

	Rural		*Suburan*		*Urban*	
Variable	*Mean*	*SD*	*Mean*	*SD*	*Mean*	*SD*
Individual and Family Background						
Black	0.076	0.266	0.053	0.223	0.251	0.434
Family income	32.441	25.854	45.163	32.563	32.213	26.866
Father attended college	0.363	0.481	0.487	0.500	0.381	0.486
Mother attended college	0.363	0.481	0.460	0.498	0.388	0.487
Family Social Capital						
Structural Attributes						
Both parent(s) in household	0.836	0.370	0.846	0.361	0.716	0.451
Number of siblings	2.272	1.448	2.103	1.359	2.347	1.508
Sibling(s) dropped out of high school	0.111	0.314	0.080	0.271	0.108	0.310
Process Attributes						
Parent(s) expect college	1.451	0.840	1.585	0.760	1.502	0.804
How often parent(s) help with homework	2.206	0.979	2.211	0.944	2.166	1.000
Discuss school matters with parent(s)	4.903	1.087	4.971	1.068	4.844	1.407
Discuss high school plans with parent(s)	2.368	1.275	2.754	1.225	2.379	1.226
Parent(s) check on homework	1.964	1.087	1.913	0.997	1.852	0.967
How often parent(s) limit TV time	2.963	1.015	2.810	1.061	2.863	1.055
Time alone after school with no adult present	1.768	1.164	1.828	1.164	1.822	1.255

Community Social Capital						
Structural Attributes						
South	0.395	0.489	0.259	0.438	0.469	0.499
Percent minority	16.802	22.720	18.393	23.000	50.910	31.730
All students from area	0.933	0.249	0.920	0.272	0.839	0.367
Process Attributes						
Number of moves since 1st grade	1.013	1.373	1.161	1.376	1.413	1.407
Know parent(s) of childs friends	3.024	1.625	2.754	1.574	2.304	1.582
Child involved in religious group	0.584	0.493	0.561	0.496	0.503	0.500
Number of other groups involved in	2.009	1.391	2.021	1.364	1.789	1.428

REFERENCES

Aplet, G., N. Johnson, J. T. Olson, and J. A. Sample, eds. 1993. *Defining Sustainable Forestry*. Washington, D.C.: Island Press.

Bales, K. B. 1979. "The Single Parent Family Aspirations and Academic Achievement." *The Southern Journal of Educational Research* 13(4):145–160.

Beatley, T. 1995a. "The Many Meanings of Sustainability: Introduction to a Special Issue of JPL." *Journal of Planning Literature* 9:339–342.

_______. 1995b. "Planning and Sustainability: The Elements of a New (Improved?) Paragidm." *Journal of Planning Literature* 9:383–395.

Beaulieu, L. J., and D. Mulkey, eds. 1995. *Investing in People: The Human Capital Needs of Rural America*. Boulder, Col.: Westview Press.

Beaulieu, L. J., and D. Mulkey. 1995. "Human Capital in Rural America: A Review of Theoretical Perspectives."In *Investing in People: The Human Capital Needs of Rural America*, edited by L. J. Beaulieu and D. Mulkey. Boulder, Col.: Westview Press, 3–21.

Becker, G. S. 1962. *Human Capital*. Chicago: University of Chicago Press.

Blake, J. 1981. "Family Size and the Quality of Children." *Demography* 18:421–442.

Blau, P. M., and O. D. Duncan. 1967. *The American Occupational Structure*. New York: John Wiley & Sons.

Bourke, L. 1994. "Economic Attitudes and Responses to Siting Hazardous Waste Facilities in Rural Utah." *Rural Sociology* 59:485–496.

Cobb, R. A., W. G. McIntyre, and P. A. Pratt. 1989. "Vocational and Educational Aspirations of High School Students: A Problem for Rural America. "*Research in Rural Education* 6(2):11–15.

Coleman, J. S. 1988a. "Social Capital in the Creation of Human Capital." *American Journal of Sociology* 94:95–120.

_______. 1988b. "The Creation and Destruction of Social Capital: Implications for the Law?" *Notre Dame Journal of Law, Ethics and Public Policy* 3:375–404.

_______. 1991. *Parental Involvement in Education*. Policy Perspective Series. Washington, D.C.: Office of Educational Research and Education, U.S. Department of Education.

Coleman, J. S., and T. Hoffer. 1987. *Public and Private High Schools: The Impact of Communities*. New York: Basic Books.

Couch, S. R., and S. Kroll-Smith. 1994. "Environmental Controversies, Interactional Resources, and Rural Communities: Siting Versus Exposure Disputes." *Rural Sociology* 59(Spring):25–44.

Ekstrom, R. B., M. E. Goertz, J. M. Pollack, and D. A. Rock. 1986. "Who Drops Out of High School and Why? Findings from a National Study." *Teacher College Record* 3:356–373.

Flora, C. B., J. L. Flora, J. D. Spears, and L. E. Swanson. 1992. *Rural Communities: Legacy and Change*. Boulder, Col.: Westview Press.

Fuguitt, G. V., D. L. Brown, and C. L. Beale. 1989. *Rural and Small Town America*. New York: Russell Sage Foundation.

Granovetter, M. 1973. "The Strength of Weak Ties." *American Journal of Sociology* 78:1360–1380.

Haller, A. O., and A. Portes. 1973. "Status Attainment Processes." *Sociology of Education* 46:51–91.

Hansen, T. D., and W. G. McIntyre. 1989. "Family Structure Variables as Predictors of Educational and Vocational Aspirations of High School Seniors." *Research in Rural Education* 6(2):39–49.

Herrnstein, R. J., and C. Murray. 1994. *The Bell Curve: Intelligence and Class Structure in American Life.* New York: Free Press.

Hobbs, D. 1991. "Rural Education."In *Rural Policies for the 1990s,* edited by C. B. Flora and J. A. Christenson. Boulder, Col.: Westview Press, 151–165.

———. 1995. "Capacity-Building: Reexamining the Role of the Rural School." In *Investing in People: The Human Capital Needs of Rural America,* edited by L. J. Beaulieu and D. Mulkey. Boulder, Col.: Westview Press, 259–284.

Humphrey, C. R., G. Berardi, M. S. Carroll, S. Fairfax, L. Fortmann, C. Geisler, T. G. Johnson, J. Kusel, R. G. Lee, S. Macinko, M. D. Schulman, and P. West. 1993. "Theories in the Study of Natural Resource Dependent Communities and Persistent Rural Poverty in the United States." In *Persistent Poverty in Rural America.* Boulder, Col.: Westview Press, 136–172.

Jensen, L., and D. K. McLaughlin. 1995. "Human Capital and Nonmetropolitan Poverty." In *Investing in People: The Human Capital Needs of Rural America,* edited by L. J. Beaulieu and D. Mulkey. Boulder, Col.: Westview Press, 111–138.

Kandel, D. B., and G. S. Lesser. 1969. "Parental and Peer Influences on Educational Plans of Adolescents." *American Sociological Review* 34:213–223

Kane, H. 1992. *Time for Change: A New Approach to Environment and Development.* Washington, D.C.: Island Press.

Kaufman, P., M. M. McMillen, and D. Bradby. 1992. *Dropout Rates in the United States: 1991.* National Center for Education Statistics NCES 92-129.Washington, D.C.: U.S. Department of Education, Office of Educational Research and Improvement.

Keeney, D. R. 1989. "Toward a Sustainable Agriculture: Need for Clarification of Concepts and Terminology." *American Journal of Alternative Agriculture* 4(3, 4).101–105.

Killian, M. S., and L. J. Beaulieu. 1995. "Current Status of Human Capital in the Rural U.S. "In *Investing in People: The Human Capital Needs of Rural America,* edited by L. J. Beaulieu and D. Mulkey. Boulder, Col.: Westview Press, 23–46.

Kline, E. 1994. "Sustainable Community Indicators." Unpublished paper.

Kretzmann, J. P., and J. L. McKnight. 1993. *Building Communities from the Inside Out: A Path Toward Finding and Mobilizing a Community's Assets.* Evanston, Ill.: Center for Urban Affairs and Policy Research.

Lee, S. 1993. "Family Structure Effects on Student Outcomes." In *Parents, Their Children, and Schools,* edited by B. Schneider and J. S. Coleman. Boulder, Col.: Westview Press, 43–75.

Lichter, D. T., and D. J. Eggebeen. 1992. "Child Poverty and the Changing Rural Family." *Rural Sociology* 57:151–172.

Lichter, D. T., G. T. Cornwell, and D. J. Eggebeen. 1993b. "Harvesting Human Capital: Family Structure and Education Among Rural Youth." *Rural Sociology* 58(1):53–75.

Lichter, D. T., L. J. Beaulieu, J. L. Findeis, and R. A. Teixeira. 1993a. "Human Capital, Labor Supply, and Poverty in Rural America." In *Persistent Poverty in Rural America.* Boulder, Col.: Westview Press, 39–67.

Lyson, T. A. 1995. "Down and Out in Rural America: The Status of Blacks and Hispanics in the 1980s." In *Investing in People: The Human Capital Needs of Rural America,* edited by L. J. Beaulieu and D. Mulkey. Boulder, Col.: Westview Press, 167–182.

Marshall, R., and V. M. Briggs Jr. 1989. *Labor Economics: Theory, Institutions, and Public Policy*. Homewood, Ill.: Richard D. Irwin, Inc.

McCrackin, B. 1984. "Education's Contribution to Productivity and Economic Growth." *Economic Review* (November):8–23.

Merton, R. K. 1968. *Social Theory and Social Structure* 3rd ed. New York: Free Press.

Natriello, G., A. Pallas, and E. McDill. 1986. "Taking Stock: Renewing Our Research Agenda on the Causes and Consequences of Dropping Out." *Teachers College Record* 87(3):430–440.

OECD/Organisation for Economic Co-operation and Development. 1995. *Developing Environmental Capacity: A Framework for Donor Involvement.* Paris, France: OECD Publications Service.

O'Hare, W. P. 1988. *The Rise of Poverty in Rural America*. Washington, D.C.: Population Reference Bureau.

Owings, J., M. McMillen, S. Ahmed, J. West, P. Quinn, E. Hausken, R. Lee, S. Ingles, L. Scott, D. Rock, and J. Pollack. 1994. *A Guide to Using NELS: 88 Data.* Washington, D.C.: National Center for Education Statistics.

Porter, K. H. 1989. *Poverty in Rural America: A National Overview.* Washington, D.C.: Center on Budget and Policy Priorities.

President's Council on Sustainable Development. 1996. *Sustainable America: A New Consensus*. Washington, D.C.: U.S. Government Printing Office.

Putnam, R. D. 1995. "Bowling Alone: America's Declining Social Capital." *Journal of Democracy* 6:65–78.

Schultz, T. W. 1961. "Investment in Human Capital." *American Economic Review* LI(March):1–17.

Semyonov, M. 1981. "Effects of Community on Status Attainment." *Sociological Quarterly* 22:359-–372.

Sewell, W. H. 1964. "Community of Residence and College Plans." *American Sociology Review* 29:24–38.

Smith, M. H. 1993. "Family Characteristics, Social Capital, and College Attendance." Ph.D. diss., Department of Sociology, University of Florida, Gainesville.

Smith, M. H., L. J. Beaulieu, and A. Seraphine. 1995. "Social Capital, Place of Residence, and College Attendance." *Rural Sociology* 60:363–380.

Stallmann, J. I., A. Mwachofi, J. L. Flora, and T. G. Johnson. 1995. "The Labor Market and Human Capital Investment. "In *Investing in People: The Human Capital Needs of Rural America,* edited by L. J. Beaulieu and D. Mulkey. Boulder, Col.: Westview Press, 333–349.

Stauber, K. N. 1994. *Rural Development Overview*. Washington, D.C.: U.S. Department of Agriculture.

U.S. Department of Education. 1994. *The Condition of Education 1994.* Washington, D.C.: National Center for Education Statistics.

Wagenaar, T. C. 1987. "What Do We Know About Dropping Out of High School?" *Research in the Sociology of Education and Socialization* 7:161–190.

Wehlage, G., and R. A. Rutter. 1986. "Dropping Out: How Much Do Schools Contribute to the Problem?" *Teachers College Record* 87:374–392.

Weidman, J. C., and R. R. Friedmann. 1984. "The School-to-Work Transition for High School Dropouts." *The Urban Review* 16:25–42.

Wilkinson, K. P. 1991. *The Community in Rural America.* Westport, Conn.: Greenwood Press.

Wimberley, R. C. 1993. "Policy Perspectives on Social, Agricultural, and Rural Sustainability." *Rural Sociology* 58:1–29.

World Commission on Environment and Development. 1987. *Our Common Future.* Oxford: Oxford University Press.

International Institute for Sustainable Development. 1995. *Youth Sourcebook on Sustainable Development.* Winnipeg: IISD.

Zajonc, R. B. 1976. "Family Configuration and Intelligence." *Science* 192(April 16):227–235.

Eleven

Rural Public Administration and Sustainability: Reasserting Governance over Government

James Seroka

INTRODUCTION

America's public imagery of rural life has shifted at least once in its history, and it is long overdue for another major paradigm shift. Traditionally, rural life was associated with a strong sense of community and a profound sense of caring for the individual in society. Beginning with the rural depression in the 1920s, however, the model for rural society shifted from the proud farmer who had tamed the frontier to the struggling and desperately poor family needing public assistance to remain on the land. Now, as we enter a new millennium, a new image of rural America must be forged. In one sense, the new image may be reminiscent of our early republic, based on democratic governance, participation, and autonomy—a society independent from urban governments for its growth and sustainability. In another sense, the new paradigm must be based on communication across rural communities, usage of technology, openness to innovation and risk-taking, and cooperation with other—not necessarily geographically contiguous—rural communities.

Rural sustainable development today cannot rest on a single legal reform. Sustainable development must be based on an awareness and willingness by local rural communities to break their long dependency on higher units of government, to reassert democratic local control over their policy-making processes and political environments, to learn to associate with other rural communities, and to replace hierarchical government with democratic local governance. Additional financial resources, although helpful, are not the full answer to the problem of rural sustainable development. Greater political and policy autonomy, greater control, greater self-confidence, and greater reliance on local democracy and self-help efforts can make a world of difference.

If national, state, and regional governments hope to assist the process of rural sustainable development, they can provide financial resources to rebuild public works, health care systems, and education infrastructure; assist in the development of home-grown administrative capacity; underwrite assistance for such technology as the information superhighway; and provide administrative and technical expertise. It is imperative, however, that higher units of government refrain from setting priorities, determining answers, forming policy parameters, and supplanting democratically derived local policies and innovations with standardized approaches and programs.

CHANGES IN THE RURAL AMERICAN MYTHOLOGY: FROM YEOMAN FARMER TO DUST-BOWL REFUGEE

Current rural public administration and policy-making have been shaped, in large part, by two mythologies that have endured for decades but have very little application today. The first and founding mythology is based on the belief that rural life is simple, pure, wholesome, and inherently democratic, and that mixture with cities and city life is necessarily corrupting the rural soul. In the republic's early years, Thomas Jefferson, Alexis de Tocqueville, and others characterized the American yeoman farmer as a virtuous individual who was committed by temperament to democracy, cared deeply for his community, and was the bedrock of stability for the emerging nation. Rural life was thus often associated with democracy, stability, and civic virtue.

The mythology of a self-contained, virtuous, caring rural community and the spirit of individualism, tolerance, and pluralism it entailed has been very important for our nation, and it helped to define government response to policy issues in America's hinterland in its early years. The struggles of the nineteenth century against the "robber barons," the railroad trusts, and the Standard Oil monopoly were framed within the rationale of the preservation of the rugged frontier individualist. Populist movements, from the time of Andrew Jackson to William Jennings Bryan, based their credibility and programs on that myth. The progressive movements of the early twentieth century also employed the imagery of the conflict between the frontier farmer and the small town businessman against the urban aggregation of power and capital.

With the closure of the frontier near the end of the nineteenth century, the social and economic bases of the myth of rural goodness and rugged individualism within a supportive community began to lose credibility. American rural settlements, with the marked exceptions of rural immigrant communities, often lacked permanence and security. Throughout the

1800s settlements were founded and collapsed with high frequency, and the landscape of rural America today is dotted with those ghostly reminders of dreams that did not materialize. Rather than adhering to a strict identification with place, as was traditional with peasant communities, the American rural resident identified with nonplace-bound occupations such as homesteader, rancher, and so forth.

For practical purposes, the rural populations, particularly along the frontier, were generally transitory, and the attachment of the settler to the village or small town was, at best, a marriage of convenience. Even the homestead itself was viewed by many settlers in the United States as a temporary speculative venture, rather than as a multigenerational commitment. The expansion of agriculture on the upper Great Plains and its equally rapid abandonment is only one example of speculative community building that turned sour (Fite 1966). The biography of Abraham Lincoln—who was born in Kentucky, spent his youth in Indiana, and lived as an adult in Illinois—reflects the lack of attachment to place that typified rural America at the time.

In reality, mobility, impermanence, and lack of community characterized much of the history of American rural life. Villages and towns were founded and floundered, and many rural residents never developed an attachment to their land or to the community in which they lived. Unlike the immigrant communities and religious groups such as the Church of Jesus Christ of Latter-day Saints, the Amish, the Shakers, and the Mennonites, most of rural America was not characterized by close social communities, shared community values, and an intricate system of interdependencies. Economic gain, speculative fever, and technological innovations, such as the steel plow, mechanized agriculture, and barbed wire, were much more significant than shared community in the creation and development of rural life in nineteenth-century America.

The mythology of rural community that transcended specific spatial attachments was important, however, particularly in a political sense. The myth provided a common political vision and served as a unifying factor in American political life. The fiery oratory of William Jennings Bryan, for example, displayed the emotional appeal of the rallying call for all rural forces to band together against the evils of urban life. It fostered the mythology of a rural solidarity against urban interests and placed the struggle in a basic conflict between good and evil as well. Bryan's message relied on the symbolism of *battle* and *struggle to* mobilize numbers of voters.

By the end of the 1800s the relative number of rural residents had declined precipitously and the battle waged by the populists was lost. From this time on, the political emphasis of the rural resident shifted from rebellion to dependency, and a new rural mythology of dependence on urban America emerged. Like the Native Americans on reservations, rural Americans were forced to the margins of American politics and society. In the course of 30 years following the closure of the frontier, the prevailing image of rural America changed from the autonomous, independent yeoman farmer to the impoverished refugee from the dust bowl or other rural disasters.

In the twentieth century the second mythology was promulgated, in which rural America became the haven for the poor, uneducated, uninspired, and unambitious. Once Jefferson's proud yeoman farmer, the rural resident has become an object of pity and a ward of the state, and in much literature and the popular media rural life is now considered alien, hostile, and often incestuous (Stein 1976).

As it became the dominant viewpoint, this second mythology of rural life exerted a destructive force on the rural citizenry and on the nation as a whole. Government in particular became part of the problem, and even today government—at both state and federal levels—continues to feed the dependency and to work to preserve a type of life, including the idealized "family farm," that is no longer viable. Rather than support programs to create economic independence and sustained viability of rural areas based on their resources and potential, government programs often serve to keep the elderly and the nonproductive population on the land. Rather than work to create a new rural American image based on the enormous productive capacity of the land, government programs often serve to perpetuate an environment of dependency and public assistance. Although the dust bowl has long since passed, the dust bowl mentality and imagery endure. As a result, the essential challenge for public administration in rural America today is to generate a spirit of self-confident renewal and trust and a system of communication that can help to break the spirit of dependency on others, obedience to hierarchical patterns of communication, and public programs that ensure waste of productive resources.

RURAL ECONOMY AND TECHNOLOGY: FROM ISOLATED SETTLEMENTS TO GLOBAL INTEGRATION

To understand the importance of adapting to a new paradigm for rural public management and governance, one must fully appreciate the importance of current changes in rural economy and technology to rural government and administration. Frank Bryan (1986), for example, argues persuasively that the inherent difference between rural and urban life is population density, and that technological change and modernization has made the density factor irrelevant for most policy considerations. There is much truth to Bryan's hypothesis, but the introduction of indigenous historical factors of rural growth should be added to the analysis to provide a more complete picture.

Contrary to the European experience, particularly that beginning in the latter half of the nineteenth century, the United States did not develop a hierarchical network of villages, market towns, and administrative centers that culminated with a single national metropolis. Our reluctance to establish a town-centered hierarchy, which has been pivotally significant for rural public administration, was encouraged by the relative abundance and low market price for land in nineteenth-century America. The open frontier, and the attendant widespread mobility it offered, permitted settlers to cultivate as much land as technically feasible (e.g., 40 acres under the Homestead Act), ignore marginally productive land, and overlook soil conservation measures. The formation of deep attachments to a specific geographic place, town, or urban center was impeded. In addition, these relatively large land holdings made it impractical for American farmers to congregate in villages, as was the norm for European peasants at the time. Finally, the subjugation of the Indian tribes and their expulsion to reservations removed the security incentives for village formation, with the result that villages lost much of their economic and administrative utility and many small towns came to be inhabited almost exclusively by merchants and service providers.

The diffusion of technology had a major impact on loosening rural America's bonds to its small towns and villages. In the second half of the nineteenth century, the railroad revolutionized the patterns of settlement and the political economy of rural America. Access to the railroad, rather than to a market town, became a determining factor of production. Grain, for example, could be shipped long distances at low prices, thereby eliminating the competitive advantage of proximity to the market and market towns. Thus, those American producers who had access to cheap land were now at a competitive advantage in the marketplace.

The expansion of the railroads also bonded rural Americans more closely to the urban market and urban services and weakened the strategic importance of villages and many small towns. Products could be shipped by rail relatively efficiently, and passenger traffic permitted much easier access to urban areas for service needs. In time, railroads dominated settlement patterns. Villages and towns off the main lines often withered, and it was the technical or commercial needs of the railroads that determined the location, size, and survival of a town. Thus, sense of community and rural resident loyalty had little to do with the formation and development of small towns throughout much of the nation in the latter half of the nineteenth century.

Other historical factors also served to discourage American rural settlers from establishing close and enduring ties to communities. Contrary to the experience of Europe, serfdom and feudalism never took hold in the United States, and the American yeoman, with the important exception of the slaves in the Southeast, had a legal guarantee of spatial mobility. Unlike that in Europe, state administration in the United States was relatively weak, and most state revenues were generated from the sale of land, rather than agricultural production. Thus, for fiscal reasons, governmental authorities in the United States benefited from the mobility of their citizens. Contrary to the order of things in Europe, and with the exception of the cotton belt in the Southeast, land in nineteenth-century America was well distributed and was not concentrated among a gentry class who employed large numbers of agricultural workers.

The one major exception to this pattern of widespread agrarian land ownership occurred among the former slaves who, for economic reasons, remained bonded to the land until the second third of the twentieth century. Nevertheless, even among this population, ties to the neighboring town and communities were weak. As a result of Jim Crow legislation, tenuous land ownership, and racial discrimination, community loyalties and bonds to the town by the black population remained weak.

Historically, land speculation has been a driving force in the settlement of the frontier, and this factor has further weakened the formation of strong rural communities. Industrial prosperity and the accumulation of wealth for many in the United States derived from the sale of land to new waves of immigrants or newly formed households. The remarkable high fertility rate of the population at this time, along with the exhaustive agricultural practices of most U.S. agricultural producers, ensured mobility and encouraged extended families to pull up roots each generation so that each successive household would have the opportunity to begin anew. For those families, sale of the family homestead provided the mechanism through which each son would get a fresh start. Contrary to popular legend, the historical attachment of the American farmer to his land was only temporary, and his attachment to the neighboring village or small town was often only for convenience (Nugent 1977).

With the closure of the frontier and the introduction of restricted immigration, the century-long speculative settlement bubble burst. Intrarural mobility no longer offered security and opportunity for prosperity, and with the beginning of the twentieth century, the focus of speculation and individual accumulation of wealth shifted permanently to the cities. Finally, the precipitous fall of commodity prices, the prohibitive rise in the cost of establishing new farms, and the disaster of the western droughts and dust storms accelerated the pace of change, so that the center of mobility and dynamism shifted permanently from the countryside to the city.

Sociological and technological changes in the first two thirds of the twentieth century were largely responsible for a major shift in the dependency relationships between rural and urban communities. These advances have fundamentally altered intrarural relationships, politics, and policymaking processes of rural areas vis-à-vis urban areas (Albrecht and Murdock 1990). Many of these changes, including rural electrical cooperatives, a burgeoning road system, free rural mail delivery, mail order houses, and school consolidation, have further eroded the economic, service, and commercial relevance of many small towns and have increasingly made their survival of marginal utility.

In the early twentieth century several technological innovations, such as home telephones, automobiles, mechanized agriculture, and governmental expansion of such services as free mail delivery and electrical cooperatives, altered the traditional relationship of the rural resident to the village and the small town. These changes largely rendered the village obsolete and unnecessary. Over time, the functions that had been performed by villages were assumed by small towns.

The most significant technological change was the availability of inexpensive and reliable private automobiles, along with the growth of the highway system to accommodate the flood of vehicles. With a single blow, transportation time to the small town was dramatically reduced, two important service functions of the village (i.e., health care and the blacksmith) were eliminated, and rural residents benefited from more favorable prices in the more distant town. The automobile, in short, eliminated the economic and political monopoly of the village, weakened the role of the small town, and ended rural residents' dependence on the railroad for transportation and shipment of goods.

A second important technological change was the introduction of mechanized agriculture, with the resultant new political economy for agricultural producers. Mechanization permitted a single producer to be more productive and efficient, reduced the need for labor, and decreased population density in nearly all agricultural regions (Berardi and Geisler 1984). The increased size of farms, decreased population density, and demise of the village changed the pattern of communication between rural residents and weakened or destroyed any prior sense of rural community.

A third significant change was the introduction of rural free delivery (RFD) throughout the nation. With this service innovation, villages ceased to provide any significant administrative or service function. In addition, the rural resident became open to the national and competitive commercial marketplace, and economic pressures on small town merchants intensified with competition from catalog sales and mail deliveries. Through RFD, rural residents could be as well informed as their urban counterparts and could participate in a much wider social, economic, cultural, and commercial marketplace. RFD did much to create a national market and weaken the economy of the small town in rural American life.

In the middle third of the twentieth century the nature of change in rural America intensified, shifting the pattern of relationships between rural residents, small towns, and urban centers. In the first third of the century, villages lost their economic and social purpose, but in the second third, small towns suffered most from the effects of technology and social change. The consolidation of schools, for example, was initially a response to the disappearance of the village and reflected the decimation of village life. During the middle third of the twentieth century nearly 100,000 school districts, accounting for 90% of the total number of in the U.S. school districts, were phased out of existence. In time, these school consolidations and the concentration of administrative and cultural life threatened the identities of small towns, eliminating a major community function and source of community identification. Heightened productivity in agriculture led to the development of larger and larger farm units and less and less population density in rural America (Albrecht and Thomas 1986).

Rural electrification brought nearly all of rural America within the electric grid, and this factor changed forever the quality of life for its citizens. An enhanced highway system, including the interstate system, and a national telephone network also made it economically feasible for rural residents to bypass small towns for most service and commercial transactions.

Combined, these changes allowed much of the surplus rural labor force to engage in part-time agriculture and to assume industrial and service occupations in urban and regional administrative centers. These changes also permitted significant numbers of urban residents to leave the city and to reside in the countryside without compromising their life-styles. Like the village in the nineteenth century, the small town in the twentieth century found itself with a steadily declining role in the lives of rural residents (Leistritz and Ekstrom 1986).

The latter third of the twentieth century is witnessing a major qualitative change in the relationship of rural America to its urban counterparts. As the national media and official analysts have indicated, many small towns are disappearing or slowly dying. Virtually all rural residents have access to national media and regional information sources, so that the role of small towns in the communication network has become redundant. Nearly every adult resident has access to personal transportation and direct access to a wider labor and commercial market. Agriculture is becoming increasingly consolidated in fewer and larger farms, and the population density in rural areas is dropping even further. In fact, in many rural areas, services or tourism has replaced the traditional base of mining, agriculture, or livestock as the major source of revenue and employment (Albrecht and Murdock 1990). Those small towns that cater to tourism or retirement communities and those communities with easy access to metropolitan areas are among the minority of small towns that have fared best in the struggle to survive.

As we enter the twenty-first century, rural America can now become technologically and electronically integrated into the modern world and its citizens can, potentially, become integrally linked with one another. Modems, facsimile machines, computers, satellite dishes, among other vehicles, are accessible to rural residents at reasonable prices, and urban and metropolitan areas no longer have a critical advantage in the analysis, coordination, and transmittal of information and data. Overall, urban America can potentially lose much of its comparative advantage in cultural and information activities, and it is becoming clearer that urban life is seriously handicapped in many aspects of the general quality of life, as well as in the cost of living. As the century has evolved, rural location has provided a progressively weaker differentiation factor, and a rural location increasingly implies a life-style choice rather than a separate cultural and economic identity. To a great extent, the United States is approaching equality of place, and the village, small town, and even the regional center has lost much of its raison d' être as a service center for a rural hinterland.

The implications for rural life in America from the emergence of the global village are enormous. At a minimum, there are profound changes in public policy decision-making and an equalization of communication patterns. For the first time in modern history, communication patterns need not follow pyramid-style hierarchical patterns, and those with common concerns and needs can communicate and interface directly without the intercession of cities or administrative centers. Rural residents are no longer forced to communicate with one another through the "hub of the wheel," and it is no longer advantageous for public policies to be externally determined and administered. Technology has made it possible for rural communities to communicate with each other along the rim of the wheel and for public policy decision-making to become more participative, innovative, decentralized, and democratic.

Over the last several years the explosive growth of the Internet and the so-called information superhighway has made locality even less relevant to the exchange of information, provision of services, and decision making. Today there is no serious economic advantage for technology-based service providers to locate in heavily urbanized areas. Indeed, the proliferation in the quality and quantity of communication technology could give an advantage to less dense areas that offer a better quality of life at a lower individual cost. In a world where time replaces distance as a significant cost factor in production, rural areas may compete on a more equal footing.

EVOLUTION OF RURAL ADMINISTRATION: FROM TOWN HALL GOVERNANCE TO ADMINISTRATIVE STEP-CHILDREN AND BACK AGAIN

Rural America's orientation to local public administration was essentially shaped by the Northwest Ordinance of 1787, arguably the most far-reaching piece of legislation introduced by the newly founded republic. This sweeping statute gave America's newly organized territories a common tradition of representative governance, a strong institutional structure of direct local democracy, and an overarching umbrella of local administrative responsibilities and obligations for essential public services, such as roads, schools, health care, justice, and law enforcement, under direct local control.

Throughout most of the nineteenth century, local administrative autonomy over nearly all activities, with the exclusion of defense, postal services, and sale of federal land, was a given. The national government played a minimal administrative role in local activities, and state government—staffed by part-time legislatures and underdeveloped administrative personnel—largely left rural administration and policy-making to their own devices. In nineteenth-century America, even large scale public works projects, a traditional governmental obligation, tended to be privatized and financed through land grants to private companies or corporations.

Figure 11.1 illustrates how rural public managerial decision making was organized and occurred in the nineteenth century. Generally, the village was not the significant policy-making institution, and rural communities communicated with one another only if they were in close proximity and faced a common problem. Inasmuch as most public functions were provided at the local level, policy decisions tended to flow from the local rural level upward. Small towns served as conduits for the administrative center (i.e., the county), and both the administrative and urban centers avoided direct interference and involvement in rural political and administrative life.

As rural America became more integrated into the national economy and into society during the first half of the twentieth century, government expanded the scope of its services and its regulative and supervisory roles in rural areas. Over time, federal and state governments have also taken a stronger leadership role vis-à-vis rural governments, and they have tended to differentiate less and less between the needs and capabilities of rural units and those of the more developed and populous urbanized political centers.

During the latter part of the twentieth century, the scope and mission of government, particularly state and local governments, grew immensely. Accompanying this growth were intergovernmental mandates and other intergovernmental demands on local units of government. Eventually, state and federal governments developed their own state or national policy agenda and imposed their demands and priorities in a standardized manner. Rarely did they take into consideration the implications of their newly standardized policies in an environment of low population density.

A major result of the growing control of state and federal governments over local administrative discretion was the appearance of a capacity gap between urban and rural administrative units. Governmental administration in rural areas tended to lack the desire to implement services that were primarily designed for urbanized populations, and they lacked the professional personnel, staff, and resources to implement these services effectively. Significantly, as state and federal involvement grew more pro-

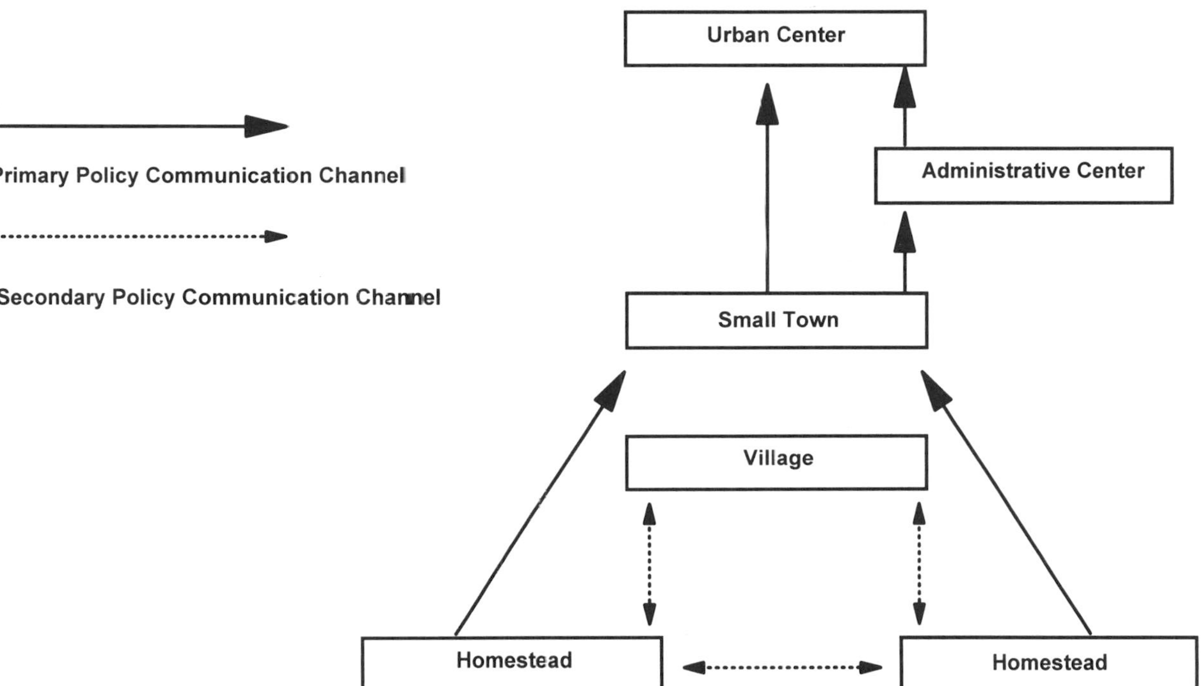

Figure 11.1 Nineteenth-century rural public policy decision-making.

nounced, local rural governments lost their political autonomy and found themselves in a stereotypical dependency relationship subject to the whims of their so-called partners at the national or state level. These higher-level governments generally argued that only they had the capacity and vision to decide what services were needed in a rural area, as well as a monopoly on knowledge of how those services could be provided. By the end of the twentieth century, therefore, much of rural public administration seemed to be without purpose, wasteful, counterproductive, and detrimental to the interests of the rural inhabitants.

Figure 11.2 outlines the rural public policy decision-making model that became prevalent in the twentieth century. The model highlights the way in which public policies and decision making were controlled by other governmental and urban centers. In this model, the rural community is dependent on the administrative and urban centers and has little capacity to alter the arrangement.

A significant outgrowth of this relationship of dependence was that rural residents often abandoned their own administrative and governmental institutions and withdrew from direct involvement in governance. By default, governance of rural areas became a state and federal responsibility. As I argue elsewhere:

> The core of the problem is not that rural America has lost its sense of direction, but rather that it has been forced to abrogate control over its future, and that it has allowed its interests to be defined by others. Depending upon the issue, rural America has been mistakenly equated to agricultural America, or to an urban America, writ small, or to a poverty stricken economic and cultural backwater in need of paternalistic care. Rural America and rural administrators need to find a way to gain control over their policy agenda and to chart their own course to the future. In brief, rural governments must assert: what needs to be done, by whom, and how. Rural American communities must take a more active role in defining their own futures (Seroka 1990a).

In the interest of equity, professionalism, and standardization, state and national governments have denied rural political and administrative units, particularly the rural counties, any significant role in the policy-making process for their citizens. Indigent health care mandates, uniform housing standards, local matches in poverty programs, and even uniform clean air and water requirements were imposed on rural government without serious consideration of the special needs and resources of their environments. Although urban local governments have had the political muscle to set the national agenda, rural local governments in most states lack that capacity. America's rural governments have no organizations with any power that

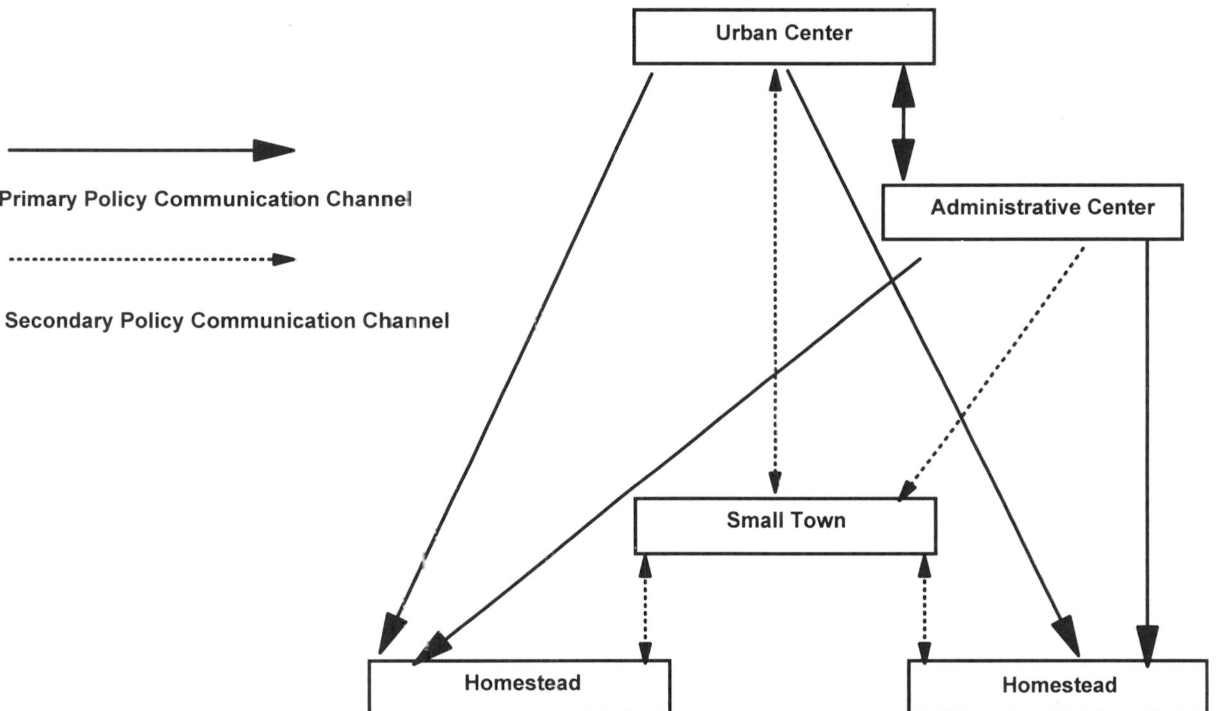

Figure 11.2 Twentieth-century rural public policy decision-making.

begins to rival the clout of such urban-dominated associations as the National League of Cities or the National Conference of Mayors. Metropolitan governments and their political leaders have the commitment and professional wherewithal to design programs and determine their visions, goals, and objectives, and national and state decision makers and legislators regularly correspond with these experts in urban settings. Meanwhile, rural governments have no significant tradition of cooperation to be able to impinge meaningfully on these significant policy debates.

The urban-based professional and bureaucratic concerns for standardization deny rural units their voice in policy implementation, and centralized hierarchical models are used to put these urban designed programs into effect within rural local communities. Finally, the tyranny of numbers in politics guarantees that the policy experiences of rural governments are considered uninteresting anomalies or the result of determined parochial resistance. The rural administrative experiences are rarely considered valid or reliable indicators of the success or failure of a program by state or national authorities.

Congress's recent attempts to devolve authority and funds to the states provide little comfort for rural communities. Federal block grants to the states have traditionally been awarded according to formulae that benefit urban conglomerates (i.e., suburbs and core cities). At the same time, mandates for services are not being lifted from rural governments. Worse yet, from the perspective of the rural administrator, the new federal devolution increases the responsibilities and obligations of rural public service providers to supply levels and ranges of services comparable to those found in metropolitan areas.

From the perspective of rural units of government, the policy involvements of national and state governments are baffling. State and national policy administrators neither understand rural governments nor do they seem to be aware that they lack significant understanding of these entities. From the perspective of state and national administrators, however, the rural disenchantment is equally baffling. They see the solution as clear—namely, rural governments must organize, cooperate, and exercise their available political muscle to take ownership or control over the policy process.

Rural policymakers and administrators are also part of the problematic environment of dependency. If rural governments continue to remain wedded to the prevailing isolationism found in many rural areas, and view themselves solely as the service delivery links in the outmoded and outdated governmental dependency pattern, rural sustainable development will not occur. If rural governments keep innovations in intrarural communication and participation at arms length, then they and their communities can lose out on the most significant leveling innovations to become available in generations.

Implications for rural public administration are also profound. Since differences in public service expectations between urban and rural residents have diminished, rural public managers are expected to provide as broad a range of services as found in urban metropolitan areas. Furthermore, they are expected to provide these services in as professional and efficient manner as possible, but without access to the necessary staff, facilities, resources, and economies of scale. If rural managers refuse to meet their constituents' new expectations for service, they will once again doom their locales to backwater and dependency status. If they continue to practice hostility toward technical proficiency, autonomous action, or intrarural cooperation and pooling of resources, they will make it enormously difficult to break the dependency relationship and to engage in solid rural sustainable development.

COOPERATIVE INTERRURAL COMMUNICATION AND ACTION

In the past, rural communities were geographically isolated from one another, but there was a semblance of community and community structure. Currently, technology has eliminated the bases for isolation, but new interrural patterns of communication and new institutions have not yet developed sufficiently to capitalize on this new environment to build a renaissance in interrural community patterns. Most communication of rural governments is still with urban and administrative centers, and very little is directed toward other rural units of government. As I argued earlier, rural governments have to forge new lateral bonds of communication to enhance their public decision making and to engage in more cooperative activities involving direct participation of citizens (Seroka 1990b). Rural-to-rural contact must supplant the long-dominant top-down, urban-to-rural communication and control.

This process of lateral interdependency is more advanced in some regions of the country than in others. Traditional bonds of rural residents to the small towns are relatively strong in New England but have weakened considerably in the Midwest. The process, however, is occurring everywhere, and it is inexorable and irreversible. Small towns must rely on national networking and cooperation with one another, and not put their hopes for survival or growth in the competitive domination of the surrounding rural area or neighboring small town network. Aggressive and cannibalistic economic development to the detriment of neighboring small towns weakens everyone. An analogous argument can also be made for urban areas, which must learn that internal long-term growth is maximized through cooperative behavior, not through the rapacious seizure of resources. Again, economic and technological competition is no longer local, regional, or even national. Competition is increasingly becoming global in character.

Figure 11.3 portrays a likely positive change in policy-making among rural communities, in which commonalities of interest replace spatial and administrative hierarchy in public policy decision making and governance. The primary line of control is through lateral interdependency on each level and the maintenance of communication across levels of government. This model for the twenty-first century is not simply a return to the Jeffersonian model. Rather, it is based on interconnectivity and interdependency, not isolation and self-sufficiency.

Even a shallow examination of the professional and trade journals reveals the presence of strong lateral commonalities of interests (e.g., public service financing and delivery, design of local government telecommunication services accessible to citizens regardless of income, distance learning opportunities for rural communities, etc.) and weakening hierarchical patterns of decision making, partly a result of federal and state fiscal retrenchment. During the last two decades, regional councils of government and regional municipal associations have begun to cooperate and act laterally rather than hierarchically. Large cities and metropolitan areas now interact more regularly with each other than in the past. State-level associations (e.g., the U.S. Governors' Conference) have taken the lead and meet regularly to discuss and act on common problems.

Among government entities, only rural governments (i.e., rural counties and townships) have stayed outside the emerging lateral decision-making trend. Rural governments, more than those at any other level, have been co-opted by the past and remain committed to the nineteenth-century pattern of hierarchy. Administrative and political institutions of counties, towns, and townships have changed little, even though their roles and responsibilities have grown enormously. Rural counties often have elected sheriffs, county clerks, finance officers, and other officials to manage complex departments and agencies, and many rural institutions also have a tendency to celebrate their isolation, rather than to set about building a climate of cooperation.

Until recently, county institutions could remain isolated because the lower level of services expected by their citizens and clients favored such isolation. Today, however, such isolation is no longer viable. Citizens in rural communities expect more from their local governments, and the national and state governments mandate more as well.

Generally, rural governments have been reluctant to utilize new technologies and to associate and cooperate with each other. Rural governments have also been slow to recognize that their problems are not unique and may be shared in other venues. Problems relating to the regulation of open pit mines, for example, may occur in rural areas as widely dispersed as southwestern West Virginia, southern Illinois, northern New Mexico, western Wyoming, and central Montana. Forest management is a concern in central Maine, western Georgia, northern Michigan, and west-

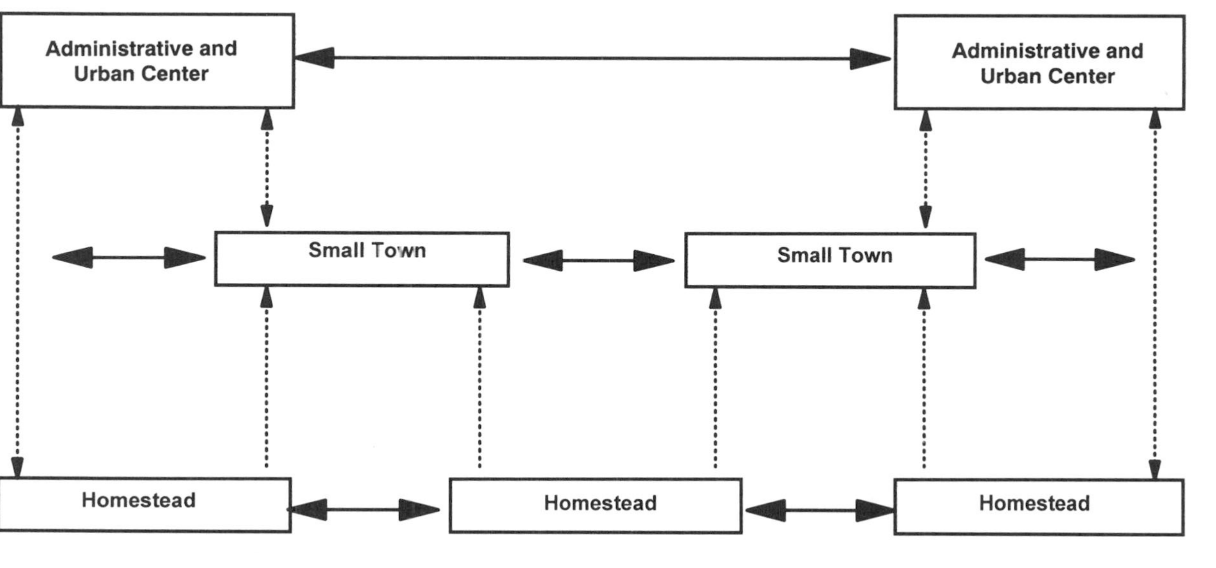

Figure 11.3 Twentieth-first-century rural public policy decision-making.

ern Oregon. Current deregulation in the telecommunication industry is critical to the needs and interests of rural residents. Negative impacts from correctional facilities are widely dispersed, but little interrural communication occurs.

The rural criminal justice community has been in the forefront of information sharing, but most of this data network is hierarchical rather than lateral. Rural governments are likely to be able to identify known felons and escaped convicts from anywhere in the nation, but there is no data network for distinctly rural concerns such as the illegal disposal of waste, misuse of herbicides, or soil runoff. Rural governments have not sufficiently communicated "along the rim of the wheel" nor pooled their resources and experiences. They have not sufficiently used the new environment to free themselves from urban dependency.

If rural governments hope to break their isolation and dependence on urban centers, they must begin to communicate and cooperate with one another. There is no realistic hope for any significant rural control over the policy process if rural governments work alone and in ignorance of one another. Rural leaders must cooperate with one another if their voices are to be heard in the national competition to set the policy agenda. Geographical contiguity no longer defines commonalities of interest and is no longer essential, or even important, in the communication process. In the twenty-first century, commonalities of interest, organization, and knowledge will determine power.

Many rural governments argue that they lack the personnel and resources to take control of their policy process. They ignore the advantages that derive from sharing and cooperation. Smaller companies, for example, use consultants to provide assistance in areas where they lack expertise. Why not rural governments? Smaller companies subcontract for services that are short-term, project-specific, or episodic. Why not rural governments? Rural governments argue that their policy issues are unique or are not shared by their neighbors. Again, the solution is communication, cooperation, and outreach, not defined by geographical proximity. Small companies form trade associations whenever commonalities of interests emerge. Why not rural governments?

One can argue that cooperative action among small rural governments is a pipe dream. The theory of collective action, for example, states that whenever there are numerous small beneficiaries of collective behavior, no collective action will be taken because no one will undertake the costs of organization to promote that collective good (Olson 1965). Further, the theory argues that the problem of free riders will overwhelm and destroy collective action unless compulsion and/or externalities are provided. The theory of collective action is persuasive in explaining why rural governments did not band together in the nineteenth century or in the early part of the twentieth. It can also explain why the populist movement disintegrated so completely after its initial defeats.

Today, however, the rural environment has changed in ways that distinctly encourage collective action. First, technology has fundamentally altered the economic calculus for potential participants. The cost of organization is much lower, and the cost of information sharing is almost negligible. Organizational costs are now low enough that even hard-pressed rural governments and policymakers may be able to participate in cooperative action. Second, just as the federal government linked the distribution of intergovernmental benefits to municipal participation in regional councils of government, the national or state governments can encourage similar associational behavior through similar incentives and requirements. Finally, rural associations can provide externalities to encourage membership and participation. Exchanges of personnel, registration of consultants, directories of experts, data collection and analysis, model grant proposals, membership-based journals, association purchase agreements, appropriate computer software, and staff training and development program packages are just some of the means to encourage membership and cooperation through the provision of externalities.

Universities, particularly land grant colleges and universities, can assist in this process. Although the efforts of these schools have been largely tied to the technical aspects of agriculture and livestock production, the basic expertise exists to assist rural governments to help themselves.

The U. S. Department of Agriculture, particularly the government assistance branch of the Economic Research Service, can be expanded to serve as a resource and consultation base for rural governmental entities. State-level organizations, such as the Department of Commerce and Community Affairs in Illinois or the Pennsylvania legislative Rural Affairs Institute, can also contribute to the effort, particularly if they are willing to share knowledge and experiences beyond the boundaries of the state.

University-based planning and public administration programs also can contribute to the cause by training personnel who can make interrural communication work. These units can also conduct research and develop and disseminate programs to meet the needs of rural entities. Finally, such programs can lend professional assistance to the efforts of rural governments to have an impact on the intergovernmental policy process.

Public managerial and political competence and capacity are necessary to develop a framework for sustainable development in rural areas. This framework must also be locally based, locally generated, and locally controlled. Its genesis, however, will not be locally driven. Just as the availability of land transformed rural America and created a mobile society in the nineteenth century, the instant availability of information via new telecommunications and the emerging equality of space may create another transformation of rural America in the twenty-first century and lead to another mobile society. There is no inherent reason that expertise cannot be shared across rural America. The political infrastructure of rural communities, however, must become more open and welcoming to outsiders with new ideas than they have generally been so far. The rural political structure must be willing to consult others, rely on local talent when available, endorse and encourage public participation, and encourage innovation and risk taking, as well as networking between the public and private sectors.

CONCLUSION

Self-sustaining development in rural areas cannot emerge so long as county, regional, state, and federal governments protect the twentieth-century administrative pattern of dependency, hierarchical control, standardization, and limited risk-taking. Most important, rural sustainable development cannot become a reality if local forces and the local population remain outside the decision-making process. We need to restore the democratic and participative essence of rural administration and governance established by the Northwest Ordinance. We need, in short, to replace intergovernmental aid with intergovernmental partnerships and closed bureaucratic government with participative governance.

Sustainable development today and in the foreseeable future is no longer based on the traditional economic triad of land, labor, and capital. Knowledge, communication, and natural resource stewardship have entered the equation, which potentially places rural communities on a more equal footing with larger urban areas. Modern development, rather than representing the accumulation of capital, can also take the form of enhanced quality of life, and the rural policy-maker's role in this process is no different from that of a policymaker in any other unit of government. To provide the level of services needed and demanded, rural governments must get access to the information and expertise that is now electronically available. To facilitate sustainable development, rural governments must become more open to their clients and citizens and must put into practice the goals of Jeffersonian democracy. In this respect, rural governments have the distinct advantage.

Sustainable rural development depends on the initiative, vision, and support of the citizenry as much as on capital accumulation. For the development of the future, qualitative growth is better than quantitative growth, more is not necessarily better—better is better. In the competition for development, rural areas no longer suffer disproportionate handicaps in comparison with areas of higher population density.

In conclusion, it must be stressed that the progress in rural sustainable development depends on the initiation of rural units of government and their willingness to associate, to communicate, and to entertain new ideas and approaches. Rural entities themselves, not the national government or the states, must set the agenda and make decisions in a democratic manner that involves the entire community. Rural entities, not governmental or university departments, must take a central role in coordination and management of the process. For the prevailing dependency relationship to be broken, the dependent must be willing to take responsibility for the future.

REFERENCES

Albrecht, D. E., and S. Murdock. 1990. *The Strategy of U.S. Agriculture: An Ecological Perspective*. Ames: Iowa State University Press.

Albrecht, D. E., and J. K. Thomas. 1986. "Farm Tenure: A Retest of Conventional Knowledge." *Rural Sociology* 51(1):18–30.

Berardi, G. M., and C. C. Geisler. 1984. *The Social Consequences and Challenges of New Agricultural Technologies*. Boulder, Col.: Westview Press.

Bryan, F. 1986. "Defining Rural: Returning to our Roots." In *Rural Public Administration: Problems and Prospects,* edited by J. Seroka. Westport, Conn.: Greenwood Press.

Fite, G. C. 1966. *The Farmers' Frontier.* Albuquerque: University of New Mexico Press.

Leistritz, F. L., and B. L. Ekstrom. 1986. *Interdependencies of Agriculture and Rural Communities: An Annotated Bibliography.* New York: Garland Press.

Nugent, W. T. 1977. *From Centennial to World War: American Society, 1876–1917.* Indianapolis: Bobbs-Merrill.

Olson, M. 1965. *The Logic of Collective Action: Public Goods and the Theory of Groups.* Cambridge, Mass.: Harvard University Press.

Seroka, J. 1990a. "Rural Administrative Capacity: What Needs to Be Done, by Whom, and How?" Paper presented at the annual meeting of the Southeastern American Society for Public Administration, Clearwater, Florida.

———. 1990b. "Inter-Rural Administrative Cooperation: Issues and Opportunities." *National Civic Review* 79(2):138–151.

Stein, B. 1976. "Whatever Happened to Small-Town America?" *The Public Interest* 44 :17–26.

Twelve

Community-Based Workshops: Building a Partnership for Community Vitality

James A. Segedy

I know no safe depository of the ultimate powers of the society but the people themselves; and if we think them not enlightened enough to exercise their control with a wholesome discretion, the remedy is not to take it from them, but to inform their discretion.

—Thomas Jefferson, Letter to William Charles Jarvis, September 28, 1820

INTRODUCTION

The sustainable development and vitality of our communities depend on an informed and involved citizenry. Although all citizens are consumers of community planning and design, they are generally uninformed about the choices available to them or how to go about getting more for their money. In the small rural town there is often a lack of synergy between citizens and planning officials. Furthermore, these communities often lack a vision for their community. The community-based workshop, or charrette, can provide local officials and concerned citizens with a set of tools that will help educate and involve the community in building a clear vision for their future. It should be noted that virtually all of the sustainable community initiatives have grown from a foundation established through some form of visioning charrette or workshop. This grass-roots involvement in the community development and planning process is the foundation of the sustainable smaller community.

The Charrette

The *charrette* (a Beaux Arts-derived term for a short, intensive design or planning activity) workshop is devised to stimulate ideas and involve the public in the planning/design process. It is a valuable tool for setting the foundation for the development of a more formal plan. Although it is not a substitute for formal planning or design, it can be integrated with such processes. By offering a hands-on and exciting community activity, a well-run community-based workshop or charrette can be a tremendous resource to the rural community (see sidebar 1). The charrette workshop establishes a platform for a free flow of information, opinion sharing and ideas. Its primary role is to provide a forum for building community consensus on a vision for the community's future through active involvement and visualization that help bring the vision to life (Segedy and Johnson 1995).

In most small communities local planning officials tolerate public hearings as a necessary evil that is of little or no help to the planning process (Creighton 1981). Too often the development and implementation of public policy is based on secondhand impressions of the public interest or is centered on a series of negotiations between local authorities and private businesses, who merely guess at the needs of the local citizenry or project their own interests as being those of the public. This approach generally has the effect of stripping the public of its right to participate in the local planning and decision-making process, often leaving the community apathetic to its own future. The impacts of community development activities are simply too far-reaching not to include all parties in the decision-making process. An educated (and involved) public is more than sufficiently equipped to make the necessary choices in a responsible manner. Experi-

SIDEBAR 1

Applications of Community Charrette/Workshops

The community charrette/workshop can be used in a multitude of applications:

- SWOT (strengths, weaknesses, opportunities, and threats) identification
- Quality-of-life assessment
- Issue identification
- Needs assessment
- Project development and identification
- Strategic planning
- Energizing the community
- Consensus building
- Visioning and visualizing
- Communication and network enhancing.

ence has shown that such an informed and involved citizenry often serves as the driving force for the implementation of the planning effort.

The Community-Based Visioning Process

Community apathy is more often a result of a lack of awareness and a feeling of disenfranchisement on the part of the community, than of citizen disinterest. For community development to be truly effective, it is imperative that communities, particularly smaller and rural communities, reach the final stage of Creighton's (1981) four stages of public involvement. In the final stage the community has been involved in issue identification, the formulation of alternatives, and their evaluation, and is ready to integrate its involvement into the decision-making and implementation stage, thereby "institutionalizing" the process. In this environment of total integration of citizen participation into the planning and decision-making process, the inadequacies of the conventional public meeting process are resolved.

It is also imperative that the involved public be an informed one. The complexity of the planning issues facing the rural community cannot be met by the public hearing process, which relies on the advocacy model and is increasingly formal. A certain level of formality is necessary in the debate over the future direction of community development. Yet the widely practiced "propose and respond" mechanism of community planning and development has little ability to inform the public of the relevant issues and choices available and often leads to confrontation rather than cooperation. What is needed to counteract this vicious cycle of apathy and disenfranchisement is the incorporation of an educational component into the public involvement process. This establishes a dialogue, as Leymann (1987) suggests, that is crucial to the development of the local citizens' objective and subjective knowledge.

An informed citizenry is instrumental in resolving the traditional challenges faced by the small town and rural planning process. When "a planning commission or task force in a small rural community meets to discuss the future of the town, to devise the plan for the year 2010, they are seeking to create a vision" (Luther 1990, 33). Hence, visioning builds on the assumption that the local citizenry has values, insights, and expertise different but no less important than those of the planning professional. The community-based visioning process becomes a conduit for the community and its individual stakeholders to share not only their concerns, but also their ideas and dreams. In Oregon "ads invited citizens to *open your mind and say, Aaahh!* And they did" (Oregon Visions Project 1993, 2).

Creating the vision for the future of a community also involves a wide array of community issues, including town planning and individual building design. This becomes more complicated when one realizes that "small towns are not [usually planned or] designed by designers. Rarely is an architect, planner, or designer involved in important design decisions. The small town designer is more likely to be a barber, garden club member, alderman, or church building committee member. These 'designers' must be equipped with a clear concept of how to develop the small town environment with a sensitivity to the impact of place and time" (Barker et al. 1981, quoted in Luther 1990, 33). Therefore, it is important not only to create a vision in the local community, but also to help citizens visualize what that future might be like. This is the heart of the public involvement and public learning process that is the foundation of sustainable community development. Seven guidelines provide support to this process (see sidebar 2).

A process that has proven successful in applying the aforementioned principles, educating the public, and stimulating the involvement of the greatest number of citizens is the presentation of multistep design and/or community development workshops, in combination with short-duration, highly intense charrettes (Costello 1987).

The Community-Based Charrette/Workshop

The direct involvement of its citizenry promotes ownership of the project, which adds immeasurably to the planning and development efforts of the rural and smaller community. The bringing together of the various stakeholders is itself an educational process. This process has been achieved through a wide variety of techniques, ranging from futuring, storefront studios, co-design, and similar dynamic group processes. The community-based workshop and charrette provides the added dimension of *hands-on visualization.*

The visualization process reinforces the mental images that are being developed and allows them to be easily shared with the community. In conventional community planning and design, and even in traditional citizen participation programs, there is a strong risk of misinterpreting ideas and concepts. One person's definition of a park, for example, may be very different from another's. In the more passive planning and development processes, these differences can lead to frustration and even remonstration against a project's implementation. Alternatively, through the hands-on visualization process that occurs in the intensive community-based charrette/workshop, community members are involved in the design, development, and evaluation of their environment by being able to quickly see words transform into pictures, thus avoiding misinterpretations of proposed projects (Wates and Knevitt 1987). In many cases the participants actually become the designers themselves.

Early examples of this process are a derivative of John Friedmann's theory of transactive planning, expanded in the R/UDAT (Regional/Urban Design Assistance Teams) approach that originated in London and has been

SIDEBAR 2

Guideline for the Community-Based Workshop Process

1. **Encourage Citizen Participation.** All segments and groups within the community should be represented throughout the process. Citizen participation should be solicited from all age groups, organizations, city/town officials, interest groups, and the general citizenry.
2. **Empathize with Participants.** Strive to fully understand the problems, perceived problems, issues, and concerns of each participant. Do not anticipate or predetermine the problems of a community nor pass judgment. Seek out recommendations and ideas and obtain lists of community assets from people throughout the community.
3. **Understand the Physical Community.** Complete a thorough inventory and assessment of the community. There is never too much information. Maps, photographs (historical and current), demographics, and other community data serve as a basis for decision making.
4. **Develop User- and Reader-Friendly Documents.** Make absolutely sure that final documents and supplemental reports can be understood by the layperson. Use whatever means necessary to make the workshop products easy to understand and follow.
5. **Get It Started.** A plan is not the final step in the process; rather, it is the beginning of the journey. Do not allow a plan to be shelved and left to collect dust. It is in the implementation stage that the majority of plans fail. To ensure completion, a group or individual must provide leadership for each project following the adoption or approval of the plan.
6. **Visualize the Vision.** Drawings and pictures are among the best tools for accurately depicting what is intended or expected. Words alone generate a different mental image for each person who reads them. Pictures provide a platform that is rarely misunderstood and that can be discussed individually, component by component.
7. **Follow-Through and Benchmark.** At a specified time following the competition of a planning project, the plan must be revisited to determine whether the projects are on track. This follow-through will also be an opportunity to check the progress to date and make any adjustments to the projects or plans.

adapted and adopted as a program of the American Institute of Architects (AIA). Recent history has seen a number of academic institutions integrate similar programs into the curricula of architecture, landscape architecture, and community planning programs. Each of these programs approaches the community planning and design process from a slightly different perspective, all build from a strong foundation of citizen involvement and participation. In this author's experience, the charrette approach has been found to be more effective than traditional "topdown" planning models at spawning community involvement and project success by virtue of its strong educational component and the public excitement generated by its short-term intensity.

A Model for Involving the Community

Before it decides to begin a charrette, a community should be informed and sensitized to what this unique planning process will ultimately accomplish, that a charrette will do the following:

- Generate strong citizen participation and motivation toward planning and community development projects;
- Enhance communication within the community and increase awareness of planning and development issues;
- Bring community groups, leaders, and citizens together to generate a common community vision and address community challenges;
- Initiate doable community development projects;
- Identify potential funding sources for community projects; and
- Give the community a starting point with specific actions steps for successful community development and quality-of-life enhancement.

The Steering Committee: Vital to the Charette/Workshop

Community participation is the main ingredient in the community-based workshop, of which its most visible form is a two- or three-day intensive charrette. Communities are not built overnight, nor is the charrette process; prior to the actual well-publicized event, a significant amount of preparation must take place. This is probably the most important step in the process. The community must first determine that it wants to get involved and is willing to do something with the results. For any community planning activity to be successful, the plan must actively involve and be supported by the community, not only by a few of its leaders. At an initial meeting with local community officials it is important to determine clearly the preliminary and secondary issues related to the project and to identify its geographic area. This process should build on solid analytical information. An economic and statistical profile of the community is a fundamental tool in the success of a visioning/public involvement process. It is also critical to the success of the project that a citizen action group, or steering committee, be established to represent a broad base of community interests. Depending on the issues, this committee should consist of 12 to 15 persons with diverse opinions and ideologies. Members of the steering committee must be actively interested and involved in their community. They can be drawn from the local chamber of commerce and business community; neighborhood, citizen, and home owner associations; elected officials; technical staff; religious organizations; service groups; public schools; state, regional, and county officials; and farmers' organizations. It is important that these people be local citizens and have time available to be active members of the process, as they will be the principal players in coordinating charrette activities. The steering committee will also be responsible for scheduling and arranging the logistics of the charrette/workshop and for organizing and managing the charrette's budget. Important to the effectiveness of the steering committee is the inclusion of an individual who possesses expertise and experience in community organization and politics. The ability to assume a leadership role and to understand and work within the system will prove invaluable throughout the process. This person should also be recognized by the community as a leader who can help build community consensus and enthusiasm.

Reaching and Soliciting Public Participation

After a number of preliminary meetings of the steering committee, a series of articles is prepared and published in the local newspaper(s) informing the citizens of the issues identified and inviting their participation in a community development and design workshop. A cooperative local newspaper provides the greatest potential for reaching the maximum audience; however, broadcast media and public access cable television systems can also be utilized. In some cases, members of the charrette team (discussed in the following paragraph) make presentations to local civic groups to heighten awareness and garner further support for the community development effort. Topical programs can be prepared for local high school civics and elementary school classes, both of which generate tremendous student interest and parental support. High school drafting and government classes can be encouraged to join the charrette team as well. Generally speaking, the more groups involved, the greater the likelihood of building local ownership of the project.

The Charrette Team

In addition to the members of the steering committee, the principal participants in the workshops are the public and the charrette team. The charrette team is usually a group of individuals with a broad range of skills and backgrounds. There are advantages and disadvantages to having both local and outside team members. Local members bring unique insights to the process, and outside members can bring fresh, objective viewpoints to the activities. Local members can also have a rather narrow view of their own community and their vision may be clouded by what they already know. It is important that the team be assembled for its skills and not just according to the interest of the individual members. The team will be primarily responsible for producing the tangible results of the workshop. In addition to the "technical" members of the team, participation in team and charette activities should be open to all interested members of the community.

In assembling the charrette team, it is important to recruit a good facilitator, one experienced in the charrette process if possible. This should be a local leader (not an elected official)[1] with good facilitation skills, or someone from an outside organization, such as a university or a professional or consultant group. The facilitator must be an objective, good communicator, trusted by the participants, comfortable with the subject matter, and able to ask and answer difficult questions. Team members should include people with skills in some or all of the following (which vary with charrette focus): urban and rural community planning, architecture, landscape architecture, marketing, civil engineering, and community and economic development.

The Community Database

An effective process begins with good information. The charrette process requires a solid base of technical data at every stage—from the preliminary issues identification meetings involving the steering committee to the actual charrette/workshops, building on public input. This is critical to having accurate information available so as to respond adequately to citizens' questions and provide support to innovative insights. Especially useful are existing plans and historical profiles. Good base maps of the study area are essential, as well as other sources helpful in documenting existing environmental, economic and social conditions (see sidebar 3).

Scheduling and Conducting a Community-Based Charrette/Workshop

As mentioned earlier, most charrette/workshops are of short duration (usually two to three days), highly organized, and very intense. This format has been found to be the most effective because it permits ample time to accomplish workshop goals while developing and maintaining public enthusiasm. Typically, each workshop involves a series of large-scale public information/education sessions, structured interviews, meetings with concerned citizens and local organizations, and informal working sessions.

SIDEBAR 3
Sources for Documenting Existing Conditions

- Identification of key players in the community
- Materials documenting existing conditions
- Aerial photos
- Maps
- Earlier environmental and physical planning documents
- Economic and statistical profiles
- Community and regional studies or reports
- Demographics and statistical information
- Videos, photographs, and sketches
- Surveys
- Historical profiles (i.e., from newspaper files, photographs, archives, historical societies, books)
- Governmental regulations and ordinances
- Reference materials and examples of related projects

Although the actual schedule must be flexible, public meeting times should be firm and adhered to. However, flexibility must be allowed so that special opportunities are not missed and that creative energy is not foregone just to keep on schedule. It should also be noted that the days do not have to be contiguous. In some cases it is better to allow several days between sessions to permit the team to catch its breath, but spreading the process over too long a period of time may cause a loss of momentum and public interest.

An effective charrette consists of eight sessions as described in sidebar 4:

SIDEBAR 4
The Charrette/Workshop Session

SESSION 1
Getting Started

The first session might be held after dinner, or at a breakfast meeting.

Goal: To develop a working relationship between the charrette team and the steering committee

This session should:

- Be held in an informal setting to make information flow more easily;

- Involve casual conversation, which is more effective than formal presentation;
- Allow members of the steering committee and the charrette team to introduce themselves (people-to-people style) with a short statement of background and interests; and
- Enable the charrette team to develop an understanding of the basic issues to be addressed in the charrette itself.

SESSION 2
Context Development and Community/Issue Orientation

This session is best held early in the morning on the day after the first session.

Goal: To give the charrette team a firsthand look at the community and provide an orientation to their background information

This session should include:

- A walking tour of the area, led by members of the historical society, neighborhood leaders, children, planning staff, etc. (an important event for both the charrette team and the steering committee);
- A summary by the steering committee of its interests and perceptions;
- An opportunity to view any available videotapes or slides of the community;
- A study of maps, aerial photos, and community database;
- A review of planning reports and other technical documents.

SESSION 3
From the Horse's Mouth: Interview and Input Sessions

Goal: To provide an opportunity for diverse citizen and public groups to discuss issues with the charrette team

In this session:

- The study team is divided into small groups to facilitate interaction and effective communication with citizens;
- The issues and goals of the community are clarified; and
- Interview times are scheduled to ensure that each citizen group will be properly heard.

SESSION 4
Impressions: Analysis and Issue Clarification

Goal: To provide an opportunity for the charrette team to assimilate and discuss observations and prepare for the public meeting

This is a critical regrouping of the charrette team to brainstorm, share ideas, and get organized for the first public workshop.

SESSION/WORKSHOP 5

Open the Doors: Community Discussion and Feedback

Goal: To summarize charrette team's initial impressions and provide the community with preliminary assessment and analysis; to obtain broader citizen input and feedback

The public is invited to this session during which:

- The team summarizes input and analysis, develops SWOT (strengths, weaknesses, opportunities, and threats) lists, and sets goals, objectives, and priorities.
- The community reacts to charrette team's initial impressions and confirms or redirects the focus.

SESSION 6

Setting the Course:
Development of Goals and Objectives: Charrette Findings

Goal: To clarify the focus of the workshop and to establish a vision for the process

Using the information obtained in session 5, the charrette team meets with steering committee to:

- Articulate a preliminary/plan;
- Revisit or investigate in greater detail key areas of concern; and
- Talk to key people in the community for additional feedback.

SESSION 7

Are We There Yet? Preliminary Review of Findings

Goal: To finalize the charrette findings with the steering committee

In this session the charrette team meets with the steering committee to review charrette findings and discuss possible courses of action and implementation strategies. At this time it is also important to involve the members of the steering committee directly in the final development of the ideas, and even the production of presentation materials, so that they begin to take ownership of the process, programs, and projects being recommended. This can help in the transition from ideation to implementation.

SESSION/WORKSHOP 8

The Big Show: Presentation of Findings to the Community

Goal: To present charrette findings to the community

In this final session the recommendations and idea pieces developed during the charrette will be orally and graphically presented at a public meeting. The session goal should be reinforced with the presence of the public media. It is in this session that the challenge of following through with the projects is presented to the community.

Where available, local broadcast media, including public access cable television systems, may carry specific workshop sessions and the large-scale public meetings. This may provide an opportunity to include still more residents in the process via live telephone hookup. Although this high-technology approach is an effective means of maximizing citizen input, it should be noted that the local newspaper is by far the most effective means of creating and maintaining awareness of community planning and development issues, particularly in smaller communities (Catanese 1984; Costello 1983). Whichever medium is used, the rapid turnaround and easy accessibility heightens enthusiasm for the project and serves to build curiosity in members of the community not thus far involved. In a recent small town charrette, more than 100 people attended the final workshop session, which was inadvertently scheduled at the same time a professional exhibition basketball game was being played in the gymnasium down the hall.

Keeping the Process Alive

Completing the charrette workshop is only part of the overall process. Most plans fail because people do not know how to begin implementing the project and its programs. A leader must be found for each project, to follow through from start to completion. Without that leader, a project has little hope of succeeding. It is often helpful to identify a single person to serve as a coordinator. The steering committee can also continue as the coordinating body.

There can be a tendency to build aspirations toward goals that are unreachable, at least in the short term. Although it is important for a process of this kind to push a community toward new and lofty goals, they must ultimately be achieveable. Start with a project that is short-term and highly visible. This type of project can get the momentum going by making small but noticeable improvements. A common mistake is to first take on a large project to create a big impact. Too often such projects fail because people lose their energy too soon. Nonvisible (behind the scenes) projects, although absolutely critical to the long-term viability of a community, should also be put off until the momentum gets going, as they are rarely recognized by the public, making them much less rewarding to the contributors.

People typically rally around successful projects and efforts. If projects and efforts fail, so too will recruitment of volunteers. Furthermore, successful projects generate more volunteers, who can then be integrated into long-term and nonvisible projects.

Benchmarking—the process of establishing dates (goals) for the completion of specific projects and tasks—should be used to revisit the process and to measure progress towards achieving project or community goals. Such revisiting should be a scheduled event, even a celebration, and offer an opportunity to keep the vision before the community. A method used to assure the successful implementation of projects is to hire an intern from a planning program (or similar field) to spearhead and coordinate the projects. This intern position can be phased out or turned over to a town manager or other local person after program momentum has developed.

A CASE STUDY: HARTFORD CITY, INDIANA

The Hartford City Community Charrette arose out of the community's participation in the Indiana Total Quality of Life Initiative (TQLI). As a result of this comprehensive needs assessment and strategic planning process, citizens, businesses, and leaders of Hartford City saw in their community a vision of who they were and what they could become. The steering committee used a charrette as an opportunity to visualize ideas and identify several specific projects that could transform goals into reality.

Hartford City, Indiana, is the county seat of Blackford County, a predominately rural county located approximately halfway between Indianapolis and Fort Wayne. Hartford City has a population of approximately 7,000, which has been declining slightly over the past two decades but is projected to stabilize. Although Hartford City has some local industry, it looks to Muncie, approximately 20 miles to the south, for much of its shopping and employment opportunities.

The Vision: Reviving the Heart of Hartford City

Throughout the Total Quality of Life Initiative (TQLI) process its citizens said time and again how much they valued the friendliness and small-town character they found in Hartford City. The importance of a close-knit and involved community was reinforced during the charrette workshop in which more than 120 people shared their vision for the future of the community. The vision included:

- The buildings and public spaces of the Blackford County Courthouse Square and Downtown Hartford City restored to the character they once enjoyed;
- Downtown Hartford City becoming a family destination, with shopping, entertainment, and restaurants;
- The entries and corridors leading into Hartford City establishing a strong identity and reflecting the pride of the community;
- A wide variety of activities for the families, youth, and senior citizens of the community;
- Hartford City supporting a strong local economy; and
- The community being a celebration of its heritage and natural features.

These concepts firmly established a collective desire for the rebirth of the heart of Hartford City.

Projects and Programs Supportive of the Vision

The heart of Hartford City is the Blackford County Courthouse square. It was recommended that downtown Hartford City begin to recapture its vitality by restoring and rehabilitating its historical buildings. Although many of these buildings have been neglected over the years, the basic character remains. The upper floors of the downtown buildings are well-suited for conversion to apartments, offices, and even community facilities. The charrette teams recommended that the false fronts that had been placed on many of the buildings be removed, and that the buildings be restored to their original character. In many cases, simple maintenance and a new coat of paint will go a long way to restore the small-town character of the heart of Hartford City (see Figure 12.1).

Building rehabilitation alone will not bring life back to the downtown. There must be a reason for people to go downtown, making a mixture of interesting stores and convenience is essential (see Figure 12.2).

Much of the historical fabric of the city remains; what is missing is the people. People not only make a place more lively and interesting, but also render it economically more viable. The charrette team suggested that the downtown should once again become a destination for the people of Hartford City and Blackford County, as well as for visitors from surrounding communities. This means that there must be a variety of things for people to do, places to shop and relax, and places for people to live—all in an attractive setting. Among the projects and programs that were identified to help spark the revitalization of downtown Hartford City are the following:

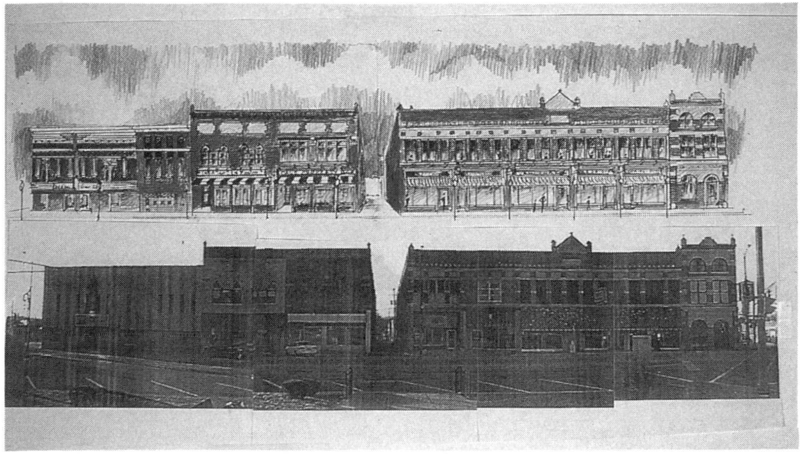

Figure 12.1 Buildings on the south side of courthouse square, as they are now (bottom) and with recommended facade restoration and streetscape improvement (top). (Drawing by Lohren Deeg)

Figure 12.2 Hotel Hartford on the courthouse square, as it is now and with recommended restoration as an inn for visiting businesspeople. (Drawing Lohren Deeg)

- Sponsorship of community activities to attract local residents and visitors, including a farmers' and artisans' market to be held Saturdays on the courthouse square (informal stands and booths or local farmers simply backing up their pickup trucks to the curb would involve minimal expense, yet attract a great many people); an arts and crafts festival; youth activities, such as bicycle races around the square, and carnivals; community gatherings and events such as pancake breakfasts and community picnics; Oktoberfest (to celebrate the strong German heritage of the area); Christmas walks; an herb festival, in conjunction with the local herb farm; a Civil War commemoration or living history; and progressive Christmas dinners.
- Upper-floor rehabilitation to provide affordable housing options for the community. This could also increase the number of destinations in the downtown while encouraging the upkeep of the properties as income-generating investments. Some potential new uses for unused space include apartments, offices, community activities (e.g., banquet hall, meeting rooms.) In rehabilitating these structures it is important that the new uses match the historical character of the building.

- Some kick-off projects, which include painting the boarded windows and restoring the supergraphics and ads on the sides of the buildings.
- Establishing downtown Hartford City as a showplace. Rehabilitation of historical buildings to their original character is important, but there must be places for people to go that serve as catalysts. The library, dressed-up storefronts, restaurants, the restored train depot, and the rebirth of the Hartford City Theater could all work to bring people and life back to the downtown area (see Figure 12.3 and 12.4).
- Promotion of a hospitable attitude. Getting people back to the downtown area means that they must be made to feel welcome. Some easy steps to this end include businesses staying open later one or two nights a week, coordinated promotional activities to make coming downtown a special and fun event, and treating customers well.

Several special projects were identified by the people of Hartford City and the charrette team as important catalysts to downtown revitalization. Some of these involved the adaptive use of old and historical buildings for new retail stores, commercial tourism, and entertainment activities.[2] Others consisted of community-based economic development efforts such as a local business support system, including a business incubator, a commu-

Figure 12.3 Downtown facade enhancement with pocket park. (Drawing by Emily Curtis)

Figure 12.4 Recommended improvements to the main downtown intersection. (Drawing by J. Rebecca Leonard)

nity-wide coordinator of business organizations, and promotional and marketing programs geared to promoting the Heart of Hartford City.[3] Furthermore, community members, through the Chamber of Commerce, the Economic Development Corporation, local banks, and civic leaders, were encouraged to provide financial and technical assistance to local entrepreneurs in establishing new businesses downtown. An incubator-without-walls, providing basic business services, training, and administering grants to local businesses, was identified as an early priority. The program required that to qualify for this assistance, the business owner and the building owner must guarantee that they would follow the redevelopment and design guidelines being developed in association with the Hartford City Community Charrette.

Promoting a Strong Town Image

Another issue identified by the charrette team and the citizens of Hartford City is the need for a strong image and identity for the city, especially for the downtown. An important part of that image is associated with the major state roads that intersect near the downtown (see Figure 12.5).

Plans are currently under way to redesign the intersection to reduce the turning and traffic-flow problems. This opportunity should be used to help create a strong invitation and link to the downtown area. More human-scale signage, lighting, plantings, banners, and distinctive paving materials would help establish a positive image for Hartford City. The de-

Figure 12.5 South gateway improvement. (Drawing by JoEllen Jacoby)

sign of this intersection should be an extension of the revitalization of the courthouse square and of the new gateways located at the city limits. As one moves from the intersection toward the courthouse square, the street furniture becomes more pedestrian oriented. Pedestrians should feel as comfortable as drivers at this intersection and along all the streets of Hartford City.

The ideas generated by the Hartford City Community Charrette are just that—ideas. Through the TQLI process and now, with the charrette, the people and businesses of Hartford City have clearly identified their goals and priorities. A few of the projects and programs visualized during the charrette/workshop sessions represent a somewhat costly and long-term vision, but most can be implemented relatively quickly and inexpensively. The key has been to get started, take a few small but highly visible steps, and build momentum. The larger projects will follow in good time.

Since the charrette, several service groups have raised money, hired a professional landscape architect, and have begun to install new gateways to Hartford City. This action was begun by one civic group, the Rotary Club, challenging the other local service groups (Lions and Kiwanis) to do the same. The projects had the money raised and invested in design and construction within two weeks of the charrette. The downtown merchants have begun a program of planting flowers in the courthouse square, and a local bank has earmarked funds for low-interest loans to local businesses that will rehabilitate their facades.

At this writing it has only been two months since the Hartford City Community Charrette ended, but enthusiasm still runs high and momentum has shifted to the remaining projects with the support of the local community.

SUMMARY AND CONCLUSION

The community-based charrette/workshop has proven to be highly successful in involving a community in the planning and development process. Many of these projects have stimulated an informed involvement of the local citizenry and have resulted in sustained action and a stronger sense of community.

It is important to note that the community-based charrette/workshop is not designed to produce miraculous results or even to generate plans. It is a process. Its strength is derived from its flexibility and its ability to stimulate the flow of ideas through visualization. It is also important to remember that the members of a charrette team serve primarily as facilitators aiding in the visualization process. It is the vision and cooperative spirit of the local community that will see the project to and through implementation.

As suggested earlier, the charrette/workshop process derives its strength from its ability to fit within a wide variety of contexts. The early projects of the R/UDAT program focused primarily on urban design issues, but recent years have seen involvement in countywide zoning plans, fairgrounds master plans, site development, and numerous downtown and neighborhood revitalization and community and economic development efforts. This approach has potential in other planning activities as well, including comprehensive planning, economic development planning, and the preparation of community sign ordinances, to name a few. Alumni of these workshops have often adapted the process to a wide range of community settings all across the nation. Recent experiences in several rural communities, such as the Hartford City Community Charrette, have shown that charrettes can also be effectively used as follow-up exercises to earlier projects or as integral components of longer-term redevelopment projects. In all of this, however, the driving force for success is an informed and involved community.

The charrette/community workshop "seeks a shared vision by building on consensus agreements at each step. Continuously, knowledge is shared, insight is gained, opinions are voiced, concerns are addressed and conflict is mitigated. In the face of rapidly changing social and economic environment, the small rural town is given an alternative to externally enforced triage and expectant death. Community design offers self-determination, self-reliance, hope and dignity" (Luther 1990, 54).

NOTES

1. It is usually not advisable to use an elected official because of possible public perceptions or political overtones.
2. The following are examples of such projects.: revitalization of the old movie theater on Washington Street into a dollar cinema, where special events can be coordinated with downtown promotional campaigns (e.g., James Dean movies during the James Dean Festival in Fairmount); opening of new establishments, such as comfortable and unique restaurants and quality retail stores that do not try to compete with the big discount and department stores; promotion of existing downtown identity by fixing up buildings, renovating historical structures, planting flowers and trees.
3. Recommendations included a local business support system that would use economic development income tax money to provide grants for local businesses to rehabilitate and/or enhance their buildings according to the development guidelines. An incubator-without-walls could provide support for startup businesses.

   Also recommended was the establishment of a Hartford City Community Roundtable, an organization of organizations to serve as an action-oriented group that coordinates the activities of the many organizations and associations of Hartford City. Because downtown Hartford City was one of the best-kept secrets of this part of the state, marketing strategies were encouraged, including the dissemination of a brochure publicizing downtown, the goods and services provided, the new expanded hours, and promotional activities, as well as shopping bags with a map of the downtown area or a new "Heart of Hartford City" logo to be used by all the downtown merchants. A walking tour of historical buildings and neighborhoods in and near downtown might also be developed. An architectural scavenger hunt of some of the unique architectural details of downtown buildings is being developed to get residents to rediscover their downtown.

REFERENCES

Catanese, A. 1984. *The Politics of Planning and Development.* Beverly Hills, Cal.: Sage Publications.

Community-Based Projects Program. 1978. *Urban Design Workshops.* Muncie, In.: Ball State University.

Costello, A. 1983. "The Powers of the Press: Communicating Preservation Concepts Through Local Newspapers." *Small Town.*

———. 1987. "The Charrette Process: University-Based Design Teams Serve Indiana's Small Towns." *Small Town* 17(6).

Creighton, J. L. 1981. *The Public Involvement Manual.* Cambridge, Mass.: Abt Books.

Leymann, H. 1987. "The Significance of the Learning Process Underlying Democratic Participation." *Economic and Industrial Democracy* 8(1).

Luther, J. 1990. "Participatory Design Vision and Choice in Small Town Planning." In *Entrepreneurial and Sustainable Rural Communities,* edited by F. Dykeman. Sackville, N.B., Canada: Rural and Small Town Research and Studies Program.

Oregon Visions Project. 1993. *A Guide to Community Visioning: Hands-On Information for Local Communities.* Salem, OR. The Oregon Visions Project, Oregon Chapter, American Planning Association.

Segedy, J. A., and B. E. Johnson. 1995. *The Small Town Charrette Handbook: Visioning and Visualizing Your Community's Future.* Muncie, Ind.: The Institute for Small Town Planning and Design.

Wates, N., and C. Knevitt. 1987. *Community Architecture: How People Are Creating Their Own Environment.* London: Penguin Books.

PART THREE

Assessing the Alternatives

Thirteen

Rural Sustainable Development: A New Regionalism?

Michael R. Boswell and Ivonne Audirac

INTRODUCTION

It has become a tradition in sustainable development writings to define sustainable development through the obligatory quoting of the Brundtland Commission's definition (WCED 1987). Once homage is paid to this wider meaning of sustainable development, the author then goes on to explain what the definition really means. Because sustainable development often involves a complex set of economic, environmental, and social/ethical interconnections for which there are no tested models—let alone complete knowledge—the author is forced to stake out a particular corner of intellectual battleground, from which sustainable development is conceptualized and defended. Those seeking novelty in presentation will not find it here; except for avoiding the WCED quote, this chapter does not defy tradition. We argue that a new regionalism that treats rural urban symbiosis as a closed loop, aims for increased regional self-reliance, and builds on a framework of collaborative participation is the backbone of rural sustainable development.

THE NEED FOR A HOLISTIC REGIONAL FRAMEWORK

Amid a diversity of intellectual approaches, several common themes—such as intergenerational equity, integration of ecology and economics, qualitative growth versus quantitative growth, renewable and nonrenewable resource maintenance, multiple spatial scales (including interregional links and local responsibility for global ecological conditions), industrial bioregionalism, community empowerment, grass-roots participatory planning, and stable human populations—pervade the sustainable development literature. Although each of these themes is important and contains a medley of environmentalist positions (O'Riordan 1995), the most relevant, and perhaps the least addressed, is the consideration of multiple spatial scales. The recognition of interconnected links between different spatial scales (global, national, regional, local, household/individual) present both difficulty and opportunity for the implementation of sustainable development. Spatial scale is rarely defined explicitly, and "sustainability may have a different definition and different measures depending on the scale of concern" (Brown et al. 1987, 717). Affecting change at a global or household level is far removed from the majority of formal planning interventions. Therefore, the familiar planning dichotomy of urban and rural, town and country, wherein the city produces industrial goods and the country food staples, persists even with increasing evidence of the blurring of these demarcations (United Nations Development Program 1996).[1] This urban-rural dichotomy fails to recognize how critical rural and urban symbiosis is for ecological sustainability. Instead of a linear open flow, where inflows and outflows are unrelated, rural-urban symbiosis recognizes a bidirectional flow of materials, energy, information, and people, on which both urban and rural areas depend for survival. It recognizes a circular flow of goods, services, energy, technology, culture, and waste, whose impacts on human and nonhuman communities may be positive *or* negative. However, to take into account this urban-rural interdependence, a holistic, regional framework is a necessary precondition.

A holistic regional framework is not new to planning. In 1923 several prominent planners, including Lewis Mumford, Benton MacKaye, and Clarence Stein, formed the Regional Planning Association of America (RPAA). The goal of this informal organization was to reduce the economic dominance of the metropolitan area in favor of a region that was more economically integrated and socially equitable—an idea that is consistent with our current conception of sustainable development. Although regional planing had strong advocates, in the postwar years the focus in planning shifted back to the cities (Sussman 1976). The regional planning project that was begun more than seven decades ago is considered "unfinished" (Hughes 1972) but still useful for solving our current economic and environmental problems (Sussman 1976).

The new regionalism espoused by sustainable development expands on the basic concepts established by the RPAA. In the last few decades, the advancement of science has made us aware of the critical role of the natural environment in supporting human life and of our ability to modify the environment at all spatial scales. In addition, we have expanded our notion of rights to include those of future generations (Nash 1989). The new regionalism is different from the old in the consideration it gives to ecological integrity, global connections, and intergenerational equity. Moreover, the new regionalism differs from the current conception of regional planning, which makes economic growth and development of the metropolitan area the privileged context of planning interventions.[2]

Although the region becomes the relevant planning scale for sustainable development, the nested spatial scales around and within the region need special attention. Around the region are national, transnational, and global perimeters to and from which materials, energy, waste, information, and people flow. However, as the spatial scale of flows increases, these larger regions are less predictable, less understandable, and less controllable. The larger the scale, the more difficult planning efforts become in the realms of environment, economy, and social equity. Therefore, planning for sustainable development—at the intersection of economy and environment—implies the management and maintenance of flows at the lowest possible local scale in order to avoid reliance on "phantom carrying capacity" (Catton 1980). Phantom carrying capacity is the illusory perception of increased local carrying capacity obtained by importing resources from and exporting wastes to external areas. This semblance of high carrying capacity for the local area disguises the fact that social and environmental degradation occurs (and carrying capacity is reduced) in the external areas. Thus, to counter the flow of resources at larger spatial scales and to avoid phantom carrying capacity, planning for self-reliance, to the greatest extent possible is a must within the spatial scales of the farm/household, the community, and the regional ecosystem.

At the intersection of social equity and environmental sustainability lies the issue of sustaining the livelihoods of the current generation vis-à-vis those of future ones. Besides the ethical and practical conflicts that the intergenerational equity principle poses to present generations, the implications of self-reliance and different property rights regimes[3] associated with programs in support of ecological stewardship are mired with social conflict at all spatial scales. Hence, sustainable development planning and implementation increasingly involve transcending conflict by promoting shared visions and common purposes. In other words, the new regionalism requires strategies that shift the focus from conflict resolution to collaborative planning (President's Council on Sustainable Development [PCSD] 1996).

SELF-RELIANCE AND BOTTOM-UP PARTICIPATION

Self-reliance is not a new concept in rural America, as Gary Green (Chapter 9), Joseph Luther (Chapter 8), and other authors in this volume discuss. It suggests that communities should seek to provide for their needs from within before turning to importing and exporting. Self-reliance is "the regeneration of community through community-controlled development of its own resources" (Sargent et al. 1991, 27). How, then, is development to occur, and who will implement this undertaking given the commitment to self-reliance? The alternative to the traditional government or corporate -directed top-down development approach is the watchword of sustainable development: *bottom-up participatory development*. From the national to the international realm, in developing and developed countries, and from the small rural village to the international community gathered at Rio's Earth

Summit, that is, at all spatial scales, bottom-up participation is the sine qua non of sustainable development implementation (Kinsley 1994; Sargent et al. 1991; Stokes and Watson 1989; PCSD 1996; Stöhr 1981; IUCN/UNEP/WWF 1991; Agenda 21 1992, Zazueta 1995). However, at the regional level—the locus of the rural-urban symbioses and environmental alliances gathered in the contributions to part three of this book—self-reliant, bottom-up development is possible only within a framework of cooperation among rural people; local, state, and federal governments; and technical and scientific organizations. This collaborative/participatory framework is essential for people and stakeholders to identify with their place in the regional ecosystem and undertake collective responsibility for protecting it.

As in many developing countries (Zazueta 1995; USAID/WRI 1994), in rural communities negatively affected by technological and global economic change as well as national fiscal retrenchment, grass-roots groups and nongovernmental organizations (NGOs) are bypassing conventional governmental channels to forge extralocal alliances and partnerships that enhance their ability to meet their social and economic needs while implementing natural resource stewardship (PCSD 1996). The sustainable development strategies tried in Richmond/Wayne County, Indiana (Chapter 16), in the Central Appalachian region (Chapter 19), and those undertaken by three Canadian indigenous communities (Chapter 20) vividly illustrate successful regional collaborative efforts.

Flaccavento (Chapter 19) describes the endeavors of the Clinch Powell Sustainable Development Initiative (CPSDI), a regional not-for-profit organization established to solve human and ecological problems in upper east Tennessee and southwestern Virginia. He begins by defining sustainable development as beneficial to humans, locally oriented, ecologically sound, based on self-reliance, and persistent. Through these principles, the debate over development in the Clinch Powell region has moved away from the "traditional jobs versus the environment" mode toward an approach that recognizes the value of both. Three specific initiatives are identified by the CPSDI: sustainable forestry and value-added wood products, diversified and sustainable agriculture, and nature-based tourism. Each of these initiatives is intended to increase self-reliance in the region, reduce the negative impact of flows and transformations, and consider global implications of local activities.

Seymoar (Chapter 20) analyzes the efforts of three indigenous Canadian communities to achieve sustainable development through community self-empowerment. The three communities were selected as recipients of the We the People: 50 Communities Awards for providing "the foundation for creating just, inclusive and sustainable communities, rooted in place and capable of cooperation with others." Each of the three communities—Walpole Island First Nation, Sanikiluaq, and Oujé-Bougoumou—faced either existing or potential environmental degradation resulting from external activities by private industry or government. These activities threaten the health and economic well-being of the communities; therefore, the communities organized to protect themselves.

Seymoar contends that there was a power imbalance between external organizations and the indigenous communities that forced the communities into a state of dependency. Eventually, a stage of conflict was reached, owing to the increasingly poor economic and environmental conditions and/or proposals for new industrial-scale projects. This stage was followed by a period of independence, in which the communities withdrew from the conflict and turned inward to envision a sustainable future. Finally, the communities reestablished partnerships within the region, but on terms that recognized the value of all communities in the region and their dependence on one another for survival. These Canadian indigenous communities were able to create a path to sustainable development by first establishing principles of self-reliance, then recognizing the symbiotic nature of the region.

Segedy and Lyons (Chapter 16) acknowledge the importance of sustainable development principles to the new regionalism. For them, urban-rural symbiosis requires a new regionalism forged out of urban-rural partnerships, alliances, and grass-roots networks. They describe the Sustainable Urban/Rural Enterprise (SURE) partnership between the city of Richmond, Indiana, and Wayne County, Richmond's rural hinterland. The rural area's impetus for the partnership was provided by successful rural communities that have showed that small-town America has been able to survive the "big box retailers" by banking on the preservation of its cultural and ecological uniqueness. Hence, SURE, a local civic venture, was initiated in 1989 with the mission to promote ecologically based economic choices. It was the nation's first broad-based local public-private partnership chartered to monitor local economic choices in terms of their ecological consequences. As a result of SURE workshops, the comprehensive plans of Richmond and Wayne County harmonized their economic development goals in support of a regional vision respectful of the unique environmental qualities of the White Water River Gorge and the interdependence of Richmond and its surrounding rural towns. These efforts forged a civic commitment to strengthening the region's economic and cultural diversity, the beautification of cityspace, and the utilization of state-of-the-art technology for transportation, communications, waste management systems, and business incubation.

Seven years after SURE's formation, the authors contend that although its influence is unmistakable, it is difficult to make tangible links between the original concept and specific policies and projects beyond Richmond and Wayne County's comprehensive plans. SURE's successful accomplishments include the Richmond Farmers' Market and the construction of wetland wastewater treatment facilities. However, they point out that the current leadership is straying from the original SURE ideas and that in order to make the process sustainable over time, it is essential that the leadership remains committed to it. They assert that although the grass roots provide strength to push the project forward, it still needs a strong advocate to steer it to fruition.

REGIONAL LANDSCAPE AND ECOSYSTEM PROTECTION

As parts of large regional ecosystems, rural landscapes cut across various jurisdictions, private and public lands, agricultural, forestry and aquatic resources, wildlife habitats, and rural villages and communities. The development of greenways and trails is another example of successful collaborative efforts between conservation NGOs, private landowners, rural and urban citizens groups, and state and local governments. These partnerships support the sustainable development goals of enhancing opportunities (e.g., ecotourism and recreation) for economic self-reliance while conserving biodiversity and historical and cultural resources. As the knowledge of landscape ecology increases and progress in ecosystem management diffuses to federal, regional, state, and county agencies, there is a growing need for inventive participatory and collaborative strategies to mediate conflicts and achieve a common ground among the region's diverse interests and stakeholders.

Starnes, Benedict, and Sexton (Chapter 15) argue that greenways and trails can contribute to the sustainability of rural areas. Greenways can be used to protect environmental features, traditional rural working landscapes (such as agriculture and forestry operations), and an area's historical and cultural heritage. Greenways offer social benefits such as recreation, social bonding, mental health, and aesthetics. They provide a link between urban and rural areas that creates economic benefits to the rural dwellers and social-psychological benefits to urban dwellers. The authors hope that rural communities will recognize the value of existing resources in a strategy for implementing sustainable development, thus upholding the traditional rural philosophy of self-reliance.

RURAL SUSTAINABILITY AND AGRICULTURE

For the last 30 years, research by agricultural economists and rural sociologists has eloquently documented the changing structure of American agriculture. This change has been characterized by higher concentrations of land and capital into fewer and larger farms, while full-time family farms, the mainstay of many rural economies in yonder years, have practically disappeared. Farm commodity programs, technological change (Cochrane 1979), and urban encroachment have been singled out as the most important causes of the decline in the number of farms and rural employment. From a regional perspective, American agriculture is increasingly metropolitan. Urban areas keep expanding over their hinterlands with detrimental effects on conservation of prime farmland, range land, and forests, and on global ecological stability as well. If the imperative for rural sustainability, as argued by William Rees (Chapter 3), involves curtailing metropolitan expansion, reducing the appropriation of natural resources from beyond the region, and increasing low-input, environmentally friendly agriculture, then the imperative also includes strengthening rural-urban symbiosis for regional self-sufficiency. This involves reorienting agricultural production to meet the food and fiber demands of immediate urban markets and the active promotion of organic and low-input agriculture.

As Heimlich and Barnard suggest in Chapter 14, however, farmland preservation programs in support of rural sustainability will have differing effects on the three types of prevailing metro farms. Adaptive farms—the most highly industrialized—which specialize in greenhouse, high-value crops, are likely to be the least favored inasmuch as they have the greatest environmental impact owing to their high per-acre use of fertilizers and pesticides. Traditional farms—the most extensive in land use, but also the most conservative in land management and agricultural practices reflecting traditional rural values—are the most threatened by urban expansion and thus likely to be favored by farm preservation programs. However, their well-known environmental impacts, such as soil erosion associated with tillage operations, odors, and nutrient runoff from livestock operations, would make them objectionable to ecological sustainability unless farmland preservation programs are accompanied by environmentally correct agricultural resource management programs. Finally, recreational-type farms—the smallest in size, primarily supported by off-farm income, and the most environmentally friendly—are likely to benefit the most if preserving a pastoral landscape is the priority and if urban demand for organic agricultural products is sufficient to support their economic viability. Thus, the authors caution that agriculture and open space preservation may be at odds with sustaining current rural livelihoods, depending on the environmental requirements of farmland preservation programs. These programs preserve open space and rural life-styles, but also could remove agriculture as a real component of the local economy.

WASTE RECYCLING: CLOSING THE RURAL-URBAN LOOP

The biorealist movement stresses that production and consumption processes must be reengineered to mimick nature's metabolism if sustainable development is to be achieved (Frenay 1995). The greening of industrial ecosystems aims at minimizing waste disposal to landfills, rivers, and oceans by closing resource and waste loops (Ayres 1994). It is a root-cause approach to reducing global and local pollution and enhancing regional self-reliance. Recycling to recover resources and the reuse of waste will undoubtedly intensify as we move from our current open industrial system, which focuses on management of point source externalities, to more holistic metabolic and closed-loop systems. The latter, ideally, seek high levels of recycling and reuse in order to approach zero-emission, sustainable, steady-state regional economies (Pauli 1995). Ayers (1994, 31) identifies three types of recyclables according to reuse potential: "(1) uses that are economically and technologically compatible with recycling under present prices and regulations; (2) uses that are not economically compatible with recycling but where recycling is technically feasible, for example, if the collection problem were solved; and (3) uses for which recycling is inherently not feasible." In the last category belong a large number of agricultural and household chemicals such as detergents, fertilizers, herbicides, pesticides, and germicides, which compellingly speak for a shift to low-input, sustainable agriculture. The first two categories offer opportunities for cooperation and synergy between urban and rural areas in closing industrial cycles. Urban yard waste, such as leaves, is clearly a type 1 recyclable with potential benefits for both farmers and municipalities (Shapek, Chapter 17). Although rural areas in different parts of the country have experimented with recycling waste (e.g., paper and poultry litter) to farms, the benefits of most rural recycling efforts remain tilted toward urban areas, given low volumes and high collection costs in rural areas. Regionalization of solid waste disposal has been a mixed blessing to rural areas. Despite generating considerable revenues, it has spawned relatively few rural jobs, while reinforcing the stigma of rural areas as dumping grounds and raising environmental justice concerns (Chapter 17).

Shapek examines the state of Florida's attempt to address the solid waste disposal problem through use of statewide mandated recycling goals. He highlights differences between the needs and abilities of urban and rural areas. Urban areas need rural areas to provide physical space for the disposal and/or processing of municipal solid waste. Rural areas need urban areas to provide the technical expertise for disposal and recycling techniques. In addition to cooperative marketing strategies for recyclable materials and subsidies for pickup in rural areas, public education about recycling and awareness campaigns offering alternatives to backyard burn barrels are still needed in rural areas.

Derr and Dhillon (Chapter 18) focus on a particular solid waste component: tree leaves. They show that leaves are a substantial component of municipal solid waste and provide a unique opportunity for urban-rural cooperation in meeting disposal needs. Leaves can be removed from the waste stream, composted, and used as mulch for agricultural operations. The urban community can pay agricultural operations a fee well below the average tipping fee for municipal solid waste, thus providing agricultural operators with a second income source and an organic mulch product. However, this opportunity for a symbiotic relationship between urban and rural areas in the recycling of solid waste is most readily available to towns and municipalities with agricultural operations in close proximity.

CONCLUSION

If Part Two of this volume made obvious the role of local communities in rural sustainable development, we argue that the works gathered in Part Three collectively revive the insights of the RPAA and reaffirm the possibility of a new regionalism based on the principles of sustainable development. Such regionalism transcends political jurisdictions, depends on bottom-up, community-based collaborative partnerships for environmental protection and resource stewardship; is supportive of self-reliant development, bolsters rural-urban symbiosis, and advances closed-loop systems of resource and waste flows within a region.

Although the above principles encapsulate one more normative perspective rather than a series of practical recommendations, we view the opportunity of a new regionalism as a challenge for implementing sustainable development in the years ahead. The following chapters are a testimony of the practical efforts put forth in this direction. Some have been more successful than others, reflecting to some degree the authors' direct engagement in the field and personal commitment to the concept, whereas in others a more detached skepticism prevails. All, however, speak to the experiments and adaptations that rural North America is undertaking in the transition to (it is hoped) a more sustainable society.

NOTES

1. Flynn and Marsden (1995) observe that the separation of urban planning and rural planning for sustainable development has received little attention. Allanson and associates (1995) question whether there is a clear distinction between urban and rural space in advanced industrial economies.
2. Sussman (1976) notes that the RPAA rejected this definition of regional planning, insisting that it was nothing more than metropolitan planning.
3. These are associated with resource management regimes that include state, private, and communal property rights.

REFERENCES

Agenda 21, Agriculture and Rural Development. 1992. "Promoting Sustainable Agriculture and Rural Development," Chap. 14 in *Report of the United Nations Conference on Environment and Development*. Rio de Janeiro, 3–14 June.

Allanson, P., J. Murdoch, G. Garrod, and P. Lowe. 1995. "Sustainability and the Rural Economy: An Evolutionary Perspective." *Environment and Planning A* 27:1791–1814.

Ayres, R. U. 1994. "Industrial Metabolism: Theory and Policy." In *The Greening of Industrial Ecosystems,* edited by D. J. Richards, B. R. Allenby, and R. A. Frosh. Washington, D.C.: National Academy Press, 23–37.

Brown, B. J., M. E. Hanson, D. M. Liverman, and R. W. Merideth Jr. 1987. "Global Sustainability: Toward Definition." *Environmental Management* 11(6):713–719.

Catton, W. R., Jr. 1980. *Overshoot: The Ecological Basis of Revolutionary Change.* Urbana: University of Illinois Press.

Cochrane W. W. 1979. *The Development of American Agriculture: A Historical Analysis.* Minneapolis: University of Minnesota Press.

Flynn, A., and T. K. Marsden. 1995. Rural Change, Regulation and Sustainability. "Guest Editorial." *Environment and Planning A* 27:1180–1192.

Frenay, R. 1995. "Biorealism: Reading Nature's Blueprints." *Audubon* 97(5):70–79, 104–106.

Hughes, M., ed. 1972. *The Letters of Lewis Mumford and Frederic J. Osborn: A Transatlantic Dialogue 1938–1970.* New York: Praeger Publishers.

IUCN/UNEP/WWF. The World Conservation Union, United Nations Environment Programme, World Wide Fund for Nature. 1991 *Caring for the Earth. A Strategy for Sustainable Living.* Gland, Switzerland.

Kinsley, M. J. 1994. *Economic Renewal Guide: How to Develop a Sustainable Economy Through Community Collaboration.* Snowmass, Col.: Rocky Mountain Institute.

Nash, R. F. 1989. *The Rights of Nature: A History of Environmental Ethics.* Madison: University of Wisconsin Press.

O'Riordan, T. 1995. "Frameworks for Choice: Core Beliefs and the Environment." *Environment* (October):4–9.

Pauli, G. 1995. "Zero Emissions: The New Industrial Clusters." *Ecodecision* (Spring):26–30.

President's Council on Sustainable Development. 1996. *Sustainable America: A New Consensus for Prosperity, Opportunity, and Healthy Environment for the Future.* Washington, D.C.: Government Printing Office.

Sargent, F. O., P. Lusk, J. A. Rivera, and M. Varela. 1991. *Rural Environmental Planning for Sustainable Communities.* Washington, D.C.: Island Press.

Stöhr, W. B. 1981. "Development from Below: The Bottom-Up and Periphery-Development Paradigm." In *Development from Above or Below?* edited by W. B. Stöhr and D. R. Fraser Taylor. New York: John Wiley & Sons, 39–72.

Stokes, S. N., and A. E. Watson. 1989. *Saving America's Countryside: A Guide to Rural Conservation.* Baltimore: Johns Hopkins University Press.

Sussman, C., ed. 1976. *Planning the Fourth Migration: The Neglected Vision of the Regional Planning Association of America.* Cambridge, Mass.: MIT Press.

United Nations Development Program. 1996. *Urban Agriculture, Food Jobs and Sustainable Cities.* Publications Series for Habitat II. Vol. 1. New York: Unitied Nations Development Program.

USAID/WRI. 1994. *New Partnerships in the Americas: The Spirit of Rio.* Washington, D.C.: New Partnerships Working Group.

WCED (World Commission on Environment and Development). 1987. *Our Common Future*. Oxford: Oxford University Press.

Zazueta, A. 1995. *Policy Hits the Ground: Participation and Equity in Environmental Policy Making*. Washington, D.C.: World Resources Institute.

Fourteen

Agricultural Adaptation to Urbanization: Farm Types and Agricultural Sustainability in U.S. Metropolitan Areas

Ralph E. Heimlich and Charles H. Barnard

INTRODUCTION

Metropolitan Statistical Areas (MSAs), defined by the Bureau of Census, now contain 16.4% of U.S. land area and 77.5% of U.S. population (Bureau of the Census 1992). Metro areas also contained one third of U.S. farms in 1991, accounting for a proportional value of agricultural products sold. A previous study showed that metro farms are generally smaller, produce more per acre, have more diverse enterprises, and are more focused on high-value production than nonmetro farms (Heimlich and Brooks 1989; Heimlich 1988). Metro farms were also found to have a generally stronger financial position than nonmetro farms (Ahearn and Banker 1988). Metro farm characterizations in these earlier studies were based on aggregate statistics, which can be misleading if several distinct types of

---

The views presented in this chapter are the authors' and do not necessarily represent policies or views of the U.S. Department of Agriculture.

The data used in this chapter are based on the Farm Costs and Returns Survey (FCRS), an annual survey conducted by the National Agricultural Statistics Service for Economic Research Service, USDA (see, for example, Bentley 1993; Morehart, Johnson, and Banker 1992). The FCRS provides detailed estimates of the expenses farmers incur operating their farms and other farm characteristics. The 1991 FCRS includes farm operators in the 48 contiguous states who sold or normally would have sold at least $1,000 worth of agricultural products in 1991. The FCRS surveys approximately 26,000 operators nationally, using a list frame of medium to large farms and a complementary area frame to ensure sample representation of smaller farms less likely to appear on lists. The 1991 FCRS had 11,988 usable responses, of which 3,831 were from farms located in the metro counties defined by the Office of Management and Budget and the Bureau of the Census in 1993. The sample is expanded to represent all 2.1 million U.S. farms.

farms coexist in the urbanizing metropolitan environment. This chapter delves beneath the metro averages, using data on individual farms classified as recreational, adaptive, and traditional. It also draws implications for farmland preservation and metro agricultural sustainability.

AGRICULTURAL ADAPTATION TO THE METRO ENVIRONMENT

The increasingly metropolitan character of the United States presents agriculture with both problems and opportunities. Metro area growth affects agriculture primarily through the markets in which farmers buy and sell and through local government institutions that exercise control over property taxes and land use (see Figure 14.1). Many of these influences simultaneously pressure farmers to adapt and offer them opportunities for change, with different kinds of adaptation resulting in different kinds of farms. The increased pressures and opportunities for change in metro areas result in greater differentiation between farm types than occurs in nonmetro areas.

How do these adaptations occur? New development to support growing populations competes with agriculture in the land market, driving up land prices. Property taxes increase, raising the cost of keeping farmland in agriculture. Landowners often seek enterprises and markets that offer returns commensurate with those derived from development. Landowners may also sell off less productive woodlots and pastureland, concentrating on more intensive production on the remaining cropland. Higher land values permit investments in new enterprises through increased equity in farmland that supports higher levels of debt.

Employment opportunities stemming from urban growth reduce available farm labor, particularly seasonal and part-time help. However, off-farm employment opportunities for the farm family help support the farming operation. Off-farm employment can also provide a transition to part-time farming, particularly if changes are undertaken that reduce full-time labor needs on the farm. In addition, urban employment opportunities can offer people working in urbanizing areas an opportunity to start recreational farms that may eventually develop into full-time, part-time, or retirement businesses.

Growing urban populations provide opportunities to grow new crops and market them in new ways. High-value and specialty crops, such as fresh fruits and vegetables and organic produce, can be sold to restaurants and gourmet grocery outlets or directly to consumers at roadside stands, farmers' markets, or U-pick operations. U-pick farms often combine produce sales with value-added products like dried herbs and flowers, jams and jellies, homemade breads and pastries, and other farm-related products. Recreational aspects of U-pick operations, such as hayrides, picnics, farm-pond fishing, and special holiday activities, also add value to urban

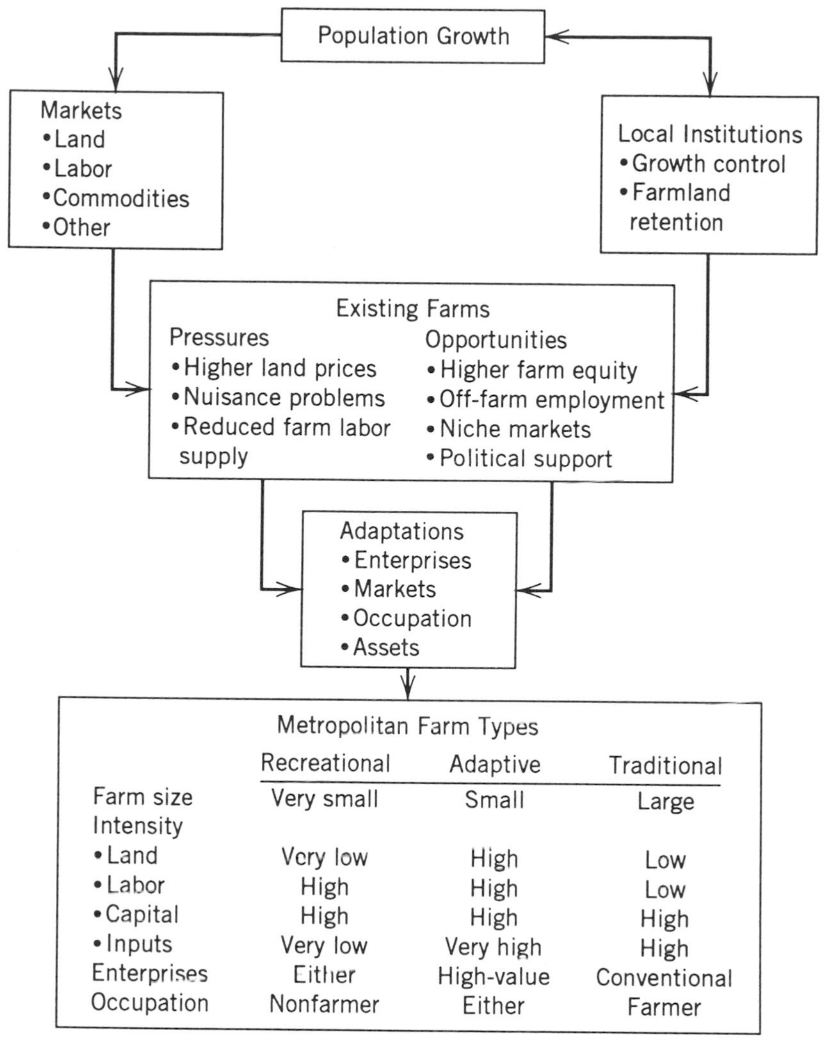

Figure 14.1 A conceptual model of agricultural adaption to urbanization. (*Source:* Hemlich and Brooks, 1989)

customers' purchases. Boarding, breeding, and training facilities for horses, cattle-breeding operations, or other specialty livestock operations may replace dairy farms and cow-calf operations.

Typically, suburban neighbors' complaints about farm odors and chemical spraying force farmers to turn to enterprises that produce fewer such negative environmental effects. Yet many suburban residents also support growth controls and farmland retention programs aimed at preserving the rural landscape to which they were initially attracted. These programs offer relief from nuisance actions, while also providing property tax reductions or financial help for metro farmers.

Urban pressures force agriculture to adapt, both to improve its economic performance and to become less environmentally objectionable. Farmland preservation programs shield farmers from such urban pressures and thus eliminate or postpone the necessity to change farm economics and operations. They can sustain farms as they are, but in doing so may delay changes that could make farming more sustainable in an urban environment that requires higher returns per acre and stricter control of environmental externalities. Nevertheless, farmland protection may be a goal of land-use planning to maintain open space, preserve rural life-styles, prevent urban sprawl, control public infrastructure costs, and preserve local agricultural economies. Despite some negative environmental externalities, farmland preservation contributes to important environmental protection goals, such as protecting watersheds, maintaining air quality, and retaining natural processes in the urban environment.

Metro Farm Types

We contend that urban pressures and opportunities result in distinct farm types that exist side by side in metro areas, with the mix changing as the area urbanizes. Others have developed conceptual or empirical classifications of metro farms. Brooks (1985) labeled farms with less than $2,500 in sales as *minifarms*, whereas those with $2,500 to $20,000 were termed *small farms*. Smith (1987) identified production and value types of urban farming adaptations. For purposes of this analysis, we present a refinement of a typology reported earlier for farms in Northeastern metro areas (Heimlich and Barnard 1992). We classify farms located in metro counties as recreational, adaptive, or traditional, based on the following criteria.

Recreational farms have an annual market value of products sold of less than $10,000. A typical recreational farmer owns and operates relatively small acreage (less than 100 acres), is engaged in enterprises that require relatively little day-to-day management (such as a cow/calf operation), and is retired or employed in another occupation that subsidizes the farm operation.

Adaptive farms sell more than $10,000 per year, with high-value products making up more than a third of all sales or with more than $500 in sales per acre of land operated. High-value products include vegetables, fruits and nuts, nursery and greenhouse crops, and specialty livestock. Inasmuch as adaptive farms cover a wide range of possible enterprises, it is difficult to describe a typical such farm. However, they tend to be relatively small (100 to 200 acres), extremely intensive in terms of both labor and other inputs, and operated as businesses, either exclusively or in combination with other enterprises.

Traditional farms have annual sales of more than $10,000, with a third or less of sales from high-value products and sales per acre of $500 or less. Farms that met the adaptive farm criteria for sales per acre, but which derive more than one third of sales from conventional livestock enterprises (beef, dairy, hogs, sheep, poultry, and related products) or had a high proportion of their acreage in conventional crops with high gross receipts per acre (cotton, rice, tobacco, sugar, and peanuts) are also classified as traditional farms. Traditional farms typically have the largest acreage (more than 200 acres), are operated more extensively than intensively, raise traditional field crops and livestock, and have relatively little off-farm employment or business.

CHARACTERISTICS OF METRO FARM TYPES

More than 696,000 farms, one third of all U.S. farms, are located in MSAs. Nationally, 54% of farms in metro counties are what we classify as recreational, 14% are adaptive, and 33% are traditional. U.S. metro farms account for 35% of all crop and livestock sales recorded in the 1991 Farm Costs and Returns Survey.

The official MSA definition used here is a proxy for areas influenced by urbanization, although not necessarily completely dominated by urban land uses. Farms outside MSAs experience some of the same urban pressures and have some of the same opportunities as farms inside metro areas. Consequently, 6% of nonmetro farms meet the adaptive farm definition However, nonmetro adaptive farms account for only 20% of nonmetro farm sales and operate only 5% of nonmetro farmland. They also average only $276 in sales per acre, less than one third that of intensively farmed adaptive metro farms. In the following paragraphs, the three metro farm types are compared and contrasted with each other and with all nonmetro farms.

Regional Distribution of Metro Farms

The proportion of metro farms varies from region to region, depending on the degree of urbanization and county land area (see Table 14.1, and Figure 14.2). Almost 60% of farms in the highly urbanized Northeast are in metro areas, although they make up only 12% of all metro farms. Forty-six percent of Southeastern farms are in metro areas, although only 10% of U.S. metro farms are found there in that region. The Corn Belt has a large proportion of all U.S. farms and the highest number of metro farms (18%), even though only 29% of Corn Belt farms are in a metro area. In the Pacific region, metro farms constitute 63% of all farms. In terms of agricultural products, the Pacific region accounts for 36% of metro farm sales, more than proportional to the number of farms because of the high value of production per acre. The Mountain and Northern Plains regions have the fewest metro farms.

The composition of farm types within metro areas also varies across regions. The Southern Plains, Southeastern, and Appalachian regions have the highest proportions of metro recreational farms (63% to 66%), possibly because of the popularity of these areas for retirement. Only 11% of adaptive metro farms are located in the Southeast, but they account for 69% of all Southeastern metro farm sales. Adaptive farms in the Pacific region account for 38% of all adaptive farms and 56% of all adaptive farm sales. Adaptive metro farms in the Pacific region account for 82% of Pacific metro farm sales. More than one quarter of all traditional farms in metro areas are in the Corn Belt, where they make up 49% of metro farms. The Pacific region has only 5% of metro traditional farms, but they account for 13% of traditional metro farm sales. Traditional metro farms account for more than 70% of metro farm sales in the Corn Belt, Southern Plains, and Northern Plains regions.

Table 14.1 Regional Farms and Sales, by Farm Type, United States, 1991

	Metro				
Farm Production Region	*Recreational*	*Adaptive*	*Traditional*	*Metro Farms*	*Nonmetro Farms*
	Farms Numbers				
Northeast	41,655	15,641	29,786	87,082	58,918
Lake States	28,287	8,038	28,909	65,234	151,673
Corn Belt	54,677	9,207	60,803	124,687	307,841
N. Plains	13,933	543	10,432	24,907	165,735
Appalachia	58,325	3,917	29,491	91,732	210,687
Southeast	47,114	10,305	14,720	72,139	84,361
Delta States	16,784	3,561	10,788	31,133	82,827
S. Plains	51,066	3,012	23,243	77,321	175,689
Mountain	10,336	6,080	6,215	22,631	95,426
Pacific	50,512	36,720	12,317	99,550	57,450
United States	372,689	97,023	226,703	696,416	1,390,607

Source: 1991 Farm Costs and Returns Survey

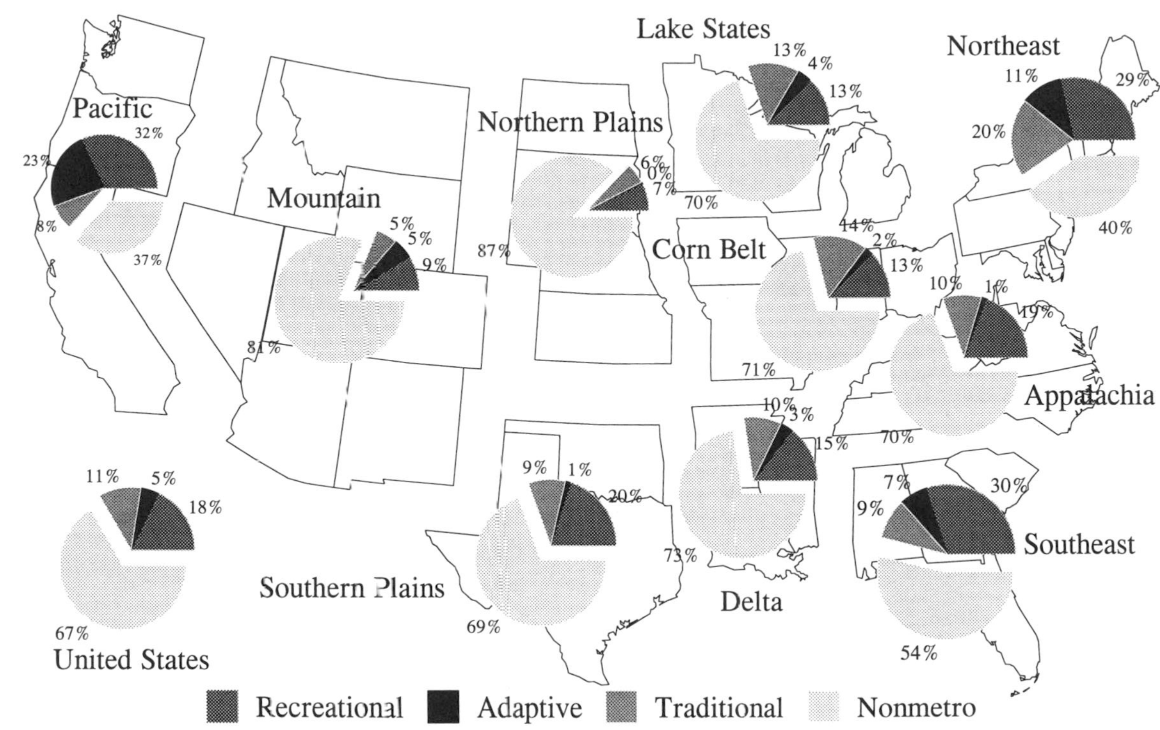

Figure 14.2 Distribution of farm types in metro and nonmetro counties, by region, 1991. (*Source:* 1991 Farm Costs and Returns Survey, USDA)

Sources of Income

Farm and Off-Farm Income

Metro farm households rely less on farming for their income than do nonmetro farm households (see Figure 14.3 and Table 14.2). Surprisingly, adaptive farms rely on farm income more than any other farm type. Traditional farms—metro and nonmetro—both depend on nonfarm activities for more than half of their household income.

Adaptive farms have large incomes from off-farm businesses, which may consist of value-added retail and wholesale ventures derived from

Table 14.2 Farm and Off-Farm Income of Farm Households, by Farm Type, United States, 1991

Operator Characteristics	*Metro* *Recreational*	*Adaptive*	*Traditional*	*Metro Farms*	*Nonmetro Farms*
	Dollars per Farm				
Farm Income					
Sales of agr. products	2,443	194,562	77,843	52,985	50,320
Government payments	286	2,943	3,307	1,632	3,579
Other farm income	631	22,236	6,518	5,466	4,835
Gross cash farm income	3,360	219,741	87,677	60,083	58,734
Net cash farm income	(4,865)	43,184	12,139	7,169	9,971
Farm income to primary operator	(4,698)	28,166	8,193	3,945	6,726
Off-Farm Income					
Wages and salaries	21,280	16,092	13,927	18,176	13,908
Off-farm business	13,276	20,792	7,189	12,296	6,284
Interest/dividends	3,155	5,062	2,949	3,344	2,603
Other off-farm income	7,121	5,097	5,205	6,222	4,748
Total off-farm income	44,832	47,043	9,271	40,038	27,543
Household income [1]	40,133	75,209	37,463	43,983	34,269

Source: 1991 Farm Costs and Returns Survey

[1] The household income figures reported here are based on a subset of the 1991 FCRS data set (see text) and consequently do not match USDA's official estimates of 1991 farm family income.

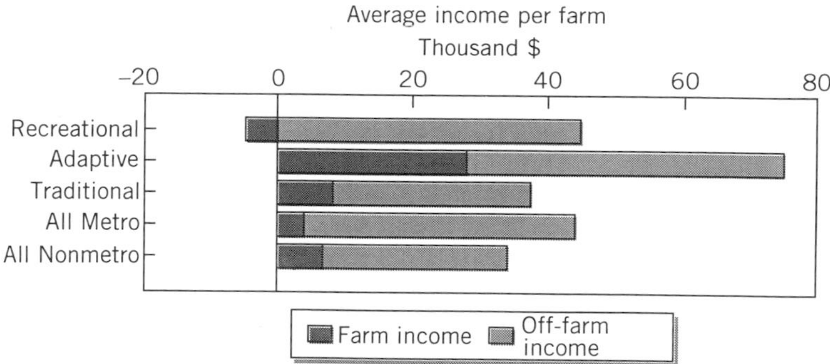

Figure 14.3 Farm and off-farm income of farm households, by farm type, 1991.

farm production. Recreational farms depend on nonfarm income sources—particularly wages and salaries—to subsidize monetary losses in their agricultural activities.

Average annual sales of agricultural products of metro farms are greater than those of nonmetro farms. However, this aggregate result is heavily influenced by the more than 372,000 recreational farms (53% of total metro farms) defined to have less than $10,000 in sales, which average less than $3,000 in sales per farm. Average sales of adaptive and traditional metro farms are 1.5 to 4.0 times greater than average nonmetro farm sales.

Government Subsidies

Averaged across all farms in each type, metro farms receive government payments amounting to only half that of nonmetro farms. Traditional metro farms receive higher government payments per farm than do adaptive and recreational metro farms and payments per farm only slightly lower than nonmetro farms. However, when adjusted for the actual number of farms that participate in government programs, these relationships shift. A larger proportion of traditional metro farms participate in programs (40%) than do adaptive (15%) or recreational (10%) farms. However, adaptive farms that participate receive payments more than twice as high as participating traditional farms ($19,435 versus $8,211).

In terms of economic sustainability, recreational farms can persist only through large infusions of off-farm wages and business income that subsidize farm operations. This is either a rather expensive form of outdoor recreation or the short-term cost of holding an appreciating land asset that will be recouped through capital gains when the farm is sold for urban development. Traditional farm types are also sustained by substantial off-farm income. Net farm incomes for traditional farms are certainly not sufficient to provide an adequate living, especially with the negative returns on assets and equity incurred on a rather substantial investment. Only adaptive farms can be judged economically sustainable based on the level of income generated by the farm operation. Even so, adaptive farms also have substantial off-farm income from wages and off-farm business enterprises. Nevertheless, it appears that adaptive farms have succeeded in meeting the economic challenges of urbanization.

FARM ENTERPRISES

In general, adaptive metro farms take advantage of their proximity to consumer markets by specializing in farm products that have high value per acre, are relatively perishable, and are more difficult to store and transport (see Figure 14.4 and Table 14.3). Although adaptive farms are defined in terms of high-value crop production, the concentration of sales in the high-value category (78%) is startling. Adaptive farms also grow field crops, but most of these crops are not supported in farm programs.

Recreational farms concentrate on livestock enterprises, especially cattle, calves, and breeding stock that can be raised by a part-time farmer. Traditional metro farms concentrate on livestock enterprises as well, but the largest sales categories are milk and dairy products and cattle and calves.

Nonmetro farms concentrate more on crops than on livestock production. Field crops, particularly those supported by government farm programs, are the dominant crop enterprises. Cattle and calves, along with hogs, sheep, and poultry, round out the nonmetro livestock enterprises.

Organic and other specialty crops may make up some percentage of product mix on adaptive farms, but it is more likely that conventional crop systems for these high-value crops are used by adaptive farmers. However, even conventional production methods for crops of this kind are less intrusive on urban neighbors than those that generate the dust, odors, and machinery noise typically associated with traditional field crops and livestock. Adaptive farm methods frequently result in environmental problems such as exposure to pesticides and nutrient runoff and leaching, but these effects are largely invisible and thus unobjectionable to urban residents until air or water quality monitoring actually detects them.

Returns per Farm

Net cash farm income per metro farm is more than 79% of net cash income per nonmetro farm (see Figure 14.5 and Table 14.4). Although net cash losses on metro recreational farms distort the metro farm average returns, adaptive net cash farm income is more than three times that of traditional metro farms and nonmetro farms. The average value of farm

Table 14.3 Farm Sales by Commodity, by Farm Type, United States, 1991

	Metro				
Operator Characteristics	*Recreational*	*Adaptive*	*Traditional*	*Metro Farms*	*Nonmetro Farms*
	Percent of Sales[1]				
Crops					
Corn	4	2	11	6	11
Soybeans	4	1	9	5	9
Wheat	1	1	3	2	5
Other crops [2]	11	9	7	8	7
Field crops	20	13	30	21	32
Vegetables	2	8	0	4	2
Fruits and nuts	4	12	0	6	3
Nursery/greenhouse	6	25	0	13	2
Contract crops	4	33	2	18	6
High-value crops	16	78	2	41	13
Total crops	36	90	32	62	45
Livestock					
Breeding stock [3]	8	0	5	3	4
Calves, cattle, hogs, and sheep	50	2	27	14	33
Poultry	0	0	5	2	1
Contract livestock	0	7	6	6	5
Other livestock	5	1	0	1	1
Livestock	63	10	43	26	44
Milk/dairy products	1	0	25	12	11
Total livestock	64	10	68	38	55
Total sales	100	100	100	100	100

[1] Does not include government payments associated with crop production.
[2] Includes barley, oats, rye, peas, sorghum, dry beans, tobacco, potatoes, hay, seeds, popcorn, and other crops not enumerated.
[3] Includes sales of beef and dairy cows, bulls and replacement heifers and sows, boars and replacement gilts.

Source: 1991 Farm Costs and Returns Survey

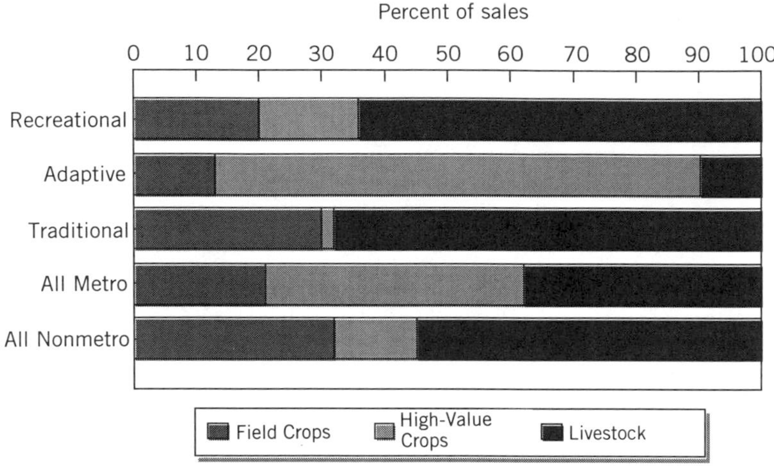

Figure 14.4 Farm sales by commodity, by farm type, 1991.

dwellings in metro areas is $62,617, versus $38,028 in nonmetro areas. The imputed gross rental value of farm dwellings substantially offsets income losses on farming operations for recreational metro farms and may be the principal economic incentive for these enterprises.

Average metro farm assets are one quarter larger than nonmetro assets, reflecting higher property values near urban centers.

Average returns on assets and equity for all metro farms are as bad as those for nonmetro farms. It must be that aggregation with recreational farms pulls down the metro average, inasmuch as adaptive farms have the only positive returns of any farm type. Even so, returns for adaptive farms are much lower than those typically earned on alternative nonagricultural investments. This raises the question of whether any form of agriculture can be truly sustainable in competition with more lucrative urban development. Despite their using less land than traditional farms, adaptive farms still hold large acreage in urbanizing areas. Thus, to the returns realized on day-to-day farming operations must be added the potential capital gains from liquidating the land investment when urban development pressures prove irresistible. Although this prospect is equally available to all three types of metro farms, it is more tenable for adaptive farms that can earn positive returns while waiting for urban land markets to fully mature.

Land Use and Value

As expected, the average metro farm operates fewer acres than the average nonmetro farm (see Table 14.5). At one extreme, recreational farms are about one tenth the size of nonmetro farms, and at the other, traditional

Table 14.4 Farm Returns, by Type, United States, 1991

Operator Characteristics	Metro: Recreational	Metro: Adaptive	Metro: Traditional	Metro Farms	Nonmetro Farms
	Dollars per Farm				
Gross cash farm income	3,355	244,697	91,770	65,760	60,891
Cash expenses	8,264	199,076	79,128	57,916	50,944
Net cash farm income	(4,908)	45,621	12,642	7,844	9,946
Net farm income[1]	(2,126)	42,653	12,629	8,916	9,703
Interest	1,855	12,797	6,304	4,840	4,890
Return on assets[2]	(6,834)	27,280	(6,549)	(1,989)	(829)
Return on equity[3]	(8,689)	14,483	(12,890)	(6,828)	(5,719)
Assets	246,925	933,140	570,875	447,982	351,957
Debt	18,178	115,524	58,282	44,795	46,830
Net worth	228,747	817,616	512,593	403,187	305,127
	Percent of Capital				
Returns[4]					
To assets	(2.8)	2.9	(1.1)	(0.4)	(0.2)
To equity	(3.8)	1.8	(2.5)	(1.7)	(1.9)

[1] Net cash farm income (or loss), plus imputed rental value of the farm residence
[2] Net farm income (or loss) less imputed charges to operator labor, management and unpaid labor, plus interest
[3] Return on assets less interest
[4] Return on assets as a percent of assets and return on equity as a percent of net worth

Source: 1991 Farm Costs and Returns Survey

metro farms operate slightly smaller acreages per farm than nonmetro farms. Adaptive metro farms own more of the land they operate (50%), whereas traditional and nonmetro farms rent more land. Grazing land rented on an animal-unit month (AUM) basis from public agencies or grazing associations is a large component of land operated by traditional metro and nonmetro farmers, particularly in the West.

Metro farmland and buildings are three times more valuable than nonmetro farmland and buildings. Moreover, adaptive metro farm property is nearly four times as valuable as that of traditional metro farms. Differences in land value by farm type may reflect differences in land quality for agri-

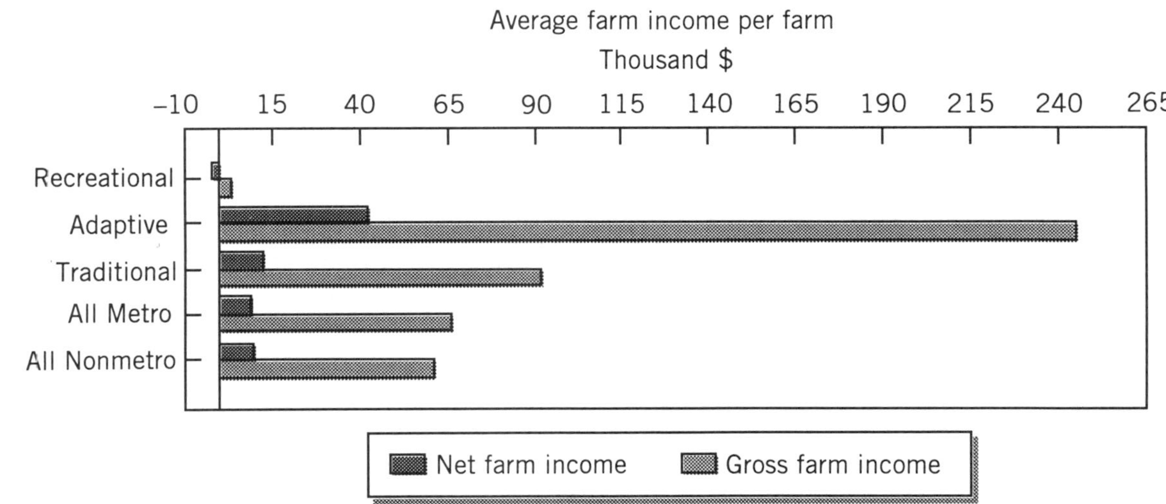

Figure 14.5 Gross cash and net farm income, by farm type, 1991)

Table 14.5 Land Ownership, Operation and Intensity, by Farm Type, United States, 1991

Operator Characteristics	Metro: Recreational	Metro: Adaptive	Metro: Traditional	Metro Farms	Nonmetro Farms
	Acres per Farm				
Acres owned	62	130	247	132	300
Cash rented in	14	93	176	78	131
Share rented in	5	37	69	30	75
AUM acres rented in	10	0	144	52	291
Rented in free	5	5	10	7	8
Total rented in	34	135	399	167	505
Rented out	6	8	19	10	23
Rented out free	0	1	0	0	1
Total rented out	6	9	19	10	24
Acres operated	90	255	628	288	784
	Dollars per Acre				
Value of land and buildings owned	3,496	5,360	1,551	2,567	771
Annual cash rental in	16	99	27	38	23
Annual cash rental out	40	104	21	34	25
Sales of agr. products	27	856	130	202	67
Return on equity	(97)	57	(21)	(24)	(7)
Capitalized @ 5%	(1,935)	1,135	(410)	(474)	(146)
Labor input	78	358	52	94	25
Nonlabor input	87	532	114	161	60
Operating expenses	165	891	167	256	85
Sales per $ of inputs	.16	.96	.78	.79	.79
Capital assets	277	547	182	243	92
Total assets	2,750	3,657	909	1,555	449
Debt	202	453	93	155	60
Debt per $ of assets	.07	.12	.10	.10	.13

Source: 1991 Farm Costs and Returns Survey

cultural production, but are more likely due to differences in proximity to urban development within metro areas.

There are interesting differences by metro farm type between rents for land that is rented and land rented out. Adaptive farms pay the most for land they rent from others and get the most for land rented to others, and they use share leasing less than traditional metro farms. It may be that adaptive farms rent better quality land, particularly in light of their intensive production. However, locational differences within metro areas and in the terms of share leases may also explain differences between adaptive and traditional farmland rents.

Productive Intensity

The value of agricultural products sold per acre on metro farms is three times that for nonmetro farms (see Figure 14.6 and Table 14.5). Traditional farms within U.S. metro areas have per acre sales almost twice those of nonmetro farms. However, adaptive farms sell seven times more per acre than traditional farms and 13 times more than nonmetro farms. This difference is partly explained by the emphasis on intensive, high-value crop production by adaptive farms, and the extensive livestock enterprises using large acreages of grazing land operated by both traditional and nonmetro farms.

Average returns on equity per acre for metro farms are about the same as for nonmetro farms. However, there is a large variation in returns on equity per acre among metro farm types, with adaptive farms earning the only positive returns per acre. Traditional farms' sales and expenses are

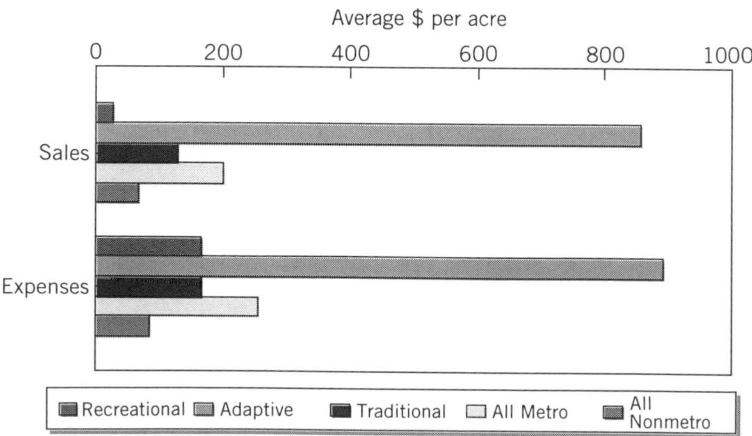

Figure 14.6 Intensity, by farm type, 1991.

twice that of their nonmetro counterparts, whereas their losses on equity per acre are larger than nonmetro traditional farms.

The capitalized returns on equity constitute one measure of the *agricultural use value* of land operated by the various farm types. In 1991 only adaptive metro farms had a positive agricultural use value equal to about 21% of the average value of the land owned. Capitalized production value on metro and nonmetro farms overall is negative.

The value of inputs used per acre is a measure of *land use intensity*. Recreational farms have lower inputs per acre operated than other metro farms. Adaptive metro farms use 10 times the input per acre of nonmetro farms and 3.5 times the inputs of the average metro farm.

It is not possible, using these data, to assess progress toward lower input use for environmentally sustainable production. Some operators in all three farm types are probably making low input adjustments, but aggregating across all farms in each type masks the effect. Low input operations can be identified only relative to more conventional operators for a particular crop or enterprise type. Thus, the input levels used by adaptive farms in metro areas may be low relative to all high-value fruit and vegetable producers, as indicated by relative input use evidence in Heimlich and Brooks (1989) and Heimlich (1988). As indicated earlier, more intensive use of pesticides and fertilizers for high-value production on adaptive farms may have greater actual environmental impacts than less intensive production methods on traditional farms, yet be less objectionable to urban neighbors. Neighbors may perceive problems with dust, odors, and machinery noise from traditional farms as greater than the actual risks from more intensive use of manufactured inputs on adaptive farms because these risks are silent and invisible.

Dividing sales per acre by operating expenses per acre, including operators' unpaid labor, yields another measure of *productive intensity*. In 1991 neither metro nor nonmetro farms produced more in sales than the value of inputs. Adaptive farms' sales and inputs are nearly equal. Traditional metro farms and nonmetro farms have nearly equal sales: input ratios, at about 80%. The excess of inputs over outputs is accounted for by including the imputed value of operator and unpaid family labor. By not paying for their own labor, these farms are, in economic terms, exploiting themselves. Recreational metro farms have very little market output per dollar of expense on each acre, clearly revealing that they are consumption units rather than production limits.

Total assets per acre and debt per acre measure *investment* in the farm operation and the extent to which operators are *leveraged*. Metro farms have 3.5 times the investment of nonmetro farms, but only 2.5 times the debt per acre. This is mainly due to higher urban land prices, inasmuch as capital assets per acre are only 2.6 times higher for metro farms. Metro farms have only 10 cents of debt per dollar of assets, as compared with 13 cents per dollar for nonmetro farms. Adaptive metro farms have the highest investment per acre of any farm type, more than four times that of traditional metro farms. Adaptive farms have 12 cents of debt per dollar of assets, slightly higher than the indebtedness of traditional farms.

CONCLUSIONS

Farms in metropolitan areas constitute an increasingly important component of U.S. agriculture, because metro areas are expanding and because agriculture in metro areas is increasing. Metro farms make up one third of all farms in the United States, control nearly 60% of all farm assets, and operate 16% of all farmland.

Within metro areas, differences between farm types blur the contributions of the more productive metro farms. As shown in this study, metropolitan agriculture is not homogeneous. Metro agriculture can be characterized by a small number of adaptive farms that are accommodating themselves to an urbanizing environment, a larger group of traditional farm operations that have increased costs and pressures without compensating increases in revenues from better-adapted enterprises, and a large group of recreational farms that use off-farm wages and unpaid family labor to subsidize their rural-like pursuits.

The future of agriculture in urbanizing areas will be a complex resultant of competitive forces, market opportunities, and public policies. It is clear that unrestricted competition for land will result in the decline of traditional farms. Adaptive agriculture's ability to replace traditional farms is limited only by farm operators' creativity in developing new market opportunities and by increasing the volume of sales and the return on each acre they farm. However, much of the land currently farmed in metro areas is not well suited to this kind of intensive agriculture. Those seeking to preserve rural landscapes and life-styles will see little benefit in the way that adaptive farms retain a place for agriculture in the metro economy. By sacrificing most of the agricultural land and all of the visual amenity and rural life-style associated with traditional farming, adaptive farms achieve economic success through a kind of agricultural industrialization with high environmental impacts.

The three farm types are clearly distinguishable in terms of agricultural inputs and environmental impacts. Most recreational farms have relatively benign impacts on the environment, with positive contributions of open space and environmental filtering likely outweighing pollution or other negative effects from farm operations. Possible exceptions are the impacts related to livestock operations, for which recreational farms may not have appropriate manure-handling facilities or barnyard runoff management. Traditional farms have well-known environmental impacts from sheet and rill and wind erosion associated with tillage operations, odors and nutrient runoff from livestock operations, and noise from machinery operation. With appropriate management, however, impacts resulting from the use of pesticides and nutrients on relatively extensive field crops can be minimal. Adaptive farms are likely to have the most substantial environmental impacts because they specialize in high-value fruit and vegetable crops that use large quantities of fertilizers and pesticides and require intensive cropping systems designed to generate high yields on small acreages. In the context of acceptability within urbanizing areas, however, it remains to be seen whether problems associated with traditional farms or with adaptive farms are more objectionable and, hence, less sustainable over time.

The three types of metro farms are also clearly distinguishable in terms of preserving rural life-styles and sustaining rural community values. Traditional farms are just what their label implies: traditional farm enterprises with roots in older life-styles and values that prevailed before urban development commenced. Whether they can sustain those life-styles and values in the face of rapid change and the dismantling of rural community infrastructure is doubtful. Recreational farms have the veneer of a rural life-style, but are imitative rather than genuine. Recreational farms may have the appearance of "real" farms to their urban neighbors, but with little of the productive substance embodied in traditional farm types. Adaptive farm types do not reflect rural life-styles and values, but embody an intensification of agriculture that is more industrial than bucolic. They often do not look like "real" farms, using plastic mulch, irrigation piping systems, and mechanized chemical blowers and pickers that most urban residents do not recognize as part of farming. Neither do they project rural values, being more in tune with urban people and urban markets.

There is considerable consensus in the sustainable development literature regarding the desirability of farmland retention programs. They have a role to play in mitigating economic competition for land to achieve public goals of balanced land use, adequate open space. containment of urban expansion, support of rural ways of life, and protection of wildlife habitats. Farmland retention programs benefit all metro farm types. Use value assessment (UVA) and purchase of development rights (PDR) programs make recreational farming less expensive. Such programs shelter traditional farms from economic competition by reducing their costs relative to urban uses and providing capital from equity holdings in farmland, thus reducing pressure to sell out for development. Finally, farmland preservation programs can ease the transition to adaptive agriculture by providing a source of financing for operational changes. Policywise, however, there is a fine line between temporarily holding off changes and undercutting the economic incentives for adaptation.

Policymakers must recognize that farmland retention programs can be directed toward saving farmland, saving farming, or attempting to do both.

Making explicit the objectives of farmland preservation program, such as maintaining open space, reducing environmental problems associated with production, using farmland to help improve water and air quality, and preserving agriculture as part of the economy and the rural life-style, will help in deciding which programs to offer and what type of farm to target. Farmland preservation programs can impact farm types differently, determining whether metro farming becomes primarily a hobby, a museum piece, or a business. If program incentives are offered primarily to owners of small parcels who live on off-farm employment and extensive farm enterprises that are economically marginal, then recreational farms will be promoted. This strategy may preserve an integrated mix of urban and rural uses with a pastoral landscape, but will likely accomplish little in regard to retaining farm businesses or traditional rural life-styles. Programs aimed at larger operating units, run by full-time farm operators in traditional farm enterprises, will benefit traditional farm types likely to preserve conventional rural life-styles and values, but could blunt economic incentives for adaptation to urban land markets. Depending on the environmental performance requirements of such programs, they could preserve open space and rural life-styles but remove traditional and industrial agriculture as a real component of the local economy. PDR programs may not be accepted by adaptive farm types who want to preserve potential capital gains, but UVA and transition loan programs may be attractive to traditional or recreational farms that want to intensify their operations and exploit urban market opportunities. Such programs could foster a renewed agricultural business presence in urbanizing areas, but would do so at the expense of conventional rural landscapes and traditional rural life-styles.

Which farm types are targeted will have a bearing on what components of farming are sustained. Preserving recreational and traditional farm types can improve the environmental sustainability of agriculture and may increase the degree to which agriculture is accepted by the larger community, but without increases in consumer demand for low-input products, it is not likely to provide for economically sustainable agriculture. On the other hand, programs for adaptive agriculture support a more economically viable agriculture, but one that does little to sustain traditional rural life-styles and values and may ultimately reduce environmental sustainability. In the absence of farmland preservation programs, urban pressures are likely to increasingly squeeze out traditional farm types in favor of a larger number of recreational farms supported by nonfarm income and a smaller, economically viable number of adaptive farms.

REFERENCES

Ahearn, M., and D. E. Banker. 1988. "Urban Farming Has Financial Advantages." *Rural Development Perspectives* (October):119–121.

Bentley, S. E. 1993. *Farm Operating and Financial Characteristics, 1990.* Washington, D.C.: U.S. Department of Agriculture, Economic Research Service, SB-860.

Brooks, N. L. 1985. *Minifarms: Farm Business or Rural Residence?* Washington, D.C.: U.S. Department of Agriculture, Economic Research Service, AIB-480.

Bureau of the Census. 1992. *Statistical Abstract of the United States, 1992.* Washington, D.C.: U.S.Department of Commerce.

Heimlich, R. E. 1988. "Metropolitan Growth and High-Value Crop Production." In *Vegetables and Specialties Situation and Outlook Report.* TVS-244. Washington D.C.: U.S. Department of Agriculture, Economic Research Service.

Heimlich, R. E., and C. H. Barnard. 1992. "Agricultural Adaptation to Urbanization: Farm Types in Northeast Metropolitan Areas." *Northeast J.* David E. Banker. 1992. *Financial Performance of U.S. Farm Businesses, 1987–90.* AER-661. Washington, D.C.: U.S. Department of Agriculture, Economic Research Service.

Heimlich, R. E., and D. H. Brooks. 1989. *Metropolitan Growth and Agriculture: Farming in the City's Shadow.* AER-619. Washington, D.C.: U.S. Department of Agriculture, Economic Research Service.

Smith, S. N. 1987. "Farming Near Cities in a Bimodal Agriculture." In *Sustaining Agriculture Near Cities,* edited by W. H. Lockeretz. Ankeny, Iowa: Soil and Water Conservation Society, 77–90.

Fifteen

Greenways, Trails, and Rural Sustainability

Earl M. Starnes, Mark Benedict, and Matthew S. Sexton

There is a tide in the affairs of men,
Which, taken at the flood, leads on to fortune . . .

—William Shakespeare, Julius Caesar, IV: iii, 49

INTRODUCTION

This chapter looks at the beneficial impacts of greenways and trails located in rural settings. It is not our intention to propose or to argue that greenways and trails are to be viewed as stand-alone social, aesthetic, environmental, and economic development projects assuring rural sustainability. Greenways and trails represent but one strategy among sets of strategies that can and do contribute to the general welfare and sustainability of rural regions; however, greenways and trails are an emerging and important facet of today's environmental movement.

Currently, the environmental movement shows signs of turning away from the more traditional site-by-site regulatory methods usually employed by the federal government and counterpart state agencies. The indications point to more holistic and comprehensive approaches and nonregulatory negotiated strategies such as ecosystem management, sustainable development, regional planning, and the creation of greenways. Attention to greenways was given national prominence in 1987 by the President's Commission on Americans Outdoors, chaired by then-governor of Tennessee Lamar Alexander. One of the Commission's recommendations was to establish "a continuous network of recreation corridors which could lead across the country" (Little 1990, 37). In the same year, Patrick Noonan, a Commission member and chair of the Arlington, Virginia–based Conservation Fund, established the American Greenways Program to promote and support the concept of greenways and greenway systems throughout the United States.

Since that time several states have developed greenways programs and allocated agency resources to continue planning and implementation of statewide greenway systems. For example, Maryland, with a greenways heritage that "has been a-building, however unconsciously, since 1906" (Little 1990, 131), initiated the first contemporary statewide greenways planning effort in 1990. This was a cooperative effort of the Conservation Fund and the Maryland Department of Natural Resources. Through the ongoing work of the Maryland Greenways Commission, its staff, and the Maryland Department of Natural Resources, the state has made considerable progress on a program of planning for a complete interconnected and comprehensive greenway and trail system. In early 1993, Governor Lawton Chiles of Florida created the Florida Greenways Commission. The commission submitted its Report to the Governor in December 1994 (Florida Greenways Commission 1994). The Greenways Commission's overriding recommendation was to create a statewide greenways system by linking existing and proposed conservation lands, trails, urban open spaces, and private working landscapes for the benefit of Florida's people and wildlife. The Commission's report contained nearly 200 specific recommendations to guide the creation of Florida's statewide greenways program and system. Maryland and Florida are but two of many states committed to creating statewide greenway systems that complement ongoing rural development and environmental preservation programs.

ORIGINS OF THE GREENWAYS MOVEMENT

The concept of greenways may have first appeared as an idea in 1865 when Frederick Law Olmsted proposed "two 'greenway' elements" (Little 1990, 9) in his plan for the College of California (now the university). One of the greenway elements was designed to link the campus with Oakland; the other was designed to preserve the Strawberry Creek Valley. Olmsted continued to design major urban parks, including Central Park in Manhattan, and parkway systems in many urban locales. Olmsted's work during the later part of the nineteenth century stimulated the demand for parkways, greenways, and connected parks and open spaces. In 1887, the Emerald Necklace was proposed in Boston. "According to Charles Birnbaum, a landscape architect involved in the restoration of the parks, the Emerald Necklace 'is considered to be the greatest greenway achievement' of the Olmsted firm" (11).

Not all early greenways were urban. Benton MacKaye, a professional forester and planner, advocated greenways as a means of controlling metropolitan sprawl along the east coast of the United States. MacKaye is best known for promoting the idea of the Appalachian Trail. "MacKaye's prescription to stem what we now call urban sprawl was 'a common public ground'" to serve as green boundaries designed to limit "metropolitanism"

(Little 1990, 18). Today the Appalachian Trail is not the greenway MacKaye envisioned. MacKaye was among others espousing regional planning and the use of greenways to connect and link urban dwellers with the hinterland and rural heritage. Recently, the Appalachian Trail Conference, a nongovernmental organization, proposed the development of MacKaye's full-fledged greenway concept. Today, with rediscovered and heightened interest in greenways, the Conference hopes the idea can be realized.

Along with MacKaye and Olmsted, Ebenezer Howard, Raymond Unwin, Clarence Stein, Rex Tugwell, and other regional planners and new town proponents of the late nineteenth and early twentieth centuries made many contributions to the greenway idea. New towns, greenbelt towns, garden cities, and Henry Ford's concept of rural factories seemed to blend the affection for rural and agricultural life with an early interest in greenways.

Early planning and support for greenways were not limited to actions benefiting only human populations. More than 60 years ago the South African Wildlife Society recognized the importance of such connections to maintaining the continent's wildlife. By the 1960s ecologists in the United States were also looking to wildlife corridors to help preserve biological diversity. Larry Harris (1985) maintains that decades of land development and increasing isolation of wildlife exacerbated the dilemma of species extinctions. Protection of wildlife corridors and linking them to large hubs of biodiversity are ways to counter the impact of habitat fragmentation. The environmental corridor concept, proposed in 1960 by Philip Lewis, offers further support for an environmentally based greenway vision: "It is stream valleys, the bluffs and ridges, the roaring and quiet waters, mellow wetlands and sandy soils that combine in elongated designs, tying the land together in regional and statewide corridors of outstanding landscape qualities" (Little 1990, 22). Whether it is for urban design, recreation, economic development, environmental protection, or a combination of these, linkage is the common principle of the greenway idea—linking natural areas, recreation lands, urban communities, and rural residents.

Greenways and Trails

The word greenway evokes many images, ranging from simple recreational footpaths to broad natural corridors linking extensive native ecosystems. The report of the Florida Greenways Commission (1994, v) offered a very inclusive working definition:

> A greenway is a corridor of protected open space that is managed for conservation and/or recreation. The common characteristic of greenways is that they all go somewhere. Greenways follow natural land or water features like abandoned railroad corridors or

canals. They link natural reserves, parks, cultural and historic sites with each other and, in some cases, with populated areas. Greenways not only protect environmentally sensitive lands and wildlife, but also can provide people with access to outdoor recreation and enjoyment close to home.

Greenway and Trail Benefits for Rural Sustainability

The planning and protection of greenways bring important benefits to rural areas. For this reason they can be considered important strategies to improve rural sustainability and economic development. Greenways can protect important rural ecological features, thereby providing diverse natural resource and environmental benefits and avoiding the serious ecological problems caused by the fragmentation of land into development parcels (Little 1990, 113). The first major consequence of such fragmentation is the loss of species requiring deep woods for breeding. The second consequence is the elimination of larger species, such as bears, which require extensive ranges and corridors to reach forage areas. Third, fragmentation can cause indigenous populations to grow beyond the sustainable landscape through human-subsidy feeding. Finally, Harris sees inbreeding within a restricted range leading to the genetic collapse of species (Little 1990). Although the areas joining water bodies and uplands often provide the best linkages for wildlife, greenways and protected trail corridors can serve as a spine to which future linkages to ranges and forest lands may connect.

Greenways provide linkages between isolated but environmentally important areas, thus creating an opportunity to preserve and enhance existing ecosystems by restoring functional natural corridors. Greenways preserve and protect existing viable agricultural and forestry operations and other working landscapes. Ecological and economic benefits accrued through the protection of rural greenways and greenway systems include protection of habitat for native wildlife and conservation of water resources for the benefit of rural residents, agricultural operations, and natural systems. In addition, greenways draw recreational users into rural communities, thereby creating a need for various services, food, shelter, and recreational equipment. Greenways can serve as recreational and transportation links between rural residents and their urban counterparts.

Greenways also benefit the preservation of America's historical and cultural heritage. An interesting example is the effort of southeastern states to identify trail segments and establish historical interpretive kiosks recognizing Hernando DeSoto's march through early America. This is but one example of efforts being launched by federal, state, and local governments and nongovernmental organizations across the land.

In addition, greenways and trails offer many social benefits. Moore and associates (1992) include "personal development, social bonding, therapeutic bonding, improved physical health, stimulation and opportunity for curiosity seeking," as well as "social interaction, mental health, and family cohesiveness." Holmes Rolston, Jr. (in Moore et al. (1992) "presents the following as ecological benefits: life support, aesthetics, scientific opportunities, natural history, habitat, and forms of philosophy and religion."

The potential for economic development of tourist markets related to the rediscovery of ecological values and the provision of recreational opportunities is also motivating rural leadership to consider greenways and trails. Recognizing the link between recreational use and rural economics, several Florida counties and local chambers of commerce along the northeast shore of the Gulf of Mexico have initiated a publicity campaign dubbing the region Florida's Nature Coast. The focus of this effort is to promote ecotourism in some of Florida's poorest rural counties. In the same region, the North Central Florida Regional Planning Council (NCFRPC) and the Suwannee River Water Management District (SRWMD) are involved in separate, but closely related, efforts designed to promote economic development coupled with environmental preservation. For five years the NCFRPC has been involved in tourist development in the predominately rural Suwannee River Valley, and for 15 years SRWMD has been buying property in the floodplains of the Suwannee River, its tributaries, and three other rivers in the region. The areas along the reaches of these rivers are emerging as a coherent system accessible for hiking, camping, hunting, fishing, and canoeing. The SRWMD is now developing a districtwide plan for greenways and trails designed to assure comprehensive linkages between areas of its nearly 100,000 acres of river land, environmentally sensitive areas, state parks, and other large holdings of both public and private lands. These examples demonstrate that greenways and trails are being recognized by public and private entities for both environmental and concomitant economic benefit to rural regions.

Greenways are often classified by function and use. Functional attributes include water management, wildlife management, and other activities related to environmental protection. Uses relate to the kind and intensity of recreation, educational opportunities, and enjoyment they provide. Some greenways primarily enhance ecological values and preserve environmentally sensitive areas. When the goal of a greenway is clearly ecological, users may have only very restricted access, limited to educational opportunities and research. Other greenways serve a variety of purposes, and human access includes a wide range of activities. Unpaved trails provide for hiking, horseback riding, and mountain biking; and paved trails can be used for walking, speed walking, jogging, bicycling, and skating. Where intense use is expected, some paved trails have been constructed with separate lanes for different uses, thus providing for the safety and well-being of users.

RURAL GREENWAYS AND TRAILS: THREE EXAMPLES

Three well-documented case studies are presented here to explore the link between greenways, trails, and rural sustainability: the St. Marks Trail in north Florida, the Heritage Trail in eastern Iowa, and the Scuppernong River Greenway in northeastern North Carolina. These studies have been the subject of considerable research in the area of greenways and rural sustainability.

The St. Marks Trail and the Heritage Trail originate in or near urban areas and pass through rural areas and small towns. The Scuppernong River Greenway, located in rural Tyrrell and Washington counties in North Carolina, passes through the small waterfront town of Columbia.

Much of the data presented here are drawn from work published by the U. S. Department of the Interior, National Park Service, in 1990 and 1992 (Moore et al. 1992; National Park Service 1990). Additional data have been gained from other cited material and the professional experience of the authors. The National Park Service work is a creditable source of information regarding the impacts of the national Rails-to-Trails Program. The national Rails-to-Trails Program originated in 1973 when Congress amended the National Trails Act of 1968 to authorize this new program. According to the Rails-to-Trails Conservancy (1993), at least 500 trails were developed in 44 states between the start of the program in 1971 (predating the federal initiative) and 1993. The number of rails-to-trails conversions continues to grow each year, and recent data (Rails-to-Trails Conservancy 1994, 3) indicate that there are now more than 600. Although most trails resulting from the conversion of abandoned railroad rights-of-way may not be considered exclusively as greenways, they certainly complement the greenways concept and indeed may serve as important components of greenways located along rivers, rural roads, and other geographic features. In this chapter, we cite the benefits of only two rural rail-to-trail cases and one rural greenway case as evidence that greenways can contribute to rural sustainability. However, the definition we offer for greenways is inclusive of corridors and linear facilities developed for protected open space and managed for conservation and/or recreation. In addition, the Scuppernong River Greenway is the centerpiece of a set of strategies being employed by rural Tyrrell County and the town of Columbia, North Carolina, to promote ecotourism, environmental education, and sustainable development.

The St. Marks Trail

The St. Marks Trail, a rail-to-trail conversion that opened in 1988, is Florida's first publicly purchased rail-to-trail project. The trail is located in the Florida Panhandle and passes through parts of Leon and Wakulla counties. It is historically significant because it follows the original overland link from Tallahassee to the town of St. Marks, which served as Tallahassee's most important Gulf port during the nineteenth century. Marine access to Tallahassee from the Gulf of Mexico was vital as the vast, undeveloped territory strove toward statehood. To provide better access to the Gulf of Mexico, a railroad was built to link Tallahassee with St. Marks. Mahon (1985, 130) reports that "a rare distinction set[s] Tallahassee apart: about twenty-six miles of railroad ran from it to St. Marks. Begun in 1834, it was one of

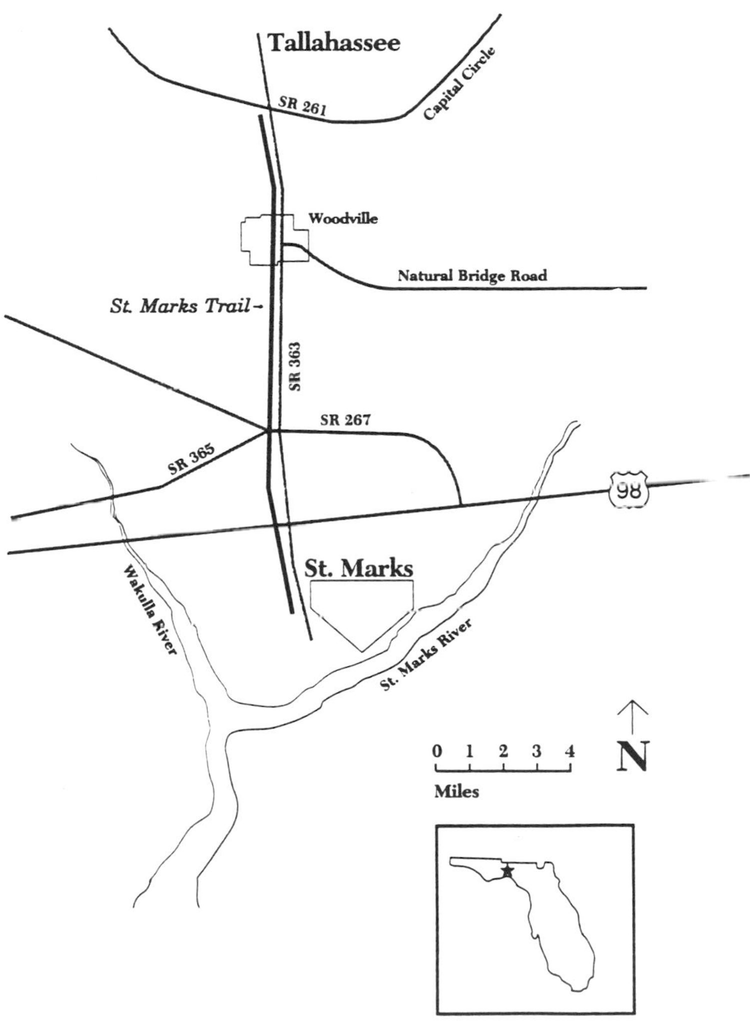

Figure 15.1 Location sketch of the St. Marks Trail and a diagrammatic map.

the early stretches of track built in the United States . . . and one of the worst." It is reported that the tracks were so poorly constructed that locomotives could not be supported by the rails, thus mules were used to pull the cars. Regardless of the railroad's technological failings, the link enabled St. Marks to serve as Tallahassee's port to the Gulf of Mexico for many years. Today, St. Marks and its tank farm act as the oil bunkering port for Tallahassee's municipal electrical utility. It is also an important commercial fishing and pleasure boat harbor.

The St. Marks Trail is 16 miles long (see Figure 15.1). Its current northern terminus is located on State Road 363, at a point about three miles south of the capitol building in Tallahassee. Work is under way to extend the trail into Tallahassee proper, with the ultimate destination near Doak Campbell Stadium on the Florida State University campus. Much of the trail parallels State Road 363 in a north-south direction. A trail user heading south through the first three or four miles finds some sparsely located mixed land uses and vacant land between the road and the trail. To the west are private lands used for silviculture. At one point the trail is touched by the Apalachicola National Forest. Other land uses spottily located along the way include a few small industrial operations, small highway-related retail stores, and very limited agricultural operations. The areas can be classified as urban fringe or simply rural. At about mile four, the trail passes through the rural community of Woodville, a bedroom community for Tallahassee with associated small businesses and schools. Woodville also provides appropriate infrastructure support for local forest operations, and forest lands dominate the trail to the south. At the intersection of State Road 267, it is only a short hike (or ride) to Wakulla Springs State Park. When the user reaches St. Marks, she or he finds a rural community whose economy depends on limited port activities, weekenders, trail users, a state museum and historical site, and a locally well-known restaurant and watering hole.

The St. Marks Trail is maintained and operated by the Florida Division of Recreation and Parks. There are no user fees, nor are use permits required. The trail is paved with asphalt concrete for its entire length and, for much of its length, has a separate, unpaved bridle path. There are several trailheads with parking areas along the way. No motorized vehicles are allowed on the trail except for the patrolling ranger's golf cart. Uses of the trail include horseback riding, walking, hiking, cycling, jogging, skating, and just simply strolling. Several annual and seasonal events are promoted by groups of cyclists and other trail users.

Benefits of the St. Marks Trail

The St. Marks Trail provides social, aesthetic, environmental, and economic benefits to its users and to the adjacent communities. Measures of the benefits are found in the work of Moore and associates (1992, Chap 2). The data are based on estimates and analyses of the overall benefits derived from the use and presence of the trail and do not support site-specific conclusions. The generalized nature of the data does permit conclusions regarding the overall benefits of the St. Marks Trail within the context of its rural setting. Moore and associates estimated that the annual number of users would be 170,000; however, the actual count made by the Florida Division of Recreation and Parks during fiscal year 1994/1995 was 184,947. The researchers reported that their survey revealed that users visited the trail 10 or fewer times the year before responding to the survey. The users, of which 81% bicycled, enjoy the beauty of both public and private forests and the sparsely settled rural places along the trail. This potential for aesthetic experience complemented users recreational goals. It is interesting to note that users in ranking the greatest benefits of the trail, chose health and fitness, aesthetic beauty, preservation of open space, and opportunities for recreation. Landowners adjacent to the St. Marks Trail tend to agree with the users. They reported the greatest benefits to be health and fitness and recreational opportunities. The lack of motor vehicles along the trail, coupled with quiet and natural surroundings, seems to heighten the aesthetic value of the user's experience (Moore et al. 1992,III–52). Understandably, some users report that the lack of rest rooms and available drinking water often interferes with their full enjoyment of the trail. Some of these amenities are being improved.

Moore and associates (1992, III–17, 18, 19) identified some interesting demographics regarding the St. Marks Trail users. Females and males use the trail about evenly, the average age is 38, 68% are college graduates, and incomes average below $40,000 a year. Very few nonwhites use the trail, and most of the users are from Leon and Wakulla counties. Only 18% come from 20 or more miles away from the trail. Of the 58 adjacent landowners surveyed, the average holding is six acres. In at least 76% of households surveyed, one or more members of the family had used the trail the previous year. On average, the trail is 1,822 feet from the landowners' actual residences. The landowners' favorable perceptions of the trail include those of 47% who are strongly supportive. Many of the problems perceived by the landowners prior to the trail's opening were not realized. Forty-three percent of landowners conclude that the trail is much better to live near than the abandoned railroad right-of-way (III–28). Moore and his associates' data are representative of the potential for social interaction, recreational benefits, aesthetic enjoyment, and concomitant social benefits to both users and landowners.

Moore and associates do not offer any data regarding the environmental impacts of the St. Marks Trail. However, the trail does provide an environmental opportunity. It serves as a central spine to which future preservation areas may be attached as such lands become available. Its proximity to the Apalachicola National Forest and Wakulla Springs State Park offers but two of many potential ecological links. Corridors even as narrow as an abandoned railroad right-of-way can provide habitat for wildlife and may be conduits for the movements of smaller animals. The St. Marks Trail offers other positive environmental impacts by promoting sustainability in its rural setting and in nearby small communities. In addition, because it is a preserved corridor and serves as a barrier to land fragmentation, it has even greater potential to contribute to the region, both environmentally and economically.

The economic impact of the St. Marks Trail on Leon and Wakulla counties may seem small in respect to their overall economic profile. More than half of the trail is located in Wakulla County, a county with a poor and rural population. In the context of sustaining natural and recreational resources, the trail's benefits are positive and viewed as worthwhile, in part because such impacts represent new income for both counties. Economic benefits result directly from the use of the trail. For instance, the average trip expenditure of a typical day-user is $11.02, or $1,873,400 per year (Moore et al. 1992, III–40). Approximately 30,600 of the annual visitors live more than 20 miles from the trail, and it is presumed that most of them are not county residents. Each of these users spends $15.18, a total annual expenditure of $464,508. Other trail expenditures include purchases of durable goods such as clothing, recreational equipment (bikes, trailers, etc.), accessories, guidebooks, and other miscellaneous durables (III–45).

There are other less direct economic impacts on adjacent landowners. Property value is often affected by development of a rural greenway and/or trail. One way of measuring the impact is to determine the potential for resale. The Moore study used two groups in its survey, landowners and real estate professionals. Landowners' perceptions of their property in such an event ranged from "much easier to sell" to "much harder to sell." The conclusion may be that the existence of the trail had little effect on perceived value. Eighty percent of the real estate professionals perceived no impact on the potential for selling adjacent properties (III–47).

The economic impacts of rails-to-trails programs are found to benefit host counties, particularly because of new money brought in by the trail users. In the case of the St. Marks Trail, no negative impacts were perceived by the majority of landowners and users clearly make a positive contribution to the local economy.

The Heritage Trail

The Heritage Trail, located in Dubuque County, Iowa, traverses the county generally from east to west. Its eastern terminus is northwest of Dubuque near U.S. 52. To the west it traverses the towns of Durango, Graf, Epworth, and Farley and ends 26 miles farther along at State Road 136 in Dyersville. The Heritage Trail is in rural surroundings throughout its length.

"Beginning in the early 1960s, as railroad companies increasingly abandoned their lines, civic leaders throughout the country, and especially in heartland states such as Iowa, sought to convert the unneeded rights-of-way to new public uses such as wildlife habitats or recreational trails" (Little 1990, 43). In the early years of the rails-to-trails movement, the pleading and support of local citizens were not often heard. However, since 1971 the national Rails-to-Trails program and state companion programs have provided financial resources and strategies to assist in the conversion of abandoned railroad rights-of-way to greenways and trails. Trail proponents, called "trailmakers" by Little, are often motivated by nostalgia for historical rural life-styles and the potential for environmental preservation. Many of today's trailmakers have rural roots, particularly in the Midwest heartland (Little 1990, 43). Development of the Heritage Trail affords the opportunity for local people and visitors to enjoy the open space, the environment, and many recreational benefits.

The Heritage Trail is built on the abandoned right-of-way of the Chicago Great Western Railroad (see Figure 15.2). The last passenger service from Chicago to St. Paul was in 1956, and in 1968 the railroad was bought by Chicago and North Western Railroad. The last freight train rumbled over the rails in 1979, and in 1981 railroad abandonment proceedings took place and the right-of-way became available (Little 1990, 46). Subsequently, a group of civic activists organized Heritage Trail, Inc., and their campaign focused attention on conversion of the former railroad to a trail. The project was approved by the Dubuque County Conservation Board and the Board of County Supervisors, despite many rancorous hearings and meetings, the vandalism of bridges along the proposed trail, and serious opposition from adjacent landowners. Money for acquisition was gathered from both public and private sources. Fund-raising was difficult, but finally successful. Doug Cheever, leader of Heritage Trail, Inc., noted that "it was an idea whose time had come" (47).

The Heritage Trail is managed by the Dubuque County Conservation Board. The board's ranger periodically patrols the trail in a small county truck. Trail-use permits are required (youngsters and seniors are exempt) and available for daily or annual use. Other than maintenance and patrol vehicles, the only motorized vehicles permitted are snowmobiles in the winter (Moore et al. 1992, II–1).

Benefits of the Heritage Trail

According to Cheever, the social, environmental, and economic benefits of the Heritage Trail are interwoven with the ethos of the region. In describing the members of the Heritage Trail board of directors, he states that "[they] grew up on a farm, and though they might now live in town, they knew what being close to the land meant to them. They wanted a chance

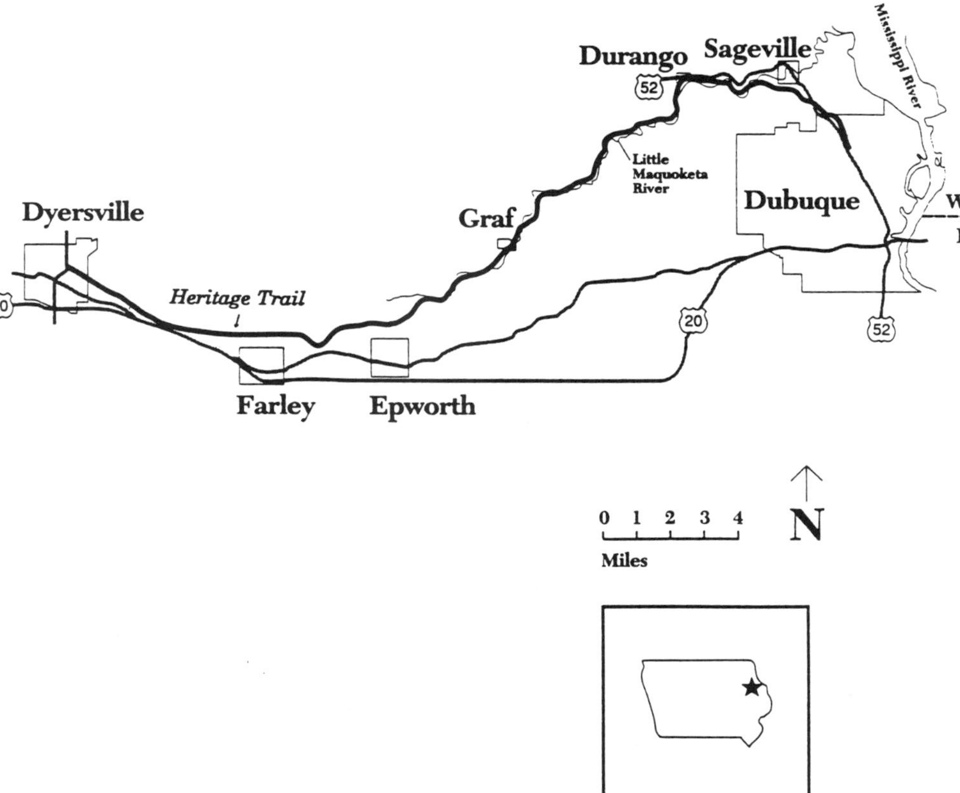

Figure 15.2 Location sketch of the Heritage Trail and a diagrammatic map.

to share that 'sense of the land' with others and saw the trail as a way to do it" (Little 1990, 48). Social, aesthetic, and environmental benefits are expressed by the images along the trail eloquently described by Little (45). The user, he says;

> encounters, in turn, wetlands; Indian burial grounds; an 1868 cast-iron truss bridge; wood duck nesting sites; deep valley with sheer limestone cliffs; lead mine boom-and-bust-town of Durango; an iron ore mine; the remnants of old stagecoach roads; a fabulous fishing spot where channel cat and lunker bass lurk in the pools; a split rock just wide enough for a 4-8-4 steam engine to negotiate; dolomite fossil beds; remnants of dry prairie with side oats gamma grass and prairie roses; a number of nineteenth-century grist mills; Paleozoic outcroppings; the birthplace of famed natural scientist W. J. McGee; remnant tall-grass prairie with big bluestem; oak savannas; and artifacts such as old railroad coaling stations, bridges, and abandoned equipment which bespeak the plains commerce of yesteryear.

Moore and his colleagues (1992, iv) calculated that 135,000 people visit the trail annually. This is more than twice the population of Dubuque. Of the trail users, 85% are bicyclists, 20% are walkers, and 3% joggers (horses are not permitted). Thirty-one percent of the users come from more than 20 miles away. The Heritage Trail is surfaced with crushed limestone, causing some users to complain that the surface is too rough (v). Regardless, whether by foot or by bicycle, it is easy to see the wonders of the surrounding prairie and behold the old towns and farms, streamsides and woods, rocks and wildlife (Little 1990, 48).

Benefits revealed by the user survey include health and fitness, aesthetic beauty, preservation of open space, and community pride (Moore et al. 1992, iv). Adjacent and nearby landowners appear to be in general agreement. They list benefits as health, fitness, recreation, open space, and research and educational opportunities. The Heritage Trail provides an economic benefit to Dubuque County in the amount of $9.21 per person per day for local users of the trail, an annual total of $1,243,350. Trail visitors from outside Dubuque County spend $13.22 per day. The data also show that $173.99 per annum is spent in the county by local trail users for durable goods similar to those listed in the St. Marks study. Expenditures by visitors from outside Dubuque County should be considered new money coming into this rural area. All of these figures indicate that use of the Heritage Trail generates benefits directly related to its existence (vi).

The perceptions of landowners and real estate professionals tend to be similar to those reported in the St. Marks surveys. In reporting economic data, it is impossible to assess the impacts of the Heritage Trail in the context of the larger economic profile of Dubuque County. It can only stand on its own as an additional contribution to the county's economy. The data do not reveal the capital or operational costs of the trail. However, the economics of the trail are clearly founded on sustainability of greenway amenities, diversity of the landscapes it pervades, and an economic activity that sustains its reason for being.

Scuppernong River Greenway

The Scuppernong River Greenway project of the Conservation Fund evolved from the coalescence of several complex land and conservation projects and the vision and creativity of the people of Tyrrell County, a small community in northeastern North Carolina. The project represents an integrated approach to dealing with social, economic, and environmental issues in a rural region where such issues are greatly intertwined. The greenway is the central feature of a regional heritage tourism strategy designed to provide economic opportunities to the region's residents by offering recreational and educational opportunities to the region's residents and visitors (see Figure 15.3).

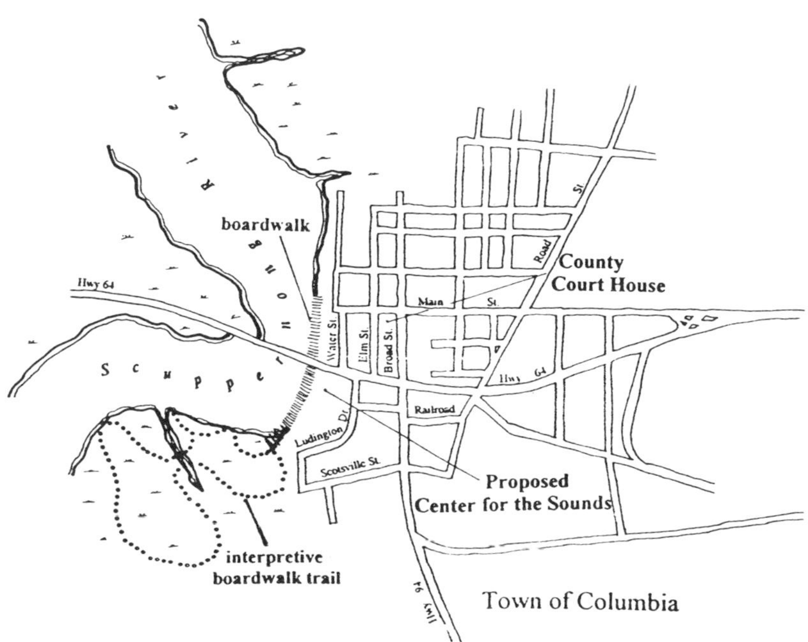

Figure 15.3 Map sketch of Columbia, North Carolina, and the Scuppernong River Greenway.

The Scuppernong River, located on the Albemarle-Pamlico Peninsula in northeastern North Carolina, is the defining landscape element of the Scuppernong River Greenway. The Scuppernong rises in Washington County, most of its 27 miles passing through Tyrrell County, and veers north at the town of Columbia on its way to Albemarle Sound.

Tyrrell County is a sparsely populated rural community that has experienced decline in both the health of its economy and the size of its population. The unsustainable industrial practices based on the region's resources drove Tyrrell County's economic and population growth prior to World War II. Such practices also led to the region's economic demise. By the 1980s the county's main industries—agriculture, forestry, and fishing—had declined sharply, along with the region's population. Giant agribusiness replaced small farms. The depletion of the majority of Atlantic white cedar led to fundamental shifts in the region's forestry industry. Decreased environmental quality, caused by point and nonpoint source pollution of the surrounding waterways, and more efficient commercial fishing practices together culminated in the decline of the area's fishery. In 1989 the county's largest agricultural employer, First Colony Farms, went out of business, further exacerbating the region's economic problems. By 1990 the U. S. Census reported that Tyrrell was North Carolina's poorest county (U.S. Bureau of the Census 1990).

The land owned by First Colony Farms represented a mix of wetlands and land previously converted from wetlands to agriculture. The wetlands included significant acreage of northeastern North Carolina's endangered swamp variety, the pocosin. The term *pocosin* is Algonquin for "swamp on a hill," the hill being an exaggeration for what in reality is only a few feet difference in elevation owing to the gradual buildup of organic materials. Pocosins provide excellent habitat for wildlife such as the black bear and the red wolf, which was recently reintroduced into the region by the U.S. Fish and Wildlife Service. Pocosins are also highly significant in terms of vegetation and include such species as the endangered pitcher plant. Most important for the region's water quality, however, is the role the pocosins in cleansing water as it passes through the system into the area's rivers and eventually into the sounds. The five sounds surrounding the Albemarle-Pamlico Peninsula at one time formed one of the East Coast's most productive fish nurseries.

Benefits of the Scuppernong River Greenway

The Scuppernong River Greenway Project grew out of the vision of the Tyrrell County community and the creation of Pocosin Lakes National Wildlife Refuge, as well as the efforts of the Conservation Fund. The region's residents saw the potential to turn the closing of First Colony Farms into an opportunity for economic gain through ecotourism and environmental education. The Richard King Mellon Foundation, through its American Lands Conservation Program, purchased the acreage owned by First Colony Farms, which included several tracts along the Scuppernong River, and transferred the land to the U.S. Fish and Wildlife Service for the creation of the Pocosin Lakes Wildlife Refuge. The Conservation Fund, in support of the emerging community efforts, effectively assisted in the coordination of the real estate, greenway planning, community development, and economic development components of the project. The result was the identification of ways to implement the community's vision of sustainable development.

One opportunity to generate economic benefits in an area rich with natural resources is provided by ecotourism. To be successful, ecotourism requires two things: access to the region's natural resources and a market. To ensure that such opportunities were available to residents and visitors required a strategy of community involvement and organization, partnership building, and a tremendous amount of learning on the part of everyone involved. To identify access opportunities and potential markets for ecotourism and to develop a community-driven process, the Conservation Fund committed a landscape architect and a community planner to work with a team of graduate students from the Department of City and Regional Planning at the University of North Carolina, Chapel Hill, and members of the Tyrrell County community. The resulting document, *Eco-Tourism in Tyrrell County, Opportunities, Constraints, and Ideas for Action* (Anton et al. 1993), identified a strategic approach to attain resource-based and sustainable economic goals.

One of the main factors shaping the county's ecotourism strategy identified by Anton and associates (1993) was the fact that more than 5,000 cars travel across the Scuppernong River and through the county seat of Columbia on their way through Tyrrell County each day. A substantial number of these cars, particularly in summer, are occupied by tourists and day trippers making their way to the beaches of the Outer Banks. Rarely did a car stop in the town. The community realized the potential for economic gain if only they could entice some of these people to stop, even if for just a short while.

The Scuppernong River provided the opportunity. The community decided to build an interpretive boardwalk along the river, visible from the road and connecting the town's main street to the natural areas only a few hundred yards away. The community also decided to pursue development of an environmental education center next to the boardwalk, along with a North Carolina Department of Transportation rest stop and visitor center.

To ensure community involvement throughout the process and in receiving the benefits of the ecotourism economic development strategy, the Tyrrell County Community Development Corporation (TCCDC) was formed. To involve the county's youths, the TCCDC formed the Tyrrell

County Youth Conservation Corps. The youth corps played a critical role in community cleanup activities and in constructing the interpretive boardwalk along the Scuppernong. The TCCDC plans to involve the youth corps in reforestation and wetland mitigation projects in conjunction with the U.S. Fish and Wildlife Service, the North Carolina Department of Transportation, and the North Carolina Department of Environment, Health and Natural Resources.

Because of its success, the Scuppernong River Greenway project became the first step in a regional effort to address social, economic, and environmental issues concurrently through nature-based tourism and environmental education in northeastern North Carolina. The project sparked creation of the Partnership for the Sounds, a regional effort to pursue sustainable development through heritage tourism and environmental education. The Partnership is building three environmental education centers on the peninsula and is actively working with East Carolina University and the U.S. Fish and Wildlife Service to rehabilitate historical Mattamuskeet Lodge as a regional laboratory for environmental science. The Scuppernong River Greenway serves as the prototype for the Conservation Fund's American Greenways and Sustainable Development projects in the Albemarle-Pamlico region and in similar rural regions across the nation. The Scuppernong River Greenway, among other strategies being employed by rural Tyrrell County and the town of Columbia, promotes ecotourism, environmental education, and economic development. It is an excellent example of social and economic benefits flowing from rural greenways.

LESSONS LEARNED FROM GREENWAYS AND TRAILS

The lessons learned from these three case studies fall into two general areas. One is the need for strong citizen advocacy and involvement throughout a greenway or trail development process. The other is the substantial social, environmental, and economic opportunities and benefits associated with successful rural greenway and trail projects.

In each of the studies, strong and persistent citizen involvement was a key factor in developing a greenway or trail and in its ongoing success. The citizen advocate plays a vital part in articulating the initial vision for a greenway or trail, and expressing the vision is vital to coalescing community interest in the project. The vision also provides a point of departure in the search for greenway partners, such as state programs, federal agencies, private foundations, and (not insignificantly) private landowners. Partnerships that grow from the beginning are important to the success of a project. The vision also has the potential to activate both opponents and proponents. Opposition to greenway proposals is often very well organized and can be fierce. This means that proponents have a special need to start early, gather solid data and information, and thoroughly analyze the proposal to understand the costs, impacts, and opportunities to facilitate public dialogue. Public dialogue is a vital component in planning and developing greenway and trail projects.

Although St. Marks Trail was originated by the state of Florida as its first rail-trail project, very active involvement and support by local citizens was vital to its success. Prior to construction of the trail, local landowners expressed grave concerns about potential negative impacts of the project. In addition to landowner resistance, the state bureaucracy set unreasonable standards for development which would make the trail far too expensive. However, a strong coalition of citizen proponents, among them representatives of local conservation and recreation organizations, succeeded in convincing the state legislature to appropriate funds for the purchase of the right-of-way and the reasonable construction of the trail. Subsequent to the trail's creation, supporters formed the St. Marks Trail Association. This group of volunteers has—through countless hours of dedicated work—sponsored trail events and cleanups; constructed benches, water fountains, and other trail amenities; and successfully advocated the extension of the trail to Florida State University's Doak Campbell Stadium.

The protracted efforts of Heritage Trail, Inc. in Dubuque County, Iowa, also represents citizen dedication and perserverence. Many serious attacks on the development of the project, both legal and physical, were overcome by the initiative of citizens. A number of citizens were directly involved in organizing support for the trail, and many others were supportive, both quietly and vocally.

The Scuppernong River Greenway, waterfront redevelopment along the Scuppernong River in Columbia, the creation of Pocosin Lakes National Wildlife Refuge, the "Center for the Sounds" Environmental Education in Columbia, and the Tyrrell County Community Development Corporation are all vital elements in Tyrrell County's economic development strategy. These strategic elements are direct results of the active support and involvement of local citizens and governmental officials, all sharing the vision for a revitalized community.

Among many researchers and commentators today, the term *ecotourism* has a great deal of circulation. Apropos of the subject of ecotourism, in its reporting of the Tyrrell County economic strategies, the *Charlotte Observer* (10 August 1992) quotes Jim Savory, manager of the Pocosin Lakes National Wildlife Refuge: "The town folks decided that the natural resources were the best thing they had, and they might as well rely on those." Broad-based citizen support was critical in the development of the three greenway and trail projects reported here. The lesson is clear. Greenway and trail projects tend to focus the motives of local citizens interested in promoting and bringing about sustainable development strategies in rural communities.

Greenways and trails afford opportunities for recreation, environmental protection, and resource-based tourism. The simple existence of a greenway or trail corridor also offers the possibility of additional linkages to environmentally sensitive lands and working landscapes. These opportunities are all found in the three case studies. Many strategies with strong community support can and do contribute to the preservation and enjoyment of rural landscapes by local citizens and visitors to the community. The existence of a greenway significantly contributes to the rural economy.

CONCLUSION

We believe that greenways and trails offer unique opportunities for economic development and sustainability. The strategy is built on a strong, active desire among rural citizens to seek out and recognize the value of existing land and historical and environmental resources. The recognition of these resources within the context of a vision of a greenway or trail is a very important discovery.

The role of community action and leadership in each of the greenway case studies is manifest in the economic benefits inured to the host communities. We recognize greenways and trails as important elements in rural strategies for sustainability. The case studies illustrate that they can be important developments among sets of more comprehensive economic and environmental strategies. We conclude that the studies reveal that each of these developments was an important focus in strengthening the economy of the rural community in which it is located and contributes to preserving and enhancing the existing ecosystem. We do not conclude that greenways and trails are stand-alone sustainable development strategies but can be—and on occasion must be—important elements in more comprehensive strategies for economic development. Economic development and environmental preservation are driven by a community's deep concern for rural livelihood and economic welfare.

Future studies of greenways and trails should include analyses of both public and private costs associated with development. In addition, communities should address greenway and trail strategies in more comprehensive planning studies, whether or not required by the state.

REFERENCES

Anton, J., C. Davis, C. Teller, E. Bergman. 1993. *Eco-tourism in Tyrrell County: Opportunities, Constraints, and Ideas for Action.* Report prepared for the Conservation Fund, U.S. Fish and Wildlife Service, and the Town of Columbia and Tyrrell County Community Development Corp. Chapel Hill: Department of City and Regional Planning, University of North Carolina.

Florida Greenways Commission. 1994. *Creating a Statewide Greenways System.* Report to the Governor. Florida Greenway Commission. Tallahassee, Fla.

Harris, L. D. 1985. "Conservation Corridors: A Highway System for Wildlife." An *ENFO* Report. Winter Park, Fla.: Florida Conservation Foundation.

Little, C. E. 1990. *Greenways for America*. Baltimore: Johns Hopkins University Press

Mahon, J. K. 1985. History of the Second Seminole War, 1835–1842, The University Press of Florida. Gainesville, Fla.

Moore, R. L., A. R. Graefe, R. J. Gitelson, and E. Porter. 1992. *The Impacts of Rails-Trails: A Study of the Users and Property Owners from Three Trails*. Washington, D.C.: Rivers, Trails, and Conservation Assistance Program, National Park Service.

National Park Service, Rivers and Trails Conservation Assistance Program. 1990. *Economic Impacts of Protecting Rivers, Trails, and Greenway Corridors: A Resource Book*. Washington, D.C.

Rails-to-Trails Conservancy. 1993. *500 Great Rails-Trails: A Directory of Multi-Use Paths Created from Abandoned Railroads*. Rails-to-Trails Annual Report. 1994. Rails-to-Trails Conservancy. Washington, D.C.

U. S. Bureau of the Census. 1990. *Statistical Abstract of the United States*. Washington, D.C.: Government Printing Office.

Sixteen

SURE (Sustainable Urban/Rural Enterprise): A Partnership for Economy and Environment

James A. Segedy and Thomas S. Lyons

INTRODUCTION

> *The city integrated successfully with its immediate hinterland is the very engine of prosperity, the basic unit of culturally diverse society (Jacobs 1984).*

More and more, small town and rural America is discovering that its traditional values of self-reliance and self-determination are critical to its cultural and economic survival. Parallel to this strong value system is a long history of sensitivity to the land and the natural environment and a culture of entrepreneurship and creative problem solving. These factors in combination form a strong and logical foundation for the development of a more sustainable approach to economic and community development.

Sargent and associates (1991, 5) assert this heritage in the basic assumptions of their Rural Environmental Planning approach:

1. Rural people place a high value on self-reliance and self-determination;
2. Rural people value cooperation as a guide to problem solving;
3. Long-term sustainability of a rural environment is achieved when citizens guide economic development according to the physical carrying capacities of the ecosystem; and
4. Increasing the self-reliance of citizens in rural communities can be the basis for sustainability.

This human-and resource-based approach to community and economic development is less concerned with the driving force of inevitable and continuous growth that is the hallmark of the economic development paradigms of our metropolitan regions. As the basic philosophies and value systems are different, so are the tools and methods that are, or should be, employed in achieving sustainable rural development.

This chapter describes the SURE (Sustainable Urban/Rural Enterprise) partnership initiated in 1989 between the city of Richmond, Indiana, and Wayne County—its rural hinterland. We perceive this partnership as offering the potential and the promise for a new ruralism. However, in assessing the experience to date, we argue that once past its initial visioning stage, the process requires continued high levels of commitment from citizens and elected and appointed leadership; that grass-roots support provides the strength to push forward, but it still needs a strong champion to steer the efforts to fruition.

THE NEED FOR A NEW RURALISM

At a time when the need for sustainable development patterns and practices is becoming ever more critical, there are still many communities adopting traditional mechanisms of community development and land use planning, following the philosophy that unbridled growth is the only means to continued economic well-being. In direct challenge to the land ethic that is the foundation of rural and agricultural culture, this growth is generally associated with the consumption of considerable energy and resource capital. History has brought us to the realization that this unrestrained growth often occurs at the expense of environmental quality and quality of life.

Furthermore, as growth continues, the community approaches and passes a series of thresholds for the provision of goods, services, and amenities adequate to fuel the needs of the expanding economy. These points of diminishing return can be directly related to physical, natural, and human (social/psychological) carrying capacities. As thresholds are surpassed, there are often significant declines in environmental and life quality, until and unless there is a considerable expenditure of energy, resource, and cultural capital. The sustainability of a community and its enterprise, therefore, can be directly associated with its relation to this series of critical thresholds. It can be further assumed that for the rural and nonmetropolitan economy to be sustainable, the strong value systems that characterize small town America will evolve from the experience gained by the generations of rural entrepreneurs who made decisions mindful of the long-term consequences of environment and economy.

Unfortunately, observation reveals that many of our communities have grown beyond the critical thresholds associated with the economies of scale and have exceeded the carrying capacity of their ecological asset base. As concern mounts as to the ability of our resource base and environment to support continuous expansion, new strategies must include demand-side considerations in the supply/demand equation. It is also important that community visions extend beyond local crossroads to address the growing reality of international agribusiness. There is growing evidence that rural sustainability lies somewhere between the traditional, self-contained farmstead and the global marketplace, the new regionalism—or, as it may become to be known, the "new ruralism."

Integral with the new ruralism is the growing importance of economic and community development that is sustainable. "Sustainable economic development [can be defined] as development which ensures that the utilization of resources and the environment today does not damage prospects for their use by future generations. Sustainable economic development does not require the preservation of the current stock of natural resources or any particular mix of human, physical and natural assets. Nor does it place artificial limits on economic growth, provided that such growth is both economically and environmentally sustainable" (Canadian Council of Resource and Environment Ministers 1989, 4).

Moreover, communities in general, and rural communities in particular, are losing their unique identities to the "expansion of the mass society" (Hobbs 1990). Efficiency of economic systems has led to a sameness in rural town character as well as a loss of services, particularly retail diversity. According to Hobbs (1991), rural communities

- Are lagging behind urban areas in the rate of economic growth;
- No longer depend on farming, forestry, mining, etc. as the basis for their economies;
- Are subject to the same economic forces as urban areas;
- Depend on access to knowledge, capital, transportation, communication, and living environments for their future well-being; and
- Are in need of a rural revitalization policy that integrates farm and nonfarm interests.

With this view of rural economies, the need for reestablishing a strong community identity and economic independence becomes critical. The strength of rural communities is derived from their unique character and asset base. Therefore, if a community is to become sustainable, it must build on its unique character and resources and develop its economy around their preservation and enhancement. This mandates a firm linkage between the uniqueness of community character, the environment, and the economy.

Despite the need for rural communities to maintain their individual character and economic self-reliance, the fact remains that many are isolated, to their detriment. Because of their remoteness from regional urban centers, their access to information, technological assistance, business support services, and sufficient economies of scale remains impaired. Nor do they enjoy the diverse interactions that typically spawn innovation. Often their economies are dominated by agriculture, or some other single industry, making economic diversification difficult (Lichtenstein and Lyons 1995).

Thus, many rural communities find themselves in a perplexing dilemma. On one hand, they must achieve economic sustainability if they are to survive. On the other, they are part of a larger regional economy upon which they are dependent. Is there a mechanism that will allow these communities to be, at once, independent economic operators and important cogs in the regional economy? The answer to this question may lie in the concept of networks and/or multicommunity collaborations. As Christaller (1933) taught us with his *Central Place Theory*, a regional economy is made up of a hierarchy of places, each with its own market area nested within the market area of a still larger community. When this theory is carried to its traditional conclusion, a smaller community can dominate its own small market area but will always be dependent on the larger market center for those goods and services the farmer's market alone cannot support. But what if this hierarchy is viewed in reverse? If the smaller market center has a good or service that is unique to that community and cannot be produced with the same quality and at the same price in the larger market center (e.g., Amish-made furniture produced in rural Pennsylvania), then the smaller community becomes the dominant market center in that region for the good or service. Thus, the smaller market center possesses something of unique value within the larger regional economy. It has something on which to build a sustainable local economy within its economic region.

"The building blocks for sustainable economic development in any community are its natural, human, and cultural resources. Careful management of these resources helps to increase local production of goods and services, replenish renewable resources, strengthen unique cultures, and broaden economic opportunity" (Sargent et al. 1991, 183). Community empowerment, integration, collaboration, and networking are being added to this foundation of the new rural economy.

When the various market centers of a region have something of value to offer one another, an opportunity for productive networking exists. Each can work to make its markets more accessible to the others. The end result is a healthier regional economy in which each component market center is taking advantage of its unique strength. For small rural communities, this ends the disadvantages caused by their isolation while allowing them to thrive in a manner that can be both economically and environmentally sustainable. A promising model for formalizing the partnerships generally sketched here is the Sustainable Urban/Rural Enterprise partnership described and analyzed in the rest of this chapter.

This concept draws its inspiration from the ideas of niche marketing, local value-added processing, and multicommunity collaboration efforts that are becoming more prevalent fixtures in successful examples of rural economic development. It is in this more sustainable approach to economic development that small-town America has also been able to survive the "big box retailers," such as Wal-Mart, K-Mart and similar retailers that typically occupy buildings larger than 80,000 square feet.

THE RICHMOND/WAYNE COUNTY, INDIANA, MODEL

A local civic venture initiated in Richmond and surrounding rural Wayne County, Indiana, exemplifies the Sustainable Urban/Rural Enterprise (SURE) process. Under the direction of the SURE, Inc. board of directors, the mission of the venture was to promote "ecologically based economic choices" (Euston 1990, 1). Beginning with a five-day workshop in October 1989 and continuing with its formal inauguration on Earth Day 1990, SURE was the nation's "first broad-based local civic organization chartered to inform local economic choices in terms of their ecological consequences" (Euston 1990, 1). It had the support of local businesses, labor unions, environmentalists, and the economically distressed neighborhoods constituting the Richmond Enterprise Zone, as well as key politicians, including the mayor, the county commissioners, Indiana's Senators Lugar and Coats, and then-Representative Phil Sharpe, a political scientist and chair of the Congressional Subcommittee on Energy and Power (Euston 1990). The SURE concept drew its strength from being a locally based economic development program, building on the unique asset base of the area and focusing on competitiveness in the marketplace integrated with environmental responsibility.

The SURE Partnership

Although community-based projects and workshops are not new to the academic arena, the Richmond endeavor represented a dramatic merging of talents, resources, and community. The foundation of the process was "a cooperative agreement between the City of Richmond, the Wayne County Commissioners, the Richmond Enterprise Zone Association, the United States Department of Housing and Urban Development (HUD), and the College of Architecture and Planning and Center for Energy Research, Education and Service at Ball State University. These signatories agreed to work together for a year to examine the principles and applicability of an integrative policy/development concept— that of Sustainable Urban/Rural Enterprise (SURE)." With the technical assistance of HUD and Ball State University, the process elicited the involvement and commitment of the communities to be on "the front lines of alternative technology application, economic development and environmental stewardship" (Koester 1990a, 2).

Sustainability: A Community Agenda

The term *sustainability* has become the buzzword used to describe everything from theoretical constructs to policy directives to planning initiatives. Its application to the environment/economy interface provides a new array of challenges. Responding to the desire for a more sustainable development pattern the communities of Richmond and Wayne County, Indiana, began to rethink the mechanisms and ideals that directed their planning and community revitalization efforts.

Richmond and Wayne County engage in a wide range of economic activities and enjoy a unique physical and cultural character. This community of approximately 40,000 persons in east central Indiana has built a strong manufacturing and technology base, yet has never lost its heritage as an agrarian community with strong ties to its Quaker traditions. In addition, Richmond maintains a distinct personality, despite its close proximity to the Indianapolis; Dayton, Ohio; and Cincinnati, Ohio, metropolitan areas. On this foundation, and with a determination to make a difference, Richmond/Wayne County has successfully renewed local public involvement and civic leadership—a basis for sustainable community development. A newly adopted Comprehensive Master Plan for the City of Richmond, and a parallel effort by Wayne County developed through active citizen participation processes, are viewed by the community as a "constitutional framework for planning activity" that allows a strategic response to a "changing inventory of resources and personnel, to yield a constantly evolving prioritization of action planning for implementing community revitalization" (Koester 1990a, 8). The five critical themes of the plan are as follows:

1. A regionalization of vision recognizing Richmond's role as the economic, political, and figurative heart of a county of small towns and productive agriculture, and the link to the metropolitan areas of Indianapolis, Dayton, and Cincinnati;
2. A sincere respect and concern for the unique physical environment offered by the natural setting, especially the White Water River Gorge, and the built infrastructure of the cities;
3. The importance of quality-of-life issues, including economic and cultural diversity, enhancement of the worker skill-base, as well as a concern for the cityscape (expressed in a beautification program that will place utility systems underground and enhance the communities' many parks and recreation facilities);
4. Utilization of state-of-the-art technology in areas of communication, information management, and distribution services (specifically in waste management systems, computer/data resources, business incubation, cultural development, and transit modes); and

5. Diversification of all community functions, including industry, arts, housing, central business district, and transportation systems.

Within this framework are nine policy areas expressed through "visions and rationale" that serve as operational goals. The 42 goals embedded in these policy areas have from one to three specific strategy objectives that guide the implementation of the plan. It is no accident that the foundation of Richmond's and Wayne County's plans so closely parallel the basic values of rural environmental planning as described by Sargent and associates (1991).

The central framework of the plan is designed to address the two challenges that became the basis for the SURE Workshops of Richmond/Wayne County: "Why should American communities be deliberately mobilizing themselves in the direction of the emerging international marketplace?" and (if so), "How might they do it?" (Koester 1990a, 5). As the dynamic linkage between economy and environment became the guide for strategic decision making, several economic development components of the framework were established:

- Ecological and economic benefit;
- Encouragement of local entrepreneurial activity;
- A future view of "planetary citizenship";
- Economic development based on import substitution, energy competitiveness, value-added resource and product exporting, resource recycling, and rural economic structuring (the regional hub);
- A strategic focus; and
- Action orientation.

From this framework the concept of Sustainable Urban/Rural Enterprise evolved. The specific objectives of this "marketplace of opportunities" include the following (adapted from Koester 1990a):

- To act on the national and global ecological crisis as a moral and practical local opportunity and obligation;
- To build on the momentum generated by citizen participation in the comprehensive planning process in Richmond/Wayne County;
- To search collectively for sustainable enterprises and to seek ideas that can be put into place within one year;
- To network with SURE pioneers, including sustainable enterprise proponents (practitioners, theorists, and policymakers) and potential sustainable enterprise *ex*ponents (Richmond/Wayne County consumers, producers, business entrepreneurs);

- To establish Richmond/Wayne County's program as a SURE prototype—an appropriate extension of its pioneering industrial and cultural history;
- To create the Sustainable Enterprise Exchange of Richmond (SEER), which will involve creating an information clearinghouse to facilitate the prioritization of options and the pursuit of all aspects of local economic development and growth that serve to reinforce the SURE objectives.

The SURE Workshop

To fortify the foundation established by the community, a five-day community-based workshop was held in October 1989, which imported 30 or more experts from around the country who shared the vision of Sustainable Urban/Rural Enterprise, and attracted several hundred members of the Richmond/Wayne County community. The workshop was structured as a hands-on, problem-solving exercise aimed at operationalizing the goals of the Richmond plan within the context of the global community. The workshop concluded with a visualization charrette[1] in which the community presented its image of the sustainable community, with the assistance of the charrette team composed of practicing architects, landscape architects, urban planners, and faculty and students from the Ball State University College of Architecture and Planning.

At the beginning of the workshop participants identified a series of nine issues that they thought were of critical importance to the future of their communities. These issues became the focus of breakout sessions for the workshop. During the workshop sessions local citizens, stakeholders, business people, and civic leaders from Richmond and Wayne County were teamed with visiting experts to develop the SURE vision in the areas that most interested them. The following paragraphs give an overview of the major topics addressed during the workshop and some of the results of the effort (adapted from Koester 1990a, b).

Sustainable Enterprise Economic Models

Building on the dynamics of the citizen participation/workshop process, one team identified the basic community resources and the opportunities that were needed for a strong, grass-roots-based economic development effort. The people of Richmond/Wayne County believe that the sustainable enterprise economic models should be analagous to plugging holes in a leaky bucket, the issues/initiatives focused on preventing the export of dollars from Richmond/Wayne County; as well as on local value-added processes, supporting existing businesses, encouraging new enterprise, identifying potential niche markets, recruiting compatible new businesses, and implementing those strategies determined to be most timely, cost-effective, and having the broadest promise for integration and networking within the local economic structure.

Wastewater Treatment and Biomass

A second workshop team focused on the application of the emerging technology of biomass treatment systems and constructed wetlands to local economic development. To the extent that such technology can be applied as a local on-site treatment system, it can be utilized in the recruitment and siting of a variety of industrial processes. This technology can allow for the decentralization of wastewater treatment, reducing the burden on centralized waste-processing systems, but requires monitoring to ensure the highest standard of water quality.

Land Use

The land use team developed a series of guidelines for the future land use policies that would influence sustainable community development patterns:

- Land use activities must be perpetuated continuously for future generations.
- Land use systems must be adaptable and resilient.
- Land use systems must not deplete the existing natural or cultural resource base.
- Planning at the microscale must include intensification of permitted uses and reduction of nuisance potential by establishing performance criteria, increased flexible zoning strategies, and creating zoning incentives to encourage redevelopment of deteriorating parcels.
- Land use decision making must involve citizen participation to develop the broadest possible base of awareness. It was determined that this could best be achieved by creating a procedure for informing citizens of land use issues, practices, and potentials; opening a "storefront" planning office; conducting civic charrettes; establishing community-wide reviews through the use of mass media; and identifying the first issues by which to test the participation/information-sharing procedure.

Energy Competitiveness

This team suggested that establishing the basis for energy competitiveness for the Richmond/Wayne County area required a wide range of technological and management resources. The locally owned and operated electric utility company had long been a proponent of conservation/demand-side management of energy resources. It was determined that this philosophy and practice must extend to the application of technology and technological systems—all components of civic function and enterprise.

Enterprise Zone Incubators

Still another workshop team evaluated the physical fabric of the Richmond Urban Enterprise Zone located within the heart of Richmond. Recommendations included the creation of nodes of "amenity discovery" within Richmond and Wayne County. These amenities included the historical manufacturing building shells, the rail depot, the old warehousing district, the historical museum, the newspaper offices, the theater, the old school, the original Richmond courthouse, and the Richmond historical area. The potential for satellite incubators in some of the smaller communities in Wayne County were also suggested, but they were not explored further during the workshop. The proposal generated by this team involved the networking of the amenities as an assets infrastructure in support of the incubation and/or accommodation of new enterprises within the city fabric.

Business incubation involves establishing a support structure for small start-up businesses, which seeks to increase their chances for survival to maturity. Such a structure is particularly useful in rural communities where entrepreneurs typically face substantial obstacles, including limited local markets, lack of industrial diversification, limited capital availability, and a general lack of support resources, to name a few (Lyons and Lichtenstein 1994).

Several industries with the potential for incubation and niche market development, locally, were identified as aquaculture (e.g., fish farming and hydroponics); composting and fertilizer substitutes; bioshelter greenhouses; organically grown brands, certification, and a marketing board; intermediate processing of recoverable wastes; clean fuel production; package-reduction design and production; commodity sales and residential consumables; environmental education through magnet school centers; regenerative input/output database creation and maintenance; environmentally benign pest control products; compressed natural gas vehicular retrofitting and service facility; product repair; cogeneration and alternative energy supply manufacturing and sales; water system purification technologies; and affordable housing production.

Agriculture

The agriculture team examined historical and present patterns of agricultural development in the Wayne County area and proposed an alternative future direction, which seeks to recapture the historical roots of the urban/rural enterprise. This would involve the use of computer/GIS (geographic information system) technology to inventory and interpret land resources for appropriate agricultural potential. The use of GIS would help analyze the potential for nontraditional agricultural systems. For example, preliminary analysis demonstrated the potential for growing peanuts and bamboo in certain parts of the county. Both of these commodities fill an immediate need in the Indianapolis and Cincinnati markets that must now be met by imports. The preliminary GIS analysis also identified certain soil and microclimate conditions that offered a potential for higher-yield farming practices. The action plan focused on the following:

- Identifying market opportunities by expanding the local food market network, expanding regional direct marketing, and developing and promoting a local farmers' market;
- Establishing a support network;
- Establishing local demonstration farms for the purpose of sharing technology and knowledge;
- Developing a program for educational outreach; and
- Developing alternative crop potentials.

Communications

Another team explored the variety of communication technologies available and their application to information exchange. The communication of ideas and technical information is critical to the success of the SURE process. Although participants in this session recognized the significance of emerging technologies and the need to integrate them into the SURE process, the primary focus was on ensuring continued open dialogue on the issues and access to information.

Quality of Life

One workshop team, focusing on quality of life, reaffirmed the criticalness of maintaining local cultural heritage and celebrating its diversity. Great emphasis was placed on establishing linkages between the unique natural environment and the built environment of the Richmond/Wayne County area. Specific programs and projects were not identified in this series of workshops, but the community's subsequent participation in the Indiana Total Quality of Life Initiative (Segedy 1995) provided a follow-up opportunity to identify, prioritize, and address these specific issues. In addition to development of the natural/cultural interface, it was emphasized that the individual citizen must never be disenfranchised by the larger community process.

PROGRESS TOWARD SUSTAINABLE URBAN/RURAL ENTERPRISE IN RICHMOND/WAYNE COUNTY

In the period since the 1989 workshop, progress toward the goals of Sustainable Urban/Rural Enterprise, like that of many visioning exercises of this type, has experienced mixed results. Initially, community leaders from all sectors often pointed to successful projects and programs and noted the impact the SURE program had had on overall community awareness (Koester 1990b). Although a reorganization of the school curriculum has precluded implementation of SURE elements into local schools, church leaders have been active in promoting the complexities of sustainability within their ministerial communications. Members of area labor organizations have been aggressively seeking ways of introducing the SURE principles into the workplace, and civic leaders and their professional staffs have retained outside experts to contribute to discussions on a wide range of SURE issues, including areawide GIS, a comprehensive master plan, and specific housing code and zoning updates.

Evidence of the application of Sustainable Urban/Rural Enterprise principles as positive tools in community revitalization can be seen in the relocation and expansion of Richmond's Farmers' Market as the first business to be situated in the Richmond Urban Enterprise Zone. The market sold out before noon on its first day of operation in its new location and has remained relatively active over the years. The community is working with Ball State University on a city/county GIS that will become the information base that will enable it to identify alternative crop potentials and develop a combined farmland profitability profile. The Purdue Cooperative Extension Service helped initiate an organic farming certification service, and the local utility company expanded its demand-side management program with a residential energy efficiency campaign that was planned with the residents of Richmond's low-income Starr Historic District. The Richmond/Wayne County Chamber of Commerce has established a committee on energy and resources, and the Richmond Main Street program has begun an initiative merging architectural, cultural, and energy conservation. The new SURE, Inc. and its clearinghouse services are viewed as pilots for a network of many more such ventures in communities across the United States. In addition, a preliminary study of the feasibility of business incubation services in support of local entrepreneurs is under way.

The Richmond/Wayne County Chamber of Commerce (renamed to reflect the linkage between the community and its rural support systems) took the initiative to hold an economic development workshop lasting two and a-half days. This charrette, an expanded version of that held during the October 1989 workshop, clearly established an agenda for the economic revitalization of the Richmond/Wayne County area. The sessions, attended by more than 150 area residents and business and civic leaders, focused on developing a more localized economy and encouraging small and "cottage industries" as the foundation for SURE development. To facilitate implementation of the SURE principles, SURE, Inc. was established to maintain community momentum and coordinate continuing projects. This ongoing effort has met with mixed success.

Because SURE is more of a conceptual foundation than a specific plan, the correlation between the SURE project and specific programs, policies, and projects is difficult, but the influence is unmistakable. Specific undertakings, such as the Richmond Farmers' Market and the development of constructed-wetland wastewater treatment facilities, can be directly linked to the recommendations of the workshop. The process and results of the Richmond and Wayne County Comprehensive Plans also demonstrate the influence of SURE principles. Other projects, policies, and programs, however, have less tangible links.

The changing political landscape in Richmond and Wayne County, as well as changes in the economic base, has probably had the most dramatic and unexpected impacts. With shifts in civic leadership there, has been a deemphasis on the goals and projects associated with the SURE program. The impacts of changes in both elected and appointed leadership suggest that although SURE is a bottom-up concept and grass-roots energy can carry the project forward, it still requires strong leadership to reach its ultimate goals.

The principles of Sustainable Urban/Rural Enterprise are firmly rooted in local initiative and grass-roots concern for the viability of natural and human systems. Quality of life for the people of Richmond/Wayne County, Indiana, means more than a rise in the GNP. Environmental stewardship, visual and cultural amenity development, and the collaboration of the urban and rural communities form the framework for the SURE process. The early successes of the SURE program in Richmond/Wayne County and parallel applications of the concept to rural business incubation; the development of an alternative grain elevator and marketing program in neighboring Farmland, Indiana; alternative agriculture demonstration projects; and countywide biomass sewage treatment systems in LaGrange County, Indiana, suggest that there is a basis for integrating demand-side resource management, grass-roots initiatives, and environmental stewardship with economic revitalization. The foundation of ideas and processes derived from the Richmond/Wayne County SURE program can, and should, serve as a model for the reintegration of the rural countryside with its urban core.

The Richmond/Wayne County example has shown that to be successful in making progress toward the goals of Sustainable Urban/Rural Enterprise, the place of community and the human factor cannot be underestimated. Simple economic and environmental forces are not enough to sustain the rural economy. There must also be a strong foundation of community involvement, support, and leadership. The economic vitality of a community cannot be removed from the civic vitality of the community. Sargent and associates (1991) suggest that a successful economic development strategy must emphasize human development, expand local control of resources, increase internal investment capacity, and change economic and social structures to increase opportunity and reduce dependency. Hawken (1993, xv) goes so far as to evoke the human spirit in suggesting that a sustainable economy must also "be fun and engaging, and strive for an aesthetic outcome." Therefore, nurtured by community involvement, civic leadership, intergovernmental cooperation, and the entrepreneurial spirit, this grass-roots localization of the economy can serve as the foundation for a sustainable future of our smaller communities and rural areas. Indeed, it is the integration of urban and rural economies within the region that makes rural economic sustainability possible.

NOTES

1. A *charrette* is a Beaux Arts-derived term for a short, intensive design or planning activity.

REFERENCES

Canadian Council of Resource and Environment Ministers. 1989. "Report of the National Task Force on Environment and Economy." CCME (Canadian Council of Ministers of the Environment), WInnipeg, Canada.

Christaller, W. 1933. *Die Zentralen Orte in Suddeutschland.* Jena: Gustav Fisher.

Euston, A. 1989. Memo to Sustainable City Project Actors. U.S. Dept. of Housing and Urban Development. (Unpublished).

_______ . 1990. "Sustainable Urban/Rural Enterprise and the Next America or When Is It Time To Shift." Presentation to the Chautauqua Institute, Chautauqua, New York, August 16.

Hawkin, P. 1993. *The Ecology of Commerce.* New York: HarperCollins.

Hobbs, D. 1990. "Keynote address." Rural Town Comprehensive Planning Conference," Mendon, Michigan, November 12.

Jacobs, J. 1984. *Cities and the Wealth of Nations.* New York: Random House.

Koester, R. 1990a. *Implementing Sustainable Urban/Rural Enterprise in the American Heartland.* Muncie, Ind.: Center for Energy Research, Education, and Service, Ball State University. (Unpublished)

_______ . 1990b. "General Observations: Status of SURE Activities and Community Momentum."summary to the Sustainability Begins at Home Conference of the Environmental and Energy Study Institute. May. Washington, D.C.

Lichtenstein, G. A., and T. S. Lyons. 1995. *Incubating New Enterprises: A Guide to Successful Practice.* Washington, D.C.: The Aspen Institute.

Lyons, T. S., and G. Lichtenstein. 1994. "New Strategies in Rural and Small Town Business Incubation: Examples of Successful Practice." *Small Town and Rural Planning* 14(1).

Sargent, F. O., P. Lusk, J. A. Rivera, and M. Varela. 1991. *Rural Environmental Planning for Sustainable Communities.* Washington, D.C.: Island Press.

Segedy, J. 1995. *The Indiana Total Quality of Life Initiative Manual and Report of Findings.* Muncie, Ind.: Department of Urban Planning, Ball State University, and Indianapolis: The Indiana Department of Commerce.

Seventeen

Solid Waste Management in Rural Areas: Lessons from Florida

Raymond A. Shapek

INTRODUCTION

The cost of solid waste disposal in the United States is increasing at an alarming rate as landfills become filled or are closed because they fail to meet the U.S. Environmental Protection Agency Resource Conservation and Recovery Act (RCRA) subtitle "D" landfill minimum standards and/or are identified as possible sources of groundwater contamination. Nearly every state has passed laws requiring local governments to initiate source reduction and recycling programs to reduce the waste stream, to lengthen landfill life, and to respond to citizens' concerns about preserving the environment. For example, Florida passed the Solid Waste Management Act of 1988, requiring all cities and counties to reduce their waste streams by 30% by 1994.

Despite the recent popularity of community-based development strategies associated with solid waste management and recycling, such as those touted by the President's Council on Sustainable Development (PCSD 1996), rural areas persistently pose unique problems and challenges to state legislative efforts to safely manage and dispose of solid waste.

THE SOLID WASTE PROBLEM

During the 1970s the costs of pollution and environmental degradation began to be factored into the prices of goods and services. The costs of waste disposal to the public rose and were related to concerns about preserving the environment. As disposal costs increased and natural resources become more costly and/or scarce, values and habits changed. The process of change involved education at the consumer level, as well as legislation providing jurisdictions with legal authority, supported by funding, to address the complex array of environmental interrelationships, especially those between urban and rural areas. Environmental problems are not simple and require increasingly costly management alternatives and innovative solutions. One area of rapidly increasing cost that is impacting rural development is solid waste disposal.

The cost of garbage disposal has increased more in the last decade than the price of oil in the 1970s. In 1960, Americans generated 2.65 pounds of solid waste per capita; by 1986 the rate was 3.58 pounds, approximately 30% more waste per capita than generated by Europeans with similar living standards (U.S. EPA 1994; Biles 1987). Florida leads the nation in the per capita generation of municipal solid waste (MSW). The 1994 statewide average generation of solid waste in Florida was 9.5 pounds per person per day, approximately 1.7 tons per person per year, or 23.6 million tons per year, all from a population of approximately 13.5 million residents (FDER 1995). Of this amount, 8.6 million tons (36.5%) were recycled (of which 1.6 million tons [6.8%] were composted); 5.3 million tons (23%), were combusted in waste-to-energy (WTE) plants; and 9.6 million tons (41%) were disposed of in permitted landfills (see Figure 17.1). These figures may be compared with national averages: 45 million tons (22%) were recycled and composted, 33 million tons (16%) combusted, and 129 million tons (62%) landfilled (U.S. EPA 1994). Since 1988, when the Solid Waste Management Act (SWMA)(Florida Statute 403.706) was passed, the amount of MSW collected has increased by 23%. Nearly 52% of all MSW is generated by the commercial sector; single-family units generated 36%; and multifamily units, including apartments, condominiums, and other units, accounted for 14%. Of this total, 23% is combusted, 36.5% is recycled, and 41% is landfilled (see Figure 17.1). Construction and demolition (C and D) debris is deposited in separate landfills (designated as Class III) that do not have to meet the rigid standards of Subtitle D, Class I landfills. The composition of this trash as compared with average U.S. generation is depicted in Figure 17.2.

Recycling rates in Florida increased at a rate of nearly 20% per year since initiation of this legislation. Between 1990 and 1994, Florida experienced an increase of 147% in the recycling of MSW (FDER 1995). Florida Statute 403.706 establishes two goals for counties with populations greater than 50,000 residents: they must meet the 30% waste stream reduction and must achieve a 50% minimum recycling rate for newspapers, glass bottles, aluminum cans, plastic bottles, and steel cans. No county has achieved the 50% rate for all five "minimum" materials, but 34% of Florida's 67 counties have met the 50% goal for at least one of the materials. Florida counties with a population of less than 50,000 are exempt from these standards as long as they provide residents with an opportunity to recycle. Nearly every county offers this opportunity, in large part because of citizen demands, even though it may be extremely costly.

Managed Municipal Solid Waste

23.6 Million Tons

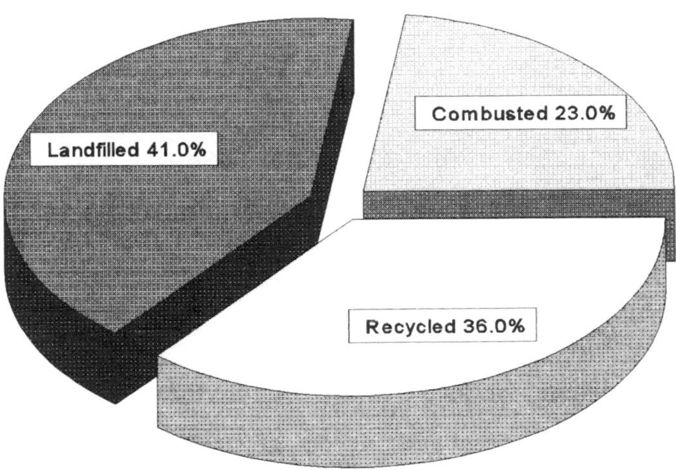

Figure 17.1 Florida municipal solid waste managed and recycled. *Source:* Bureau of Solid and Hazardous Waste. 1995. *Solid Waste Management in Florida*. Tallahassee: FDER, 21.

In many ways, Florida exemplifies what other states can do to better manage solid waste. Florida's larger per capita waste generation may be attributed to more accurate reporting of waste generation, but is also the result of waste generated and left by the more than 41 million annual visitors to the state. The annual solid waste stream is made up of roughly the same percentage of materials, however, regional, seasonal, and other variables, such as the concentration of certain industries, varies the composition of what is discarded. Although much of this waste is recoverable and recyclable, a significant portion consists of materials that pose special management problems, especially in view of Florida's fragile ecosystem and underlying aquifer. Florida collected and recycled 70.5 thousand tons of waste tires, 1,567,346 gallons of used oil from a network of 1,350 public used-oil collection centers, 173.1 thousand tons of white goods (household appliances such as refrigerators, washers, etc.), 1,594 thousand tons of yard trash, and 6,160 thousand tons of C and D debris (FDER 1995). Hazardous waste must be transported to Alabama or South Carolina by licensed carriers, because Florida has no hazardous waste landfill. Biohazardous waste is

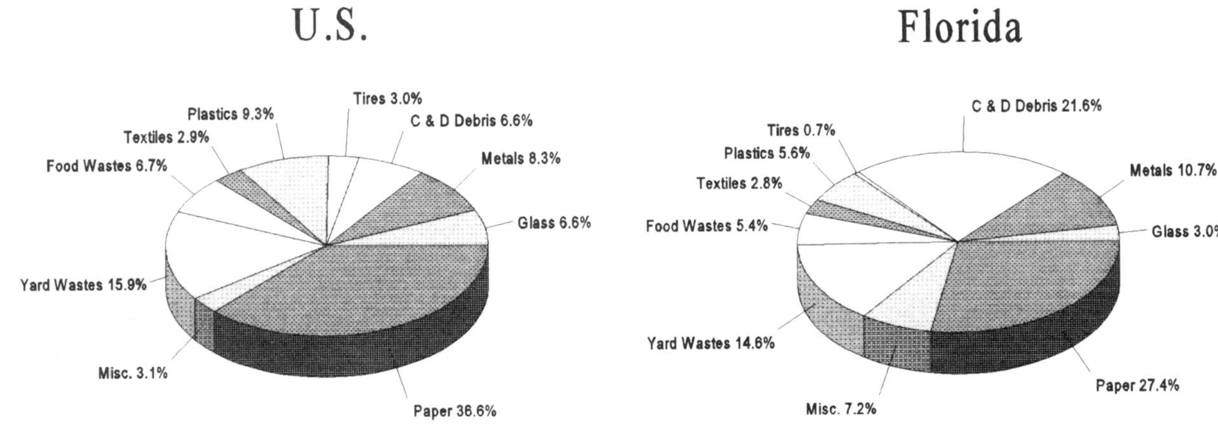

Figure 17.2 Solid waste generations: Comparison of United States and Florida. *Source:* Bureau of Solid and Hazardous Waste. 1995. *Solid Waste Management in Florida*. Tallahassee: FDER, 21.

managed through incineration, steam treatment, chemical disinfection, and microwave shredding at permitted facilities prior to disposal.

Trash Disposal

Landfilling is the most common form of solid waste disposal and accounts for 72% of all waste in the United States. However, since 1991 the number of landfills has decreased by 8%, from 5,812 to 5,386 in 1992, a decrease of 48% since the end of 1988. Approximately 22% of existing landfills were projected to close by the end of 1993 (Steuteville et al. 1993). EPA's Resource and Conservation Recovery Act (RCRA) Subtitle D requirements went into effect on October 9, 1993, imposing stringent mandates for management of landfills. Consequently, as urban landfills close, the distance to those still open increases. To offset the increased costs of transportation to these remote locations, transfer stations are built to accumulate larger amounts of solid waste for consolidation and loading into larger trucks. Once transfer stations or long-distance hauling is involved, recycling and composting programs usually become attractive or costly mass-burn facilities are considered. Tipping fees increase with the reduction of available landfills and the increased costs of landfill upgrades, and to pay for the bonds needed to construct transfer stations or to build new landfills or incinerators.

The national *incineration* rate—both for mass-burn and waste-to-energy or refuse-derived fuel—in the nation's 171 incinerators was 11% in 1993 (U.S. EPA 1994). Florida incinerates more waste than any other state (14,496 tons per day); New York is second, with 14,000 tons per day; and Massachusetts, Pennsylvania, and Virginia rank third, fourth, and fifth, respectively. Florida's 13 waste-to-energy facilities are located predominantly near large metropolitan complexes, in part because of population concentrations and in larger part because there is little or no space in the region for further landfill expansion. Seven of these 13 facilities are located near the five counties with the greatest population. Only two relatively rural counties have waste-to-energy facilities; this situation is due to political decisions as well as to the extent of wetlands in the counties that may prohibit development of new landfills. Although combustion appears to be an easy solution to solid waste disposal, air pollution control and other health issues make this form of disposal extremely expensive. Exploration of new sites for mass-burn facilities in more remote rural areas raises questions about the discriminatory effects of such siting (Vittes and Pollock 1994).

Composting yard waste (primarily leaves, grass, and brush) is becoming increasingly popular as a means of disposal. There are nearly 3,000 composting facilities in 23 states, and the number is increasing each year (Steuteville and Goldstein 1993; Steuteville et al. 1993). Most states do not have home composting programs, but several provide education programs, materials, or subsidies to home owners to purchase composting bins, or give bins directly to home owners. Twenty-two of Florida's 67 counties have extensive composting programs; 13 of these are near the 20 most populated counties, and only 2 are located in more rural counties. Composting is a relatively low-cost means of waste disposal when combined with marketing or soil amendment programs for disposal of the final product. Compost is also used as cover material at landfills, thus saving valuable soil. In Florida, yard waste has been banned from landfills without composting programs. The problems of urban composting do not affect rural areas where composting is common or where open burning of all trash is practiced (even when prohibited by state law). State regulators are now reviewing air pollution prevention standards to prohibit open burning of trash and requiring mandatory trash collection, but recycling is generally voluntary in rural areas.[1]

Recycling

As markets for processed waste develop and landfill or incineration costs increase, recycling programs also increase. The number of people receiving curbside recycling service reached nearly 90 million in 1993 (Steuteville et al. 1993). Curbside or source-separated recycling is in part a result of waste disposal laws, many of which mandate recycling and/or set recycling goals as a means of waste reduction and landfill preservation. Each year, new laws expand the range of groups they regulate, from commercial recycling enterprises and multifamily housing to public facilities, private recreation areas, and rural ones.

Recycling is a complex process involving three different activities: collecting secondary materials, marketing, and recycling materials into new products. Local governments have become involved in the first two phases, but residential recycling has commanded the most public atttention and participation. Local residential recycling programs are predicated on policy decisions made within a complex matrix of options governed by political as well as economic decisions (see Figure 17.3). Each choice, such as materials to be collected in the recycling program, has cost implications limiting further decisions. Once commitments are made (e.g., to make collection or recycling mandatory, and as to who collects, type of routing to employ, types of vehicles to use, number of collections per week, types and number of containers to use, which recyclables to collect, who sorts, where sorting and consolidation take place, and where to sell commodities and at what prices), there are significantly different cost implications.

Rural areas are generally paid scant attention. As part of a county, a rural area of low population density finds itself treated differently from more populated unincorporated areas or cities. Choices as to the types of recycling and waste management are affected by cost factors that become

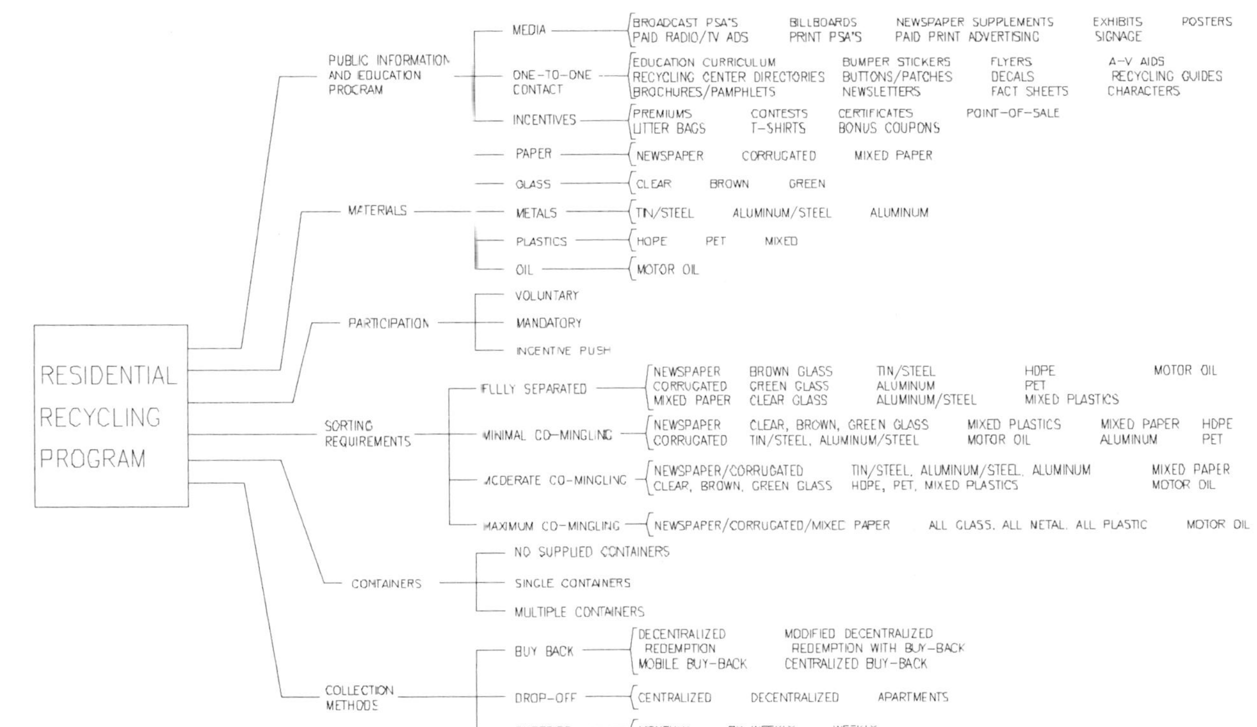

Figure 17.3 Residential recycling policy decisions. *Source:* American Public Works Association. Designing and Implementing a Recycle Program. 1990. Los Angeles:

more compelling in sparsely settled areas. Thus, state or urban recycling policies are rendered more costly to implement, less relevant, less significant, and easier to ignore. Although recycling is fundamental to sustainable development, ironically, in rural areas it elicits considerably more resistance from long-time residents than from those who migrated there, particularly among retirees from urban areas, probably because such programs were in existence in the areas from which they relocated. With few exceptions, Florida's rural counties recycle less and landfill more because, to a large extent, of high collection/processing costs (FDER 1993, 1994, 1995). Florida's inattention to rural areas was apparent in its initial Solid Waste Management legislation, but rural disposal practices have become an increasingly important segment of the state's efforts as it strives to achieve higher waste reduction goals, and as rural residents have come to accept solid waste management as a necessity.

SOLID WASTE MANAGEMENT IN FLORIDA'S RURAL AREAS

The impetus for Florida's Solid Waste Management Act (SWMA) was, in part, heightened environmental consciousness, but, more directly, a decreasing number of landfills resulting from attempts to comply with the EPA's Resource Conservation and Recovery Act (RCRA). RCRA's subtitles C and D require substantial changes in waste disposal practices (U.S. EPA 1991). Because of their traditional "open" dumping practices, their limited financial resources of these areas, and the prevalence of illegal dumps, the RCRA has had a substantial impact on rural areas.[2] Recently, however, rural counties became eligible for financial assistance in closing old landfills, which provides some incentive to initiate recycling programs (Gandi 1994). However such incentives face particular barriers in sparsely populated regions.

Florida counties quickly learned the difference between city/urban and unincorporated/rural solid waste management plans, programs, and problems. Persons living in rural areas tend to be more independent and self-sufficient than city dwellers. Many rural residents have enough property that they can burn or bury their solid waste, or they may haul it to local dump sites where tipping fees may be nonexistent or minimal. Others compost regularly or dispose of trash through illegal dumping or disposal in a dumpster at a nearby convenience store. Still others dispose of trash by contracting individually with haulers. Recycling activity is generally voluntary, with items deposited at drop-off stations. However, the amount deposited in such cases is generally extremely small (Cosper et al. 1994). The rise in environmental awareness, state laws regarding solid waste, and the pressures of increasing populations are forcing changes in public attitudes and practices, some of which meet considerable rural resistance (*Orlando Sentinel* 16 July 1994).

The terms *rural* and *unincorporated* generally denote areas of low total population and density (Goldman and Keller 1992). Rural areas generally lack significant business and industrial development. A single large industry, a unique characteristic (such as a tourist attraction), or the development of other features that draw significant numbers of travelers may affect the nature of solid waste management in rural areas, thus creating individualized and unique problems. Residential and urban areas, on the other hand, are relatively homogeneous in their solid waste management practices, and their inhabitants bring their attitudes and expectations with them when they relocate to rural settings. However, rural and nonrural areas also have commonalities. Collection or recycling programs may be optional or mandatory. Curbside collection may be optional. Yard waste may be collected as part of the trash, and composting may be encouraged. In many urban and rural areas, environmental awareness may be similar.

However, collection and processing of recycled materials pose additional problems and accentuate urban/rural differences. Recycling may not be feasible or cost-effective in rural areas. Long distances between residences or to authorized dump sites, reduced volumes, lack of professional or trained county staff or funds, lack of community support, negative attitudes, entrenchment of established practices, and inadequate knowledge of recycled commodities marketing make recycling in rural areas more difficult and less feasible than in urbanized areas. Yet many rural areas overcome these obstacles (Cosper et al. 1994). Some rural areas maintain drop-off centers or collection points, utilizing volunteers and community or church groups to collect recyclables, but the amount of material collected is generally negligible. Thus, many counties provide operating grants to encourage recycling (Snyder 1993).

The establishment of drop-off or recycling centers is a low-cost alternative to expensive recycling collection programs. Staff personnel can pick up commodities on a call basis or set up regular stops at schools, churches, or grocery stores. These services can be more cost-effective if they target specific commodities, such as old corrugated cardboard (OCC), aluminum cans, old newspaper (ONP), or glass and plastic, where local markets exist.

A recent study, sponsored by Resource Integration Systems Ltd., of 26 drop-off programs in Ontario, Nova Scotia, the United States, and Europe identified five practices that were most successful in recycling collection programs (Flemington 1994). The following recommendations are derived from this study:

- Collect a wide range of materials, such as ONP, OCC, plastics, glass, telephone directories, textiles, and even materials that may be banned from landfills (such as scrap tires and automobile batteries) to maximize the amount of material collected per household. Household hazardous waste and used oil pose special collection problems and should not be collected at unmanned sites or at sites where operators are not trained to receive these materials.

- Staffed drop-off centers increase local interest and participation. The Resource Integrations Systems' study indicated that household participation rates were higher at staffed drop-off centers than at those that were unattended.
- Continuing promotion and public education programs are essential to any recycling program. Local areas must determine which types of media have the greatest impact on public participation (Shapek 1991).
- Convenience is also essential in encouraging and promoting participation. Recycling drop-off sites should be placed to conform to local traffic patterns, and/or at strategic locations around the county, such as at fire stations and major intersections. Another option is to schedule pickups at grouped mailbox locations. Of course, the more sites that are available, the greater the convenience for the populous.
- Drop-off sites should be well designed and well maintained. Sites should be clean and free of debris, as well as visible and easily accessible from the roadway. A mix of roll-off containers, recycling igloos, or 95-gallon cans can be used to emphasize separation. A four- or five-material collection system is the norm among most drop-off programs.

Despite sparsely located populations and limited resources, rural areas can employ successful recycling strategies. The first rule is *convenience and simplicity*. Drop-off and salvage/reuse opportunities should be available to residents as well as to commercial/institutional producers of waste. Curbside collection of recyclables should be provided for population clusters. Existing, low-tech, small-scale facilities can be utilized to aggregate or sort collected materials; trash-collection vehicles can be used to collect recyclables, or for the co-collection of trash and separated recyclables. Opportunities for volume-based rates for trash disposal with no cost for sorted recyclables, can be provided. Finally, participation in recycling can be mandated (Platt 1993). Stinnett (1995a) identifies 10 steps for planning a rural collection system: (1) define the service area, (2) analyze the service area, (3) use MSW inventory data, (4) determine the town's collection frequency, (5) determine the types of separate services, (6) evaluate co-collection, existing equipment, (7) price equipment options, (8) price alternatives, (9) brief elected officials, and (10) procure equipment (Stinnett 1995a, b).

Marketing small quantities and low volumes is a particularly difficult problem for rural areas. Generally, market prices are higher for larger volumes of materials, as well as for guaranteed amounts. Few rural drop-off or collection centers can offer these minimums and are thus subject to the lowest prices and fluctuating demand. One answer is to utilize cooperative marketing. Champaign County in Champaign-Urbana, Illinois, uses a cooperative scheme that is similar to agricultural extention co-op programs (Snyder 1993). Forty-one cooperative marketing programs are currently operating in the United States and Canada, and 17 more are in the planning stages (Schoenrich and Kohrell 1995). In cooperative marketing, a number of local communities pool their collected materials to obtain volume prices. Information sharing and improved economies of scale are the primary benefits of these arrangements. Organizations participating in such cooperatives also share the costs of buying equipment and of processing, storing, and shipping the collected materials. An alternative to this single-center approach is a network of processing and collection centers, serving a maximum number of residents while combining collectibles in one location to maximize brokering for a single, higher price. Among the goals of cooperative marketing are the following (Schoenrich and Kohrell 1995):

- To consolidate smaller amounts of recyclables in order to gain access to long-term, reliable markets and improve prices through economies of scale
- To provide technical assistance, access to markets, and collection and/or processing services to emerging recycling communities
- To increase regional market prices for recycled materials and attract more buyers and end users

Another alternative is provided by a government corporation. The Tennessee Valley Authority initiated the Center for Rural Waste Management in 1992 as a waste management resource for rural communities in Tennessee and surrounding states. The Center's goals are as follows (Cosper et al. 1994):

- To serve as a clearinghouse of current, practical information for rural residents;
- To disseminate waste management information;
- To facilitate discussion of rural waste management issues and, as an information exchange, to share knowledge and skills; and
- To initiate and cooperate in developing new rural waste management information, research, programs, and skills training opportunities.

The U.S. Department of Agriculture operates some 250 Resource Conservation and Development Councils (RC&D) across the United States as part of the Soil Conservation Service, mostly in rural areas. These RC&Ds have been involved in land conservation, water management, community development, and improving fish and wildlife habitats. Many council members are now becoming involved in composting and recycling projects.

In Sweetwater, Texas, the Big Country RC&D is recycling diverted wastes into cropland (Goldstein 1993). It is now in the process of testing how well waste paper can reduce soil loss resulting from wind erosion and boost crop yields. Gainesville, Georgia, has initiated a poultry litter composting venture on six farms. The RC&D assists farmers to begin on-farm composting by buying equipment and testing to determine optimum application rates.

Waste Hauling, Landfill Siting, and Regionalization: Growing Problems in Rural Areas

Hauling

Refuse haulers generally avoid areas of low population density, preferring areas of high concentration. Logistics and unfavorable public attitudes (although they appear to be changing) also make recycling programs difficult and costly in rural areas. Less populated areas do not generate enough recyclables to encourage private operations without costly subsidies, even if the population wants such a program. Public education about recycling is also needed but costly in rural areas (see Shapek 1993). Many people think that recycling is unnecessary and inconvenient or that solid waste disposal is not a problem. Many rural residents utilize backyard burn barrels, which also raises health-related and other environmental issues. From an industry perspective, these problems can be summarized as follows:

- Not enough material (trash or recyclables) is generated to finance collection or sorting.
- The public is nonsupportive and may be uncooperative or even hostile. Many people continue to feel that recycling is a money-making business, rather than one that generates only a portion of the revenue needed to conduct the activity.
- Small unit haulers may not be able to afford the necessary equipment, and expenses may make the cost of services so high as to discourage voluntary participation.

Landfills

Landfills are the most common locations for waste disposal. Approximately 60% (41% in Florida) of the nations's solid waste is landfilled (U.S. EPA 1994). Most landfills are located in rural areas adjacent to urban centers. With the passage of the RCRA, older landfills that were near capacity or too expensive to update were closed. The problem of siting new landfills was compounded by cost and public opposition. Siting restrictions apply to airports, floodplains, wetlands, fault areas, seismic impact zones, and other unstable areas. In addition, the NIMBY (not-in-my-backyard) response pushed new landfills farther and farther into the countryside, increasing truck traffic and decreasing the land values of surrounding properties. As a result of new regulatory requirements, landfill development costs rose to $80 to $400,000 per acre, including expensive bonding and financial commitments. These costs have increased tipping and dumping fees from a few dollars to as much as $200 per ton (in New Jersey and several other northeastern states). Disposal restrictions accompanied increased operating costs. Materials such as tires, white goods, motor oil, chemicals, and lead-acid batteries may be restricted. In Florida, yard waste may be separated from normal trash and composted at landfills or other sites, depending on the county's program. Most counties have no yard waste collection program, and, with few exceptions, those that do have not made yard waste separation mandatory. It is easy to see why rural residents refuse to pay the high costs of disposal when burning or illegal disposal are alternatives.

Another issue is related to the question of environmental equity and racism. Although academic research on exclusionary zoning or locational decisions related to airports, urban highways, low-income housing, and large industrial facilities are abundant, the question of greater health risks to low-income minority communities as a result of decisions on landfill or the siting of hazardous waste disposal facilities has recently attracted national attention (Vittes and Pollock 1994). The Vittes and Pollock draft report based on EPA's Toxic Release Inventory (TRI) data indicates that siting decisions may, in effect, discriminate against poorer, more rural areas, many of which are inhabited by minority groups. The question of who came first, the landfill or the poorer or minority residents, has not been resolved. However, because land values are generally higher closer to urbanized areas, landfills tend to be located increasingly farther from the more desired path of urban development. Yet, as development spreads closer to existing landfills, residents begin to complain and seek government relief, even though they purchased homes knowing that they were near a landfill. Officials are now exploring the option of regionalized landfills served by rail transport (Fraser and Titterton 1995).

Regionalization

As small landfills close and the costs of landfills located close to urban areas increases, the concept of regionalizing landfills has become a reality. To reduce trip miles and concomitant fuel costs, waste managers have developed transfer stations at which waste or recyclables can be accumulated and transferred to larger vehicles or railroad cars. The number of transfer stations located in rural areas has increased dramatically in the past 10 years. Rural transfer stations typically include solid waste services and some recycling and/or household hazardous waste collection. Customer vehicles are generally not weighed and not attended by full-time operators. A rural transfer station typically handles 200 tons of waste per day, serving approximately 60,000 people, whereas a typical suburban transfer station handles 60 tons per day and serves approximately 20,000 people (Colville and McFeron 1994). Residents may dump garbage or deposit recyclables at a minimal cost.

Rural transfer stations are typically of two types: roll-off and drop-box operations. Roll-off containers are typically 24 feet long, 8 feet wide, and 6 feet tall, with a capacity of 40 cubic yards, or seven tons, of garbage. A roll-off container is transported on a straight-body truck equipped with a tilting frame and winch for loading and unloading the container. Transfer trailers are typically 40 feet long, 8 feet wide, and 14 feet tall, with a capacity of 109 cubic yards, or 16 tons, of trash. Smaller communities generally use one roll-off container; larger stations use two or more, which are picked up at varying intervals, generally daily but sometimes more often. Costs may be covered by an annual waste disposal fee charged to all residents of the unincorporated areas and are not generally collected at the container, because the stations are usually unmanned. Transfer stations may also accommodate one or more trailers, and many are not enclosed, surrounded by walls, or covered with a roof. In these facilities, customers generally drop their waste directly into the transfer trailer. Rural transfer stations are less expensive to construct, operate, and maintain than urban transfer stations. Because of the high cost of landfills located near urban areas, the EPA is encouraging the concept of multistate regionalized landfills. Regionalized transfer stations may be the first step in making that concept a reality.

Although the EPA encourages regionalization for solid waste disposal, there are many barriers to its development (see Lynch 1995). Permits are difficult to obtain, and citizens and politicians dislike the stigma of their community serving as some other community's garbage disposal site. Multistate regionalism is still in the infant stage, yet in many cases, there are arrangements for communities with large landfills to import solid waste from distant cities (e.g., northeast Ohio imports trash from New Jersey and New York). Revenue is lucrative, as local tipping fees may be as low as $15 per ton; tipping fees in the northeast are more than $200 per ton, but the fee charged to transshippers in northeast Ohio is only $90 to $100 (not including shipping costs). Despite the potential revenue for rural areas, these arrangements generate few jobs, with the associated disadvantage that the receiving area will gain the reputation of being a dumping ground with potential health risks. There is no current data on the effect of regionalization of solid waste disposal on rural economic development, but most reports indicate a neutral or minimal promotion of growth. However, for the regionalization and/or cooperative marketing of recyclables, new industries have emerged that may provide a different perspective. Regionalization and state/local cooperative efforts in waste stream management have resulted in new jobs and businesses in collection processing and the manufacture of recovered materials (Manfood 1995).

LESSONS FROM FLORIDA

In view of the high initial cost of start-up efforts in collecting, sorting, and marketing recyclables, and in light of declining tax revenues, most states will not continue to invest in these high cost ventures at the same rate they have in the past. The initial movement to reach the high goals set by the states for recycling has slowed (U.S. EPA 1994). States are now fine-tuning existing programs and incorporating new recycling, source reduction, and waste management initiatives that include previously neglected segments of the community. States that did not have the more comprehensive programs are expanding legislation to incorporate additional waste management provisions. All states are taking advantage of the lessons learned from pilot programs or legislation in other states so as to avoid mistakes and added costs.

State governments have a number of opportunities and options to capitalize on the growing body of research and experience of the federal government, as well as that of other states. As creatures of the state, urban and rural local governments can help shape state legislation to fit their individual needs and requirements. A number of lessons can be learned from Florida's six-year experience under the SWMA.

Lesson 1: Strategic Planning and Goal Setting

- Establish comprehensive statewide planning efforts with consensus leading to realistic and attainable goals. Involve as many segments of the population as possible in this effort: citizens, environmental groups, the business community, waste haulers, landfill operators, commodity marketing specialists and brokers, power generation officials, and elected officials of state and local governments. Include government waste management, transportation, and environmental representatives. Once initial goals are established, incorporate annual reviews to adjust or modify goals or to set new targets. Think strategically.

One of the greatest errors is for a state legislature to unilaterally develop a program and mandate goals. There is a wealth of knowledge readily available in other state programs. Poor initial decisions or planning may lead to millions of dollars in wasted or inefficient/ineffective efforts. Florida initiated the SWMA following several years of public debate with substantial input from the public, the state City and County Manager's Association, the Florida League of Cities, and many other organizations. This input led to public and media support. Initial legislation set ambitious but attainable goals. Many of the first provisions were subsequently modified to fit the changing conditions and the new knowledge gained following passage of the Act. The 30% waste-reduction goal was the "strategic" aspect of the act, which required substantial supporting legislation and program funding to become attainable and realistic. As the larger counties initiated programs, the legislature was able to fine-tune and adjust the initial legislation, permitting increased attention to rural counties and/or those with low-density propulations..

Lesson 2: Information and Education

- Recycling/waste reduction programs, as well as public support and participation in resource conservation and recovery programs, should be initiated only after extensive public information and education campaigns by all levels of government (Shapek 1991,1992,1993). Initial funding for these programs is substantial, and those dealing with public education may be constantly refined to target specific segments of the population. Increasingly sophisticated information can be incorporated to encourage higher participation in recycling programs and to generate a greater amount of collected materials. Do not neglect K–12 environmental education.

State governments can establish councils or information clearinghouses, provide technical assistance, advertise directly, and establish grant programs for local governments' public education efforts. They provide information and technical assistance on markets and market trends to manufacturers. Educational programs raise environmental consciousness and help consumers identify materials and products that can be recycled. Programs aimed at elementary schools provide long-term benefits by instilling in youngsters resource conservation values that last a lifetime. Public education is a state and local function that offers long-term dividends in resource conservation and recovery, as well as an enhanced quality of life for citizens.

Recycling can be enhanced through community development programs. The Institute for Local Self-Reliance, a Washington-based not-for-profit organization matches communities with companies that use recycled materials. The American Recycling Market, Inc., a private corporation in Ogdensburg, New York, provides a comprehensive database on the recycling industry in North America and Canada. The *Recycled Products Guide* includes more than 35,000 companies and agencies involved in the scrap market and more than 1,500 listings of manufacturers and distributors of recycled products in 250 classifications. Regional waste and resource exchange services provide market information that addresses a growing portion of the recycled commodities markets (see: http://es.inel.gov/program/iniative/waste/wexrev.html for a review of industrial waste exchange information provided by the United States Protection Agency)

Through the solid waste management trust fund, Florida has provided more than $136 million to assist local governments in establishing recycling programs, the major portion of which is designated for public education. In addition, in 1989-1990, the Florida Department of Education, in cooperation with the Florida Department of Environmental Protection, initiated a special waste management curriculum project (called "the 4Rs Curriculum") to be used in grades K–12 throughout the state. The state, in cooperation with the Florida Federation of Women's Clubs, also developed a kindergarten-first grade program and added used-oil recycling concepts to an existing environmental education program.

Lesson 3: Research and Development

- Provide direct research and development information to enterprises and/or grants and contracts to stimulate relevant research on recycling. Utilize the research capabilities of universities and the business community to refine approaches to recovering more materials, to substitute nontoxic for toxic substances, and to remediate problems with particular commodities, such as scrap tire piles.

Research and development, as well as technological innovation, are essential to find new ways to deal with increasing waste disposal problems. Many industries conduct research to enhance reprocessing capabilities, yet technologies do not yet exist to adequately or cost-effectively reprocess some materials and products. Inadequate technology also inhibits the use of some secondary materials in the manufacturing process. Although some improvements are occurring, products are not usually designed with a recycling criterion. Government sponsored or funded research on improvements in product design and recyclability, such as Environmental Conscious Design and Manufacturing (ECDAD) accelerate the pace of change (see: http://iv.cee.tufts.edu:8000/cee/Berger/devices/proposal.html).

In 1988,Florida established the Florida Center for Solid and Hazardous Waste Management (FCFSHWM) at the University of Florida, with an ongoing $500,000 research budget to fund investigations of interest throughout the state. In addition research funds are available to counties through the Solid Waste Management Trust Fund and the Waste Tire Trust Fund.

Lesson 4: Standardized Product Labeling

- Develop product standards, testing procedures, and labeling systems that show post-consumer recycled content. Initiate legislation mandating fair recycled content standards for products such as newsprint and other paper products, rubberized asphalt, glass and plastic containers, and other products.

There is an absence of standardized language for recycled products, which confuses consumers, reprocessors, and manufacturers. Definitions and standardized vocabulary for product labeling to reflect recycled content, or "recyclability," would help consumers make better decisions about products (Bennett 1988). In conjunction with public education programs, standardized labeling would enhance the recognition of recycled and recyclable products. National legislation on labeling and recycled content standards is needed to provide uniformity nationwide. In the absence of federal initiatives, several states have developed minimum standards for a number of commodities, such as dry-cell batteries, newsprint content, and packaging.

In early 1993, Florida established the Florida Packaging Council (FPC) (Section 34, Chapter 93-207, Florida Statutes) "to ensure that the recycled material content goals specified by law are technically sound and are achievable through a diligent effort by manufacturers." Another state goal is to reduce the amount of packaging materials going to final disposal by December 31, 1996.

Lesson 5: Market Development

- Stimulate markets through procurement programs, direct subsidies to industry, grants, and tax incentives for local economic development for recycled products, and help to develop export markets. Assist rural areas in developing recycling programs and cooperative marketing efforts.

Government intervention in commodity markets is difficult and complex. However, governments can influence these markets through programs of procurement, subsidies, grants, tax incentives, and other economic development initiatives, and can assist in developing local and, sometimes, export markets. There is already a trend in the development of public-private cooperative marketing programs in small towns and rural areas (Lynch 1993; Schoenrich and Kohrell 1995). Government agencies should be required to procure products made with post-consumer recycled materials with reasonable price advantages. Section 287.045 of the Florida Statutes permits a price preference of 10% for recycled products and provides for an additional 5% price preference for products created from recycled goods produced from recovered Florida materials. In addition, in 1993 the Florida legislature required a minimum 10% post-consumer recovered materials content in all printing and fine writing grades of paper purchased by the government.

Subsidies or tax incentives could be provided for industries using recycled materials. Markets can be encouraged by linking local economic development with secondary-materials-processing facilities. States can promote business development in desired industries through low-interest loans, loan guarantees, and direct grants.

At least 30 states have laws providing preferential treatment to recycled goods. Texas authorizes state agencies to pay up to 15% more for asphalt using recycled tires. Louisiana requires that at least 5% of all state purchases use recycled materials. In 1993, Connecticut mandated that all newspapers published in the state must be on recycled newsprint. In California, 25% of the paper used in state agencies, newspapers, publishing companies, and printing businesses must be of recycled stock. This percentage will rise to 50% by the year 2000 (Franklin Associates, Ltd. 1994).

In 1993, Florida established the Recycling Markets Advisory Committee (RMAC), funded from the Solid Waste Management Trust Fund, to work with the private sector and state agencies to coordinate policy and overall strategic planning on recycling market development, to encourage and develop new end-use markets for recovered materials, and to expand and enhance existing end-use markets for recovered materials.

Lesson 6: Revenue Fees and Pricing Policies

- Establish reasonable fee and pricing policies to promote the desired waste management behavior.
- Use regulatory powers to require secondary materials recovery, ban certain materials from landfills, and require deposits on containers that are not recycled at desired percentages.

One of the temptations of advanced-disposal-fee or fee-for-disposal programs is to set fees high enough to generate revenues that exceed the costs of the program, with the excess reverting to the general fund. In addition, many jurisdictions use these funds to establish reserves, and others accrue millions of dollars of debt by selling revenue bonds based on future projections of needed landfill space that may never become a reality, thus putting the burden on today's taxpayer. Fundamental to any government program is the need for cost-benefit analysis linked to the programs' goals. The greatest incentive to reduce trash is not voluntary public participation but, rather, fees based on the actual weight or volume generated. States and municipalities with weight/volume-based fee programs consistently demonstrate the highest recycling and waste reduction rates (Brown 1993).

Fees or charges may be imposed on different segments of the solid waste generation system to stimulate recycling. These include disposal fees and product charges, and fees to increase the costs of landfilling, incineration, and hauling trash. Existing subsidies of virgin materials by the federal government should also be ended. Ultimately, the solid waste management programs of the states must explore all facets of the solid waste problem: source reduction, incineration, landfilling, composting, and recycling. Each program has unique requirements and problems. Incentive mechanisms should be designed to create, build, and develop markets or collection patterns to meet the individual requirements of the program strategy. Separate but coordinated efforts by all levels of government should be undertaken to explore all aspects of collection, transportation, sorting, distribution, marketing, and market development.

CONCLUSIONS

Rural solid waste management efforts are a product of progressive state programs. Florida's Solid Waste Management Act, requiring all jurisdictions to reduce their waste streams by 30%, is an example of such efforts. Most jurisdictions responded by initiating public education, recycling, waste-to-energy or mass-burn programs, and/or composting. In many cases, cities initiated one series of efforts, and counties began others. However, unincorporated areas and rural counties lagged behind cities in these efforts.

Rural areas and those of low population density constitute an important and largely overlooked segment of the solid waste problem and are often treated as an afterthought in state legislation. As state programs become more sophisticated, rural areas are receiving more attention, but, ultimately, they are likely to remain a low priority of state waste management programs and initiatives. A solution to the waste disposal problem is not possible without consumer support and, at this stage, strong government participation and encouragement. Florida's SWMA, which offers a comprehensive state solid waste management policy that is incrementally including more and more of the solid waste stream, provides a good model for emulation by other states. Whereas landfills are predominantly located adjacent to urban concentrations, regionalization and long-hauling of trash, with intermediate collection points at transfer stations, appears to be the next stage of evolution. These trends will take place in rural areas.

As declared in *Sustainable America: A New Consensus*, by the President's Council on Sustainable Development, solid waste management—particularly reuse, recycling, and waste reduction—has become a fundamental component of sustainable development objectives. Life-cycle approaches to reduce downstream environmental impacts emphasize improved recycling, resource recovery, and extended product responsibility by businesses, governments, and consumers (PCSD 1996). Moreover, because urban areas' need for more disposal options can be met only through the use of rural land, opportunities to use the solid waste stream to develop community-based recycling businesses are emerging as a rural development alternative. However, the impact is not yet clear. In the meantime, the lessons learned from this macro-to-micro experience in Florida may be applied to rural areas everywhere.

NOTES

1. Florida's 1991 Pollution Prevention Act defined pollution prevention as steps taken by a potential generator of contamination or pollution to reduce the contamination or pollution before it is discharged into the environment.
2. In 1980, Florida had 500 open dumps. By 1994, this figure had decreased to 105 "Class 1" permitted landfills, most of which were lined, 276 C and D debris facilities, 13 waste-to-energy plants, and a substantial statewide recycling program.

REFERENCES

Bennett, R. A. 1988. "Market Research on the Plastics Recycling Industry," Center for Plastics Recycling Research Technical Report #17, Piscataway, NJ: Rutgers, the State University of New Jersey.

Biles, S. 1987. *A Review of Municipal and Hazardous Waste Management Practices and Facilities in Seven European Countries*. Portland, Ore.: The German Marshall Fund of the United States.

Brown, K. 1993. *Source Reduction Now*. Minnesota Office of Waste Management. St. Paul, Min

Colville, E. E., and N. J. McFeron. 1994. "Rural Transfer Stations: Built for the Long Haul." *World Wastes* 94(2):28–32.

Cosper, S. D., W. H. Hallenbect, and G. R. Brenniman. *1994. Case Studies in Rural Recycling*. Public Service Report OSWM-14. Chicago: University of Illinois.

Florida Department of Environmental Regulation (FDER). 1992. *Solid Waste Management in Florida*. Tallahassee: Bureau of Solid and Hazardous Waste Management.

_______. 1993. *Solid Waste Management in Florida*. Tallahassee: Bureau of Solid and Hazardous Waste Management.

_______. 1994. *Solid Waste Management in Florida*. Tallahassee: Bureau of Solid and Hazardous Waste Management.

_______. 1995. *Solid Waste Management in Florida*. Tallahassee: Bureau of Solid and Hazardous Waste Management.

Flemington, R. A. 1994. "Drop-off Recycling: Factors for Success." *Resource Recycling* XIII(1):41–44.

Franklin Associates, Ltd. 1994. *The Role of Recycling in Integrated Solid Waste Management to the Year 2000*. Keep America Beautiful, Inc., Prairie Village, KS.

Fraser, S. H., and J. Titterton. 1995. "So, You Wanna Do Rail Haul, Bunky?" *Solid Waste Technologies* 9(9):36–42.

Gandi, P. 1994. "Small Counties Can Apply for Financial Help to Close Outdated Landfills." *Florida Specifier* 16(7):10.

Goldman, M., and R. Keller. 1992. "Rural Recycling: Not Just Another Curbside Program." *Resource Recycling* XI(7):62–69.

Goldstein, J. 1993. "New Ground Rules." *BioCycle* 34(9):4.

Lynch, M. 1993. "Markets, Markets, Who's Got the Markets?" *Resource Recycling* XII(6):87–92.

———. 1995. "On the Road Again: Trends in Transporting Recyclables." *Resource Recycling.* XIV(3):46–49.

Manfood, S. 1995. "Missouri Regionalizes Recycling—and Puts Its Money Where Its Mount Is." *Resource Recycling* XIV(8):38–43.

Platt, B. 1993. "Four Small Communities Recover More Than 40 Percent of Their Household Wastes." *Resource Recycling* XII(6):47–52.

Schoenrich, L., and M. Kohrell. 1995. "Cooperative Marketing Helps Rural Recycling." *BioCycle.* 36(2):61–63.

Shapek, R. A. 1991. *Creating Public Education and Information Programs for Recycling: A Manual and Guide.* Gainesville: Florida Center for Solid and Hazardous Waste Management.

———. 1992. "Measuring the Effect of Media Use in Recycling Education/Information Programs." *Proceedings of the Second U.S. Conference on Municipal Solid Waste Management.* Washington, D.C.

———. 1993. "Data Collection and Analysis to Improve the Quality and Effectiveness of Recycling Education Programs." *Resources, Conservation and Recycling* 0921-3449/93:3–12.

Snyder, M. 1993. "Cooperative Marketing of Recyclables in Illinois." *Illinois Recycling Association Newsletter* 13(8):2.

Steuteville, R., N. Goldstein, and K. Grotz. 1993. "The State of Garbage in America." *BioCycle* 34(6):32–40.

Steuteville, R., N. Goldstein. 1993. "The State of Garbage in America." *BioCycle* 34(5):42–50.

Stinnett, D. S. 1995a. "How to Prepare a Rural Waste Plan." *World Wastes* 38(7):29–38.

———. 1995b. "10 Steps for Planning a Rural Collection System." *World Wastes.* 38(10):60–64.

U.S. EPA. 1991. *Federal Register* 56(196, October 9); 50978.

———. 1992. *Characterization of Municipal Solid Waste in the United States: 1992 Update.* EPA/530-R-92-019. Washington, D.C.: Office of Solid Waste and Emergency Response.

———. 1994. *Characterization of Municipal Solid Waste in the United States: 1994* Up-*date.* EPA/530-S-94-042. Washington, D.C.: Office of Solid Waste and Emergency Response.

Vittes, M. E., and P. H. Pollock III. 1994. *Race and Ethnicity, Income and Potential Pollution Exposure: An Analysis of Environmental Equity.* A report to the White House on Environmental Policy. Orlando: University of Central Florida.

Eighteen

Economic Analysis of Leaf Management Alternatives for Local Government

Donn A. Derr and Pritam S. Dhillon

INTRODUCTION

The initial interest in leaf recycling (in the early 1970s) was driven by the implementation of air pollution laws for a cleaner environment. The fall ritual of burning leaves at the curbside had to be abandoned in favor of cleaner air. From this point on, leaves were picked up from residences by municipalities for disposal at a landfill or were transported by home owners to a storage site or, in a few situations, to a leaf composting site. The increase in the municipal solid waste stream from several sources, including leaves from residences, commercial/industrial parks, and public areas, shortened the life of existing landfills, and stringent environmental regulations placed limitations on siting new facilities. The not-in-my-backyard attitude also placed restrictions on future sites. The result was a rapid escalation of tipping fees for disposal of municipal solid waste (MSW). For instance, in New Jersey tipping fees increased from $15 per ton in 1984 to $40 in 1987, to $80 and then to $100 by 1994. At transfer stations, where municipal solid waste is shipped out of state, tipping fees are in excess of $125 per ton. These tipping fees provided the economic incentive for recycling several MSW components, including leaves, food waste, paper, glass, cardboard, and sewage sludge (Derr et al. 1988).

Today states have implemented mandatory recycling laws that require communities to separate newspapers, plastic, and glass, and prohibit leaves being placed in landfills. On average, leaves alone can account for 10% of the municipal solid waste stream. Composting leaves not only can reduce the volume of waste going into landfills, but also offers a real opportunity to exploit a symbiotic relationship between urban areas that must recycle their leaves and agricultural operators who have the land to utilize the end product. This, in turn, can provide supplemental income for agricultural operators and give them a chance to utilize excess labor and equipment. Not only can this process help sustain the agricultural operator, but it can help sustain the land resource through adding organic matter, replacing macro nutrients from chemical fertilizers, providing protection against soil erosion, enhancing water retention capacity, and increasing plant nutrient retention. It can help diversify the rural economic base and contributes to the formation of a rural-urban partnership. Furthermore, by integrating the agricultural community into the waste management process early on, this approach provides greater opportunity for on-farm utilization of the end product.

The need to develop on-farm composting and mulching is further reinforced by the fact that composting in general has expanded so much that supplies have greatly exceeded their use and are continuing to accumulate at municipal and regional sites (Buhr et al. 1993; Kashmanian 1993). A case in point is the composting of sewage sludge with little or no market research/development. As a result, the potential revenue to be derived from composted products is basically limited to nationally known companies that have extensive marketing experience with large-volume garden and lawn products that enjoy brand-name recognition. The uniformity of the physical and chemical properties of these products and a cosist-ent and reliable supply are important to businesses that use them as potting mixes for annual and perennial plants.

This chapter focuses on leaf composting and leaf mulching because of the potential benefits to both Urban and Agricultural Communities. The opportunity for on-farm leaf composting and mulching is enhanced for highly built-up areas that have limited space for municipal composting sites yet have operating farms within a 20- to 30-mile radius. In New Jersey, about 7% of the total amount of leaves is utilized on 70 farms producing mainly corn, soybeans, and vegetables. But farm operators indicate that they could use twice this amount and, given the problems incurred by municipalities with their own composting sites, at least 30% of the 5 million cubic yards of leaves collected annually could potentially be utilized on cropland (Derr et al. 1995b).

THE COMPOSTING PROCESS

The composting of leaves is a waste-treatment process in which the volume of waste is reduced by at least 80% and the material is transformed into a useful end product through microbial decomposition. Primarily, leaf compost can be incorporated into soil or used as a mulch (Patterson 1985; Flannery and Flower 1988). It is, however, considered a soil conditioner rather than a fertilizer because of the low N-P-K (Nitrogen, Phosphate, Potash) content. Using composted leaves is preferable to adding raw or uncomposted leaves to the soil because the latter are bulkier, need to be integrated into field operations, and need soil nitrogen in order to decompose. The pH of compost is near neutral, therefore reducing the need for liming.

It is a well-established fact that the addition of organic matter, including composted leaves, improves the soil's tilth (ease of cultivation), facilitates root development, increases water-holding capacity, reduces water runoff, serves as a buffer, and helps reduce the loss of plant nutrients. In effect, compost is considered good for both sandy and heavy soils. It can also be used as a mulch around plants and shrubs to control weeds, enhance water infiltration, and reduce evaporation.

From among the various MSW fractions, recycling leaves via composting offers several advantages (Derr 1988): (1) because of the seasonality, leaf collection can be limited to a few pickups, (2) leaves are already separated from other waste, (3) odor problems associated with leaf collection are minimal, (4) transportation costs can be relatively low, especially for local municipal composting facilities, (5) several levels of composting technology are available to meet specific local conditions, (6) existing municipal equipment can be used for composting, and special-purpose equipment is available, (7) compost is considered a useful end product that can be utilized by local residents, landscapers, and agricultural operators, and (8) the composting process is cost-effective, especially in comparison to high tipping fees at landfills.

Strom and Finstein (1989) have identified four levels of leaf composting technology: minimal, low level, intermediate, and high. Each option reflects the rate of decomposition (time required to produce usable compost), depending on the amount of oxygen added to the leaf windrows. With the minimal-level technology leaf piles are turned once each year, and it takes three years to complete the process. For low-level technology leaves are assembled in windrows that are turned over three times, normally with a front-end loader, which includes the formation of the curing pile (see Figure 18.1). This technique takes about 18 months to complete and is recommended for municipal leaf-composting operations. Maintaining a minimal water content in the windrows facilitates the microbial process.

The intermediate-level technology involves use of specialized equipment such as compost turners, which are either self-propelled or powered by large farm tractors/front-end loaders. Technology of this type is used extensively by commercial compost operations and by some municipalities and on-farm composting. There is frequent turning of the windrows,

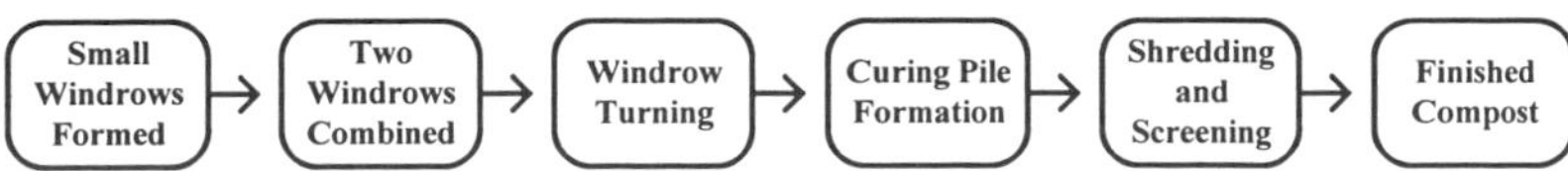

Figure 18.1 Low-level compost technology.

which is determined largely by temperature, and composting can be completed in less than 150 days. This approach requires more land area, in order to maneuver the equipment and form smaller windrows, as compared with the low-level technology. This level is applicable to on-farm composting where there are several windrow composters available (see Figure 18.2).

The high-level technology involves close control of temperature and oxygen level in the windrows. Optimum conditions for decomposition are maintained through the use of blowers that regulate the temperature, oxygen level, and moisture. Currently, this technology is used little for leaves.

On-farm composting activity, however, extends well beyond leaves. A major problem for livestock operations is processing and utilizing animal wastes. Most livestock operations are larger today than in the past, and in close proximity to neighbors, and animals are confined more, largely owing to economics. Animal waste combined with bedding material, like wood shavings or straw, forms a good mix for composting. The animal waste supplies nitrogen, and the bedding material provides the bulking agent. Fabian and associates, in their study of the feasibility of composting on New York farms, reported that farm operators were using off-farm bulking agents, including sawdust, cardboard, and shredded paper (in some cases already in the bedding). The mixing of poultry waste with leaves also showed promise in early trials. At a county landfill facility in upstate New York, leaves, animal waste, and food wastes were being composted. In another case, lake weeds, chicken manure, and moldy hay were combined and composted (Fabian et al. 1992).

Cranberry hulls have been successfully composted at a site in New Jersey. This facility also experimented successfully with composting leaves and grass clippings, but the supply of these components could not be matched up.

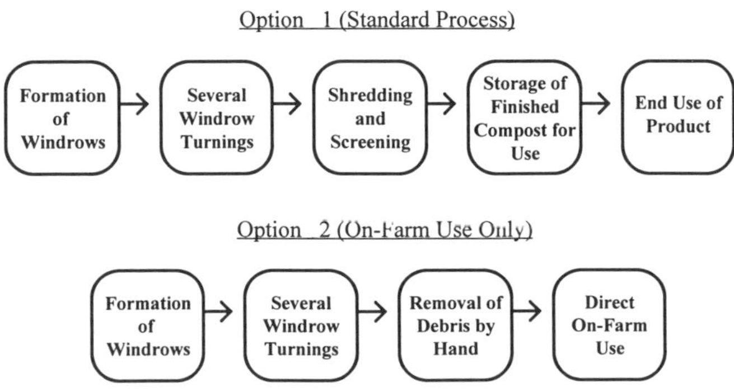

Figure 18.2 Intermediate-level compost technology.

Damaged and overripe fruit, dated retail food items, and postconsumption food waste from cafeterias, restaurants, and similar businesses lend themselves to on-farm composting. For waste with a high water content, some type of bulking agent is needed to create air voids and absorb water. However, variations in the bulking agent used and materials being composted (animal waste, food waste, vegetable waste, leaves, grass clippings, wood chips, etc.) affect the physical and chemical properties of the end product. This limits marketing opportunities, suggesting that on-farm processing and land application with the potential of revenue from tipping fees is the most feasible option.

ON-FARM LEAF MULCHING

Leaf mulching involves the direct application of collected leaves to the soil and their subsequent incorporation. New Jersey promulgated leaf mulching regulations in 1988 whereby up to six inches of unbagged leaves could be applied by farm operators, but the leaves must be incorporated no later than the next tillage season (State of New Jersey 1988). Typically, farm operators load the leaves with a tractor/front-end loader and spread them with a manure spreader or a loader. The leaves are usually incorporated into the soil with a mold board plow or a disc harrow, although research suggests that a good leaf/soil mixing can be accomplished with a chisel plow. Corn, soybeans, and vegetables are typically grown on fields where leaves have been applied. The major advantages of on-farm leaf mulching are that it reduces the management burden for local government, is more cost-effective than municipal composting, offers a source of off-farm income and organic matter to agricultural operators, and has the potential for reducing soil erosion (Kelly 1994; Derr et al. 1995c). The disadvantages of on-farm leaf mulching include the risk of adding nonbiodegradable materials (cans, bottles, paper, branches, etc.) to fields, difficulties in scheduling leaf deliveries around inclement weather and field activities, intermittent acceptance of leaves by an agricultural operator, lack of uniform application of leaves prior to soil incorporation, and tipping fees that are insufficient to encourage a long-term public-private partnership (Derr et al. 1995). The steps in the on-farm leaf-mulching process are shown in Figure 18.3.

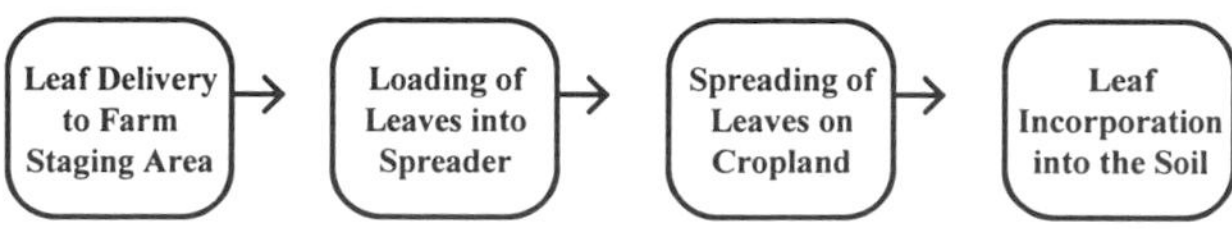

Figure 18.3 On-farm leaf mulching.

ECONOMIC ANALYSIS OF ALTERNATIVES

Because of the individuality of different communities, one would expect the optimum leaf composting/mulching option to vary from municipality to municipality (Derr and Dhillon 1989). Characteristics that influence the selection of a cost-effective compost/mulching option include the volume of leaves generated, which, in turn, depends on the size of the community; geographic distribution of development (i.e., whether concentrated or scattered); pattern of the major land uses (i.e., whether primarily residential, residential and commercial, or residential with agricultural land); availability of suitable sites for composting facilities, particularly with buffer zones to safeguard residences, hospitals, retirement communities, and other environmentally sensitive areas; availability of adequate cropland within a 20- to 30-mile radius; and the development of effective public-private partnership arrangements (Derr et al. 1993).

In line with these characteristics, four basic types of leaf management alternatives involving five options are analyzed here: small-scale on-farm composting, large-scale on-farm composting, municipal composting, private commercial composting, and on-farm leaf mulching (see Table 18.1 for a comparison of facilities).

Small-Scale On-Farm Composting

The intent of small-scale on-farm composting is to recycle leaves near their source at a very low cost. Although the leaves will be completely composted, it is not necessary to pass them through a shredder/screener because of hand separation of most of the nonbiodegradables and direct application of the finished product to the fields.

This alternative is consistent with the sustainable agricultural thrust of the United States Department of Agriculture (Madden 1988; USDA 1988). In recent years environmental concerns have prompted a reexamination of agricultural production technology. There is a common belief that the existing technology is relying too heavily on modern inputs such as chemical pesticides, herbicides, and fertilizers, which are contaminating the soil and polluting the general environment. It is feared that with the continuation of the present technology, soils will become so degraded that in the future they may not sustain agricultural production. Leaching of nitrates and nutrient runoff into rivers and lakes are causing serious damage to the underground and aboveground water supply in major agricultural regions of the country. As a result, the USDA is currently collaborating with state experiment stations and other interested organizations to explore the feasibility of adopting low-input sustainable agriculture (LISA). Leaf composting can play a complementary role in achieving the objectives of LISA, because the application of organic matter to the soil can reduce the need for chemical fertilizers.

Table 18.1 Characteristics of the Five Types of Composting/Mulching Facilities

	Type of Facility				
Characterixtics	*Small On-Farm Composting*	*Large On-Farm Composting*	*Municipal Composting*	*Private Composting*	*On-Farm Leaf Mulching*
Size (Cubic Yards)	5,000–10,000	10,000+	20,000–30,000	80,000+	5,000+
Level of Technology	Intermediate	Intermediate	Low	Intermediate	Low
Purpose	Low cost near source	Low cost near source	Service to community	Composting with resale	Low cost near source
Type of Facilities	"Add-on"	"Add-on"	"Add-on"	New	"Add-on"
Basic Equipment	Windrow composter	Windrow composter	Front-end loader	Windrow composter	Tractor/loader manure spreader
Screening Equipment	None	None	Yes	Yes	NA
Marketing	None	None	Local use	Sale of product	NA
Source of Leaves	Local	Local	Local	Off-site	Local
Investment Level	Low	Low	Moderate	High	Low
Land Area Needed (Acres)	3	6	15	42	At least 5 acres

The major investment for this approach would be the purchase of a compost turner that could be powered by an existing farm tractor. Screening equipment is not required inasmuch as the compost can be worked into the soil directly. The compost turner would keep the leaves loose, breaking up any clumps while nonbiodegradable items fall to the edge of the windrows for easy pickup. Prior to composting, some manual separation of nonbiodegradables would be required. Site improvements, such as grading and reinforcing access roads to the compost area, would also be necessary. No additional structures would be required, and land costs charged to composting would be negligible. A major advantage of this option is that complaints about odors or noise would be minimal, as compared with those incurred by municipal facilities.

This type of on-farm composting provides an opportunity for agricultural operators to earn additional income by using excess family labor and agricultural equipment during off-peak seasons. It is ideal for crop farms, because leaves would be picked up and delivered during the time when field activity would be slowing down. The only disadvantage to the agricultural producer would be the risk of adding some nonbiodegradable material to the soil. However, the risk can be minimized by specifying how leaves are to be placed at curbside and the frequency of collection.

The cost of composting 5,000 cubic yards of leaves on a farm site is estimated to be $21,232 annually, or $4.25 per cubic yard (see Table 18.2). This includes charges for all resources used in composting (labor, fringes, capital, management, land, fuel, supplies, approval fees, etc.). Depreciation and interest costs are determined by using a capital recovery factor based on a 10% interest rate. Fringe benefits for hourly labor are 11.95% of the base wage for part-time labor and 17.5% for the equipment operator. A downtime of at least 20% is assumed (downtime refers to time that must be paid for, but is not being utilized because of equipment maintenance, moving equipment to and from the work area, and waiting for delivery of leaves to the site).

To complete the compost process and generate a stabilized product, 165 hours of labor, at a total value of $1,491, are budgeted. In addition, the owner would receive a 10% rate of return on his or her investment ($2,713), a management fee of 7% of all costs ($1,284), and $589 for overhead. A total income of $6,077 for labor, management, and return on investment would flow to the agricultural operation annually. In addition, 1,000 cubic yards of finished compost (unscreened) would be available for field application. This has an estimated net value of $3 per cubic yard, or $3,000 (screened compost has a market value of $4 to $8 per cubic yard [FOB]). In total, the composting operation would generate $9,077 in total benefits. Although disposal costs to the municipality would be minimized with the use of this alternative, it may involve additional administrative costs to contract with more than one agricultural operator. Moreover, the

Table 18.2 On-Farm Leaf Composting Costs for a 5,000-Cubic-Yard Operation, July 1994

Component	*Cost ($)*
Tractor(windrow construction)[1]	1,350
Planning/Approval Fee[2]	1,620
Site Improvements[3]	3,060
Tractor (windrow composting)[1]	780
Composter[4]	4,995
Maintenance of Composter	299
Pickup Truck	780
Labor[5]	1,359
Water Supply (including labor)[6]	1,525
Disposal of Debris	1,080
Land Rent for the Site[7]	600
Insurance	900
Contingencies[8]	1,011
Return to Management[9]	1,284
Overhead[10]	589
Total	21,232
Cubic Yard[11]	4.25

1. Based on a 130 hp tractor. An hourly rate of $27.00 was used for the tractor to cover the cost of depreciation, interest, maintenance; and fuel.
2. Fees and engineering costs prorated over five years.
3. Improvements, including grading and road reinforcement, prorated over five years.
4. Based on a seven-year life, 10% interest rate, and a salvage value of 20%.
5. A rate of $12.00 per hour including fringes is used for the equipment operator and $5.65 per hour including fringes for seasonal labor. Most labor costs are included in this component.
6. Charge for water application and fire protection.
7. Land rent for three acres.
8. Based on 5% of accumulated costs. This includes the incremental costs associated with variations in weather conditions and costs otherwise not allocated to specific categories.
9. Based on 7% of accumulated costs.
10. Based on 3% of accumulated costs.
11. Based on 5,000 cubic yards.

agricultural producer may insist on a multiyear contract to justify the investment of about $49,000 to establish an on-farm composting facility.

Large-Scale On-Farm Composting

For a 10,000-cubic-yard on-farm composting operation, the cost is estimated at $3.19 per cubic yard. It would generate $17,206 for labor, management, investment, and organic matter. The equipment is similar to that used in a 5,000-cubic-yard operation. However, unit costs are lower because the fixed costs are spread over a larger output. Given that there is currently an excess supply of compost, and that wholesale and/or retail sales would require shredding and screening, the agricultural operator is better off utilizing the product on the farm.

Municipal Compost Operation

Municipal composting is the most prevalent among the various approaches to composting, at least in the northeastern United States. Normally, between 20,000 and 30,000 cubic yards of leaves are processed each year. The site must be graded so that water or leachate does not accumulate. The advantages of this alternative include minimal transportation costs and the provision of a convenient supply of compost to the municipality and its residents. The persistent and potential problems include odors that can develop because of the concentration of a large volume of leaves, the burden of management of the site, lack of room to expand, and health hazards caused by airborne spores from the fungus *Aspergillus fumigatus* (Strom and Finstein 1989). The fungus is a problem primarily because of the close proximity of residential areas to municipal compost sites.

For optimum composting, Strom and Finstein (1989) recommend that leaves be brought to a staging area where two relatively small windrows are constructed side by side. Then, about four weeks later, the windrows are combined into a larger single windrow. In the spring (three to four months later) the windrow is turned over for aeration. Moisture content is critical to the process, and water may have to be added while the windrows are turned. Approximately five to six months later the windrows are broken down and a curing pile is formed. About seven months later the compost is ready for use, and screening can take place at this time. The compost is utilized in three ways: through small quantity pickup by home owners/gardeners, in municipal landscaping, and in bulk sales to vendors such as landscapers. Using a 15% yield rate for the screened product, 30,000 cubic yards of leaves would generate 4,500 cubic yards of finished compost.

In estimating costs, an interest rate of 9% is charged on the investment, equipment is depreciated over seven years, and site improvements are depreciated over ten years. Labor costs are inclusive of fringe benefits.

Because the front-end loader is shared with other municipal departments, costs are charged on an hourly basis, whereas the screening/shredding equipment is charged exclusively to the composting operation.

The total annual cost for this alternative is budgeted at $173,889, or $5.80 per cubic yard (see Table 18.3). The screening costs alone account for 27% of the total costs. Screening costs (including labor, interest, and depreciation) constitute a major component of the total costs inasmuch as this item it is an expensive, specialized, and labor-intensive procedure requiring three people. Screening is considered important, however, because it improves the quality of the compost by breaking up clumps, removing non-biodegradable materials, and eliminating uncomposted leaves that must be recycled through the compost process a second time. Land charges, overhead/engineering, and site improvements are the next three most important cost components.

For a municipal facility, special attention must be paid to the buffer area surrounding the site. As mentioned previously, odors, airborne spores, and noise from the equipment can be problems for sites that are in close proximity to residences. For the 30,000 cubic-yard facility budgeted in Table 18.3, ten acres were used for windrowing/composting leaves, four acres as a 60-foot buffer, and one acre for the work area. For a facility of this size, each additional 15 feet of buffer zone requires one more acre of land, which would increase the total annual cost by $2,070.

The New Jersey Department of Environmental Protection recommends a buffer zone of 500 feet for all compost sites or a total buffer area of at least 34 acres for a 30,000-cubic-yard facility (State of New Jersey 1990). The total cost would thereby be increased by $62,100 annually, or $2.07 per cubic yard, resulting in a total cost of $7.87 per cubic yard. This would make the municipal facility more expensive than the other alternatives considered here. Thus, the buffer zone requirement is a major factor in determining compost costs and the siting of compost facilities.

Commercial Composting Operation

A commercial composting operation is designed to accept leaves from several sources; its emphasis is on marketing screened compost. Municipalities that are highly built up and have difficulty locating a suitable composting site, or do not see contracting with local agricultural operators as feasible, may consider this alternative. This type of facility is capital-intensive, that is, it has a large investment in equipment and facilities relative to labor. Once leaves are brought to the site, windrows are immediately constructed to avoid odors and a backlog of piled leaves. The windrow composter is the principal equipment for this alternative. Because leaves are constantly turned, high-volume equipment is used. There are several sizes of composter available, so an owner can always find an appropriate size to meet a

Table 18.3 Annual Composting Costs for a 30,000-Cubic-Yard Municipal Facility, July 1994[1]

Component	*Cost ($)*
Land[2]	31,050
Site Improvements[3]	13,496
Site Maintenance	6,930
Initial Windrow Formation[4]	6,752
Combining Windrows[4]	4,068
Water Supply/Wetting Leaves	9,462
1st Turning[4]	3,043
2nd Turning[4]	2,029
Curing Pile Formation[4]	2,948
Shredding/Screening[5]	46,519
Disposal of Debris	3,360
Insurance	8,600
Supplies	5,500
Contingencies[6]	12,881
Overhead/Engineering7	17,251
Total	173,889
Cubic Yard[8]	5.80

1. An interest-rate charge of 9% is applied to resources used lasting for more than one year.
2. Includes a total area of 15 acres at a market value of $23,000 per acre.
3. Depreciated over a 10-year period.
4. Includes equipment and labor costs. An hourly rate is used for the front-end loader and its operator.
5. Includes interest, depreciation, labor, maintenance, and fuel. Equipment depreciated over 7 years. Based on a gross yield rate of 20% or 6,000 cubic yards of compost.
6. Refers to nondistributed costs. Reflects potential variations in costs because of weather conditions, breakdown in equipment requiring equipment rental, periodic security, distribution of screened compost, temporary labor for peak labor periods, and general maintenance. Based on 8% of accumulated costs.
7. Overhead is 9% and engineering is 3% of accumulated costs.
8. Based on 30,000 cubic yards.

given situation. The other major equipment component is the compost screener/shredder. Screening greatly improves the quality of the compost by removing nonbiodegradables and leaves that are not completely composted. Odors and health hazards are less problematic because of the site's distance from populated areas and better site management. Compost selling for $7 per cubic yard (FOB) with a 15% net yield could potentially offset costs by $1.05 per cubic yard, or 14%. However, market development has been difficult because of excess supplies of compost (both from sludge and leaves) and changes in regulations governing compost operations.

For a composting operation of this type, buildings for office space, personnel, and equipment maintenance are needed. Scales are also required to weigh the finished product. More full-time people are needed, and significant resources are devoted to marketing the product and soliciting business.

Costs are estimated at $7.57 per cubic yard, with labor (including fringes), equipment (depreciation and interest), land, and maintenance (site, buildings, and equipment) as the major cost items (see Table 18.4). An interest rate of 10% is used to reflect a higher cost of capital in the private sector. An 8% return to management is used to reflect additional risk associated with government regulations.

ON-FARM LEAF MULCHING

In an on-farm leaf mulching operation, leaves are transported directly from a municipality to a farm operation. Weather permitting, leaves are discharged at several locations to facilitate spreading. For the cost analysis presented here, it is assumed that the farm operator uses a tractor with front-end loader and a large-capacity manure spreader, and that leaves are incorporated with a chisel plow. Uniform application of leaves to the soil is important in an operation of this kind. An hourly rate is used for equipment and wages, including relevant fringe costs. Costs also include a return to management, overhead, contingencies (nonallocated costs) for exceptional weather, and additional labor for handling bagged leaves and debris. The cost is $3.50 per cubic yard to apply 5,000 cubic yards (see Table 18.5). This would generate $5,119 for 546 hours of labor, $1,091 for management, $500 for overhead, and $3,000 for organic matter. On-farm leaf mulching is a viable option, particularly for municipalities with their own composting sites, inasmuch as they report a high incidence of odors and space problems.

COST COMPARISON

In comparing costs, large-scale on-farm composting (10,000 cubic yards) ranks first, with the lowest cost; on-farm leaf mulching ranks second, small-scale on-farm composting (5,000 cubic yards) is third; followed by

Table 18.4 Composting Costs for a 90,000-Cubic-Yard Commercial Facility, 1994

Component	*Cost ($)*
Land[1]	84,000
Site Improvements[2]	20,295
Buildings[3]	30,571
Equipment[4]	122,834
Engineering/Design[5]	8,534
Maintenance[6]	51,200
Utilities, Fuel, etc.	45,973
Insurance	23,040
Labor[7]	202,518
Disposal of Debris	6,480
Contingencies[8]	38,585
Return to Management[9]	47,636
Total	681,666
Per Cubic Yard[10]	7.57

1. Based on 42 acres at $20,000 per acre.
2. Site improvement costs are 20% of land costs prorated over 10 years at a 10% interest rate.
3. Includes office, lunchroom, and equipment for building repairs. Prorated over 20 years.
4. Equipment costs depreciated over 7 years with a 10% salvage value and a 10% interest rate.
5. Based on 5% of the capital costs for site improvements, buildings, and equipment prorated over 10 years.
6. Maintenance costs for site improvements, buildings, and equipment.
7. Includes base wage and fringe benefits.
8. Based on 6% of accumulated costs. Refers to undistributed costs. Reflects potential variations in costs because of weather conditions and regulatory fees.
9. Based on 8% of accumulated costs.
10. Based on 90,000 cubic yards.

municipal operation and the private commercial operation (see Table 18.6). It should be recognized that this comparison deals with on-site composting/mulching costs alone and does not account for differentials in transportation costs. Normally, on-farm and private commercial operations would be farther away from the source of leaves than the municipal operation and would entail somewhat higher transport costs. However, transport

Table 18.5 On-Farm Leaf Mulching Costs for a 5,000-Cubic-Yard Facility, 1994

Component	*Cost ($)*
Loading (Equipment)[1]	3,550.50
Spreading (Equipment)[1]	5,786.00
Soil Incorporation (Equipment)[1]	524.16
Labor[2]	5,718.90
Return to Management[3]	1,090.57
Overhead[4]	500.10
Contingencies[5]	343.40
Total	17,513.63
Per Cubic Yard[6]	3.50

1. An hourly rate is used for the equipment that covers all fixed and operating costs.
2. Labor costs include relevant fringes. Labor is needed for loading leaves into the spreader, spreading the leaves, and soil incorporation.
3. Based on 7% of accumulated costs. This charge covers the owner's time for supervision, coordination and scheduling of leaf deliveries to the farm, and contractual arrangements.
4. Based on 3% of accumulated costs. This charge covers the cost of farm lanes and maintenance, insurance, and office costs (telephone, supplies).
5. Based on 2% of accumulated costs. This pertains to nonallocated costs for exceptional weather conditions, compacted leaves, removal of excess debris including trash bags.
6. Based on 5,000 cubic yards.

Table 18.6 Summary of Costs for Five Leaf Management Options, 1994

Option	*Cost Per Cubic Yard ($)*		
	On Site Processing	*Transport*[1]	*Total*
On-Farm Leaf Composting(10,000)	3.19	.64	3.83
On-Farm Leaf Mulching (5,000)	3.50	.64	4.14
On-Farm Leaf Composting (5,000)	4.25	.64	4.89
Municipal Composting (30,000)	5.80	0	5.80
Commercial Composting (90,000)	7.57	.92	8.49

1. Incremental costs associated with transporting leaves beyond the municipal site to nearby farms and commercial composting facility.

costs are not expected to differ much, and their inclusion is not expected to change this ranking. A major portion of transport costs involves loading and unloading leaves, and this would remain unaffected by the distance to the compost site. Only travel costs would undergo a slight increase. As compared with costs to municipal facilities, the additional transport costs to farms are estimated at $0.64 per cubic yard and to private commercial sites at $0.92 per cubic yard.

SUMMARY

Rapidly rising tipping fees and lack of adequate landfill space that is environmentally safe have forced communities to recycle a significant portion of their municipal solid waste. Leaves, estimated to account for 10% of the MSW, ideally lend themselves to recycling via composting and mulching.

The breakdown of leaves by microbial action is a slow, natural process. However, with the addition of oxygen, achieved by turning leaves that are assembled in windrows and keeping them wet, the process can be accelerated severalfold. If leaves are piled and turned once a year (minimal technology), the process takes three years to complete. However, if leaves are assembled in windrows and are turned three times (low-level technology), composting time is reduced to 18 months. By turning the leaves at more frequent intervals with a compost turner (intermediate-level technology), the time can be reduced to less than 150 days. After shredding and screening, an end product is produced that can be incorporated into the soil or used as a mulch.

Four basic alternatives for leaf management—on-farm composting, municipal facility, commercial operation, and on-farm mulching—were considered here. Each has its own advantages and disadvantages for a community seeking an option. On-farm composting or mulching can minimize costs for the municipalities and contribute to the symbiotic relationship between municipalities and agricultural operators. This approach can provide an additional source of income for local agricultural operators who can use their excess labor and equipment. It also helps the long-term sustainability of land with the addition of organic matter, which in turn reduces water and nutrient runoff and can be a source of macronutrients (N, P, K). It encourages the local community to develop a public-private partnership between rural and urban areas. The on-farm alternatives reduce the burden of operating a municipal facility and can effectively deal with odors and the problem of future site expansion. However, the municipality may have to contract with several local agricultural producers, thereby increasing administrative costs.

A municipal facility has the option of sharing existing equipment and personnel (which may or may not be an advantage) with other departments, but its composting costs are relatively high. This alternative is also more service-oriented, because the compost is given away free of charge to local residents. The availability of suitable compost sites, however, places a real constraint on this alternative. Increasing the buffer zone around the site can reduce the problem of odors, noise, possible airborne spores, and space for future expansion, but can also increase composting costs.

The remaining alternative for a municipality is to employ a private commercial facility that is located at some distance. This is a capital-intensive approach and transport and composting costs will be higher. Costs can be reduced somewhat by marketing the end product. Yet, although more costly, this approach minimizes the responsibilities of the municipality. The best alternative is one that minimizes administrative, transport, and composting/mulching costs and that varies from municipality to municipality, depending on the community's characteristics.

REFERENCES

Buhr, A. R., T. McClure, D. C. Slivka, and R. Albrecht. 1993. "Compost Supply and Demand." *Biocycle* 34(1):54–58.

Derr, D. 1988. "The Economics of Turning Over an Old Leaf." *Waste Age*. National Solid Wastes Management Association, 94–99.

Derr, D. A., and P. S. Dhillon. 1989. "Minimizing the Cost of Leaf Composting." *Journal of Waste Recycling* 30(4):45–48.

Derr, D. A., D. Kluchinski, and J. Morgan. 1993. "Developing Successful Public/Private Partnerships—the Case of Leaf Mulching." Paper presented at the 48th annual meeting of the Soil and Water Conservation Society, Fort Worth, Texas.

Derr, D. A., D. Kluchinski, and W. R. Preston. 1995a. "The Potential for On-Farm Leaf Mulching." *Biocycle* 36(3):67–69.

———. 1995b. "On-Farm Leaf Mulching Opportunities and Constraints for Local Government." *Proceedings of the Eleventh International Conference on Solid Waste Technology and Management.* Philadelphia: University of Pennsylvania Press.

Derr, D. A., D. Kluchinski, W. R. Preston, and R. H. Bruch. 1995c. "On-Farm Leaf Mulching Can Reduce Municipal Cost." *New Jersey Municipalities* 72(4):14–15.

Derr, D. A., A. Price, J. Suhr, and A. Higgins. 1988. "Design and Economics of a Statewide Food Waste Recycling System." *Journal of Waste Recycling* 29(5):58–63.

Fabian, E. E., T. L. Richard, D. Kay, D. Allee, and J. Regenstein. 1992. *Agricultural Composting: A Feasibility Study of New York Farms.* Final Report Prepared for the New York State Department of Agriculture and Markets. Staff Report; Departments of Agricultural and Biological Engineering, Agricultural Economics and Food Science. Ithaca: Cornell University Press.

Flannery, R. L., and F.B. Flower. 1988. *Using Leaf Compost: Rutgers Cooperative Extension Service.* Leaflet FS 117. New Brunswick, N.J.: Rutgers University Press.

Kashmanian, R. 1993. *Markets for Compost.* (EPA/530-SW-073A). Washington, D.C.: Office of Policy, Planning, Evaluation, Office of Solid Waste and Emergency Response, US EPA.

Kelly, F. 1994. *Leaf Mulching on Cropland.* NJ Job Information Sheet No.16. Somerset, N.J.: Natural Resources Conservation Service.

Madden, P. 1988. "Policy Options for a More Sustainable Agriculture." In *Proceedings, 38th National Public Policy Education Conference. Farm Foundation*. Cincinnati, Ohio:

Patterson, J. E. 1985. "Enrichment of Urban Soil with Composted Sludge and Leaf Mold—Constitution Gardens." *Compost Science* 16(3):18–21.

State of New Jersey. 1988. *Statement of Imminent Peril to Public Health, Safety, and Welfare Mandating Adoption of Amendment at N.J.A.C. 7:26-1.7(g) and New Rule at N.J.A.C. 7:26-1.11 and 1.12 by Emergency Proceedings.* Trenton, N.J.: Department of Environmental Protection.

_______ . 1990. *Composting Facility Buffer Zone Recommendations*. Trenton, N.J.: Department of Environmental Protection.

Strom, P. F., and M. S. Finstein. 1989. *Leaf Composting Manual for New Jersey Municipalities.* New Brunswick, N.J.: Department of Environmental Science, Rutgers University.

USDA. 1988. *Low-Input/Sustainable Agriculture*. Research and Education Program. Washington, D.C.: Cooperative State Research Service and Extension Service.

Nineteen

Regenerating a Regional Economy from Within: Some Preliminary Lessons from Central Appalachia

Anthony Flaccavento

INTRODUCTION

Hancock County is a sparsely populated community nestled in the mountains of northeast Tennessee. The poverty rate among the 6,000 residents is one of the highest in the state, in excess of 30%. In February 1991 the county schools were in crisis. A revenue shortfall of more than $300,000 threatened to close the entire school system before the end of the term.

At the same time, an out-of-state firm was negotiating with county officials to build a landfill to receive garbage from other states. The landfill firm promised the county annual revenues in excess of $300,000. To some, the landfill seemed to be an answer to their prayers: a revenue source that could keep the county schools operating.

Assisted by members of a community organization in neighboring Lee County, Virginia, however, local citizens organized in opposition to the landfill. In spite of the promised financial benefits, these residents decided that the risks were too great. They believed that Hancock County's economic future should be built on something other than the importation of garbage. The landfill proposal was resoundingly defeated, and the school system was kept afloat through the year with assistance from the state.

The story does not end here. Two years after the landfill struggle, a grass-roots community organization, Jubilee Project, was formed in hopes of rejuvenating the county's economic and cultural life. Jubilee Project quickly set to work on its two main priorities: providing training and employment opportunities for young people and rebuilding the local economy from the bottom up. An entrepreneurial literacy program was developed, adapting a microenterprise training program developed in nearby Abingdon, Virginia, to meet the needs of people with no business experience and limited education. Youth groups soon formed, followed by business training classes linked with the vocational school. A number of small youth-run businesses were started, including one in graphic design and others in wood crafts.

In 1987 a wafer board factory was under construction in the tiny town of Dungannon, Virginia, on the beautiful Clinch River. Dungannon's 350 residents derived their incomes primarily from growing small plots of tobacco and/or part-time work in sewing factories or convenience stores. Most people with full-time employment traveled an hour or more to Kingsport, Tennessee, to work. Under the circumstances, the wafer board factory's promise of 100 jobs seemed a boon, not only to the town's residents, but to loggers throughout the county.

A Dungannon woman, who had lived in other towns where this company had done business, raised questions about its worker-safety and environmental records. She organized a small group of citizens to advocate with company officials for improved in-plant safety procedures and for the installation of available pollution control equipment. Company officials said they could not afford the equipment and did not need additional safety procedures. They threatened to close up shop and leave town if forced into taking these measures.

Over the next few years, a divisive and sometimes frightening battle simmered in this tiny town. Company officials and many area loggers were pitted against a small group of environmentalists and, as the saga developed, against those organizing for a union. The local woman who had first organized the citizens' group found herself intimidated, harassed, and threatened. Even though she documented systematic misrepresentation of air pollution emissions by the company, she gained only marginal public support in the community. Many who privately supported her felt unable to speak out because their sons, brothers, or sisters had jobs at the plant.

It was the classic conflict between economic growth and environmental health faced by much of Central Appalachia. However, as with the citizens of Hancock County, local people in Dungannon and throughout Scott County were able to transcend the conventional impasse of jobs versus the environment.

Seven years after the wafer board fight began, the 20-year-old son of the local woman who had raised concerns, completed training and began his work as an environmentally sensitive logger, using horse power to replace conventional heavy machinery. He is now one of a small but growing number of horse loggers in the region.

Building on both its cultural heritage and the beauty and rarity of species in the Clinch River, a local citizens' group began work to find ways to diversify Dungannon's economy, thereby reducing dependence on the wafer board factory or any other single employer. This local group, the Dungannon Development Commission, has initiated an ecotourism strategy to create jobs based on preservation of the region's ecological and cultural heritage.

This chapter examines community-based efforts under way in the Central Appalachian regions of northeast Tennessee and southwest Virginia. The jobs-or-environment conflicts described earlier are commonplace throughout the mountains. A regional not-for-profit organization, the Clinch Powell Sustainable Development Initiative, has provided the venue wherein these local groups can learn from and reinforce each other's work. It has also been the primary catalyst in the region for some innovative responses to the root problems of human and ecological well-being.

The first two sections provide a short summary of the economic and ecological issues facing this region and describe the origin of the Clinch Powell effort. The third section delineates the working definition of sustainability developed by the Clinch Powell Initiative and explains how this understanding was forged among a diverse set of participants. The fourth section explores strategies employed by Clinch Powell to foster sustainable development and some of the significant accomplishments to date. The final section provides an insider's evaluation of the progress and problems we have experienced, unexpected opportunities and obstacles, and what we have learned about the practice of sustainability.

ECONOMIC AND ECOLOGICAL OVERVIEW

In spite of relatively strong employment and continued economic growth in Virginia and Tennessee, the Appalachian sections of these states continue to stagnate economically. Although the Commonwealth of Virginia's January 1996 unemployment rate was 5%, rugged Dickenson County endured a rate of 19%, and neighboring counties fared only slightly better (Virginia Employment Commission 1996). Perhaps even more troubling has been the growing underemployment rate, with an increasing number of workers employed in part-time, low-wage jobs in either the service/retail sector or cut-and-sew factories. Many of these workers, a very large proportion of whom are women, work 25 to 30 hours per week, yet are still unable to make ends meet. Poverty rates are extremely high in some of the counties served by the Clinch Powell Initiative, including a rate of 29% in Lee County, Virginia,[1] and a rate of nearly 40% in Hancock County, Tennessee[2] (United States Census 1990).

Much of the region maintains good air quality and, in most cases, clean waterways. This relatively clean environment, along with its beautiful landscape and cultural heritage, contributes to the increasing popularity of the region among visitors and tourists.[3]

With coal mining employment steadily declining,[4] many local and regional officials are looking to the forests as a primary source of industry in the region. Enormous chip board factories are locating in Central Appalachia, facilities so large that they dwarf the wafer board plant in Dungannon. Most of these industries extract timber at a rapid rate, potentially well beyond the capacity of the forest to regenerate this renewable resource. Moreover, this type of timber production creates minimal economic value in the region, because much of the wood is used for low-value products.

Tobacco has been the mainstay of agriculture for generations in the mountains of Virginia and Tennessee. Beef and dairy operations have also been important. Green County is the largest producer of burley tobacco in Tennessee, and Washington County leads Virginia in burley production. With a decrease in smoking in the United States and increased importation of tobacco, tobacco production in the region has declined by almost 30% in the past 20 years (Purcell 1994). Apple orchards, once plentiful in the region, have now all but vanished.

One could argue that southwestern Virginia and northeast Tennessee have faced a jobs-or-environment dilemma for the past 50 years and have usually lost on both counts.

ORIGINS OF THE CLINCH POWELL INITIATIVE

For years, many public officials and private not-for-profit agencies have been working to overcome poverty in Central Appalachia. For much of that same time, grass-roots citizen groups and environmental coalitions have been fighting flagrant or dangerous abuses of the forest, soil, and waterways. More often than not these groups have been at odds with each other, and they have almost never worked cooperatively to achieve common goals.

In 1993 the situation began to change. A loose network of public agencies, private not-for-profits, entrepreneurs, and grass-roots organizations came together around a common issue: How can we build healthier local economies in the region and begin to regenerate and restore the natural ecosystems that are our home? United by this simple but formidable goal, state and federal agencies, planning and development districts, chambers of commerce, grass-roots citizen groups, conservation organizations, and some innovative entrepreneurs began a nine-month planning process in the hope of finding viable strategies and solutions. The conveners of the process were the Coalition for Jobs and the Environment, a regional citizens' coalition spanning eastern Tennessee and southwestern Virginia and the Appalachian Office of Justice and Peace, a community development office of the Richmond, Virginia, Catholic Diocese. With financial and technical assistance from Virginia's Center on Rural Development, these two organizations coordinated the planning process which culminated in the hands-on strategic plan *Sustainable Development in Northeast Tennessee and Southwest Virginia*. The plan, published in 1994, suggested seven directions for more sustainable and community-based economic development.[5]

Calling themselves the Clinch Powell Sustainable Development Forum, the original actors remained a voluntary association until October, 1995, when the organization incorporated as the Clinch Powell Sustainable Development Initiative (CPSDI)—a not for profit organization dedicated to the implementation of the strategies outlined in the Forum's original plan. To this date, the CPSDI has launched three major programs: sustainable forestry and value-added wood products, diversified and sustainable agriculture, and nature-based tourism. [The major activities associated with these programs are detailed later in this chapter.]

TOWARD A DEFINITION OF SUSTAINABILITY

Although the members of the Clinch Powell Sustainable Development Forum were aware and appreciative of the conceptual work under way regarding sustainability, their purpose from the outset was to define such development *in practice*. Because sustainability involves community and culture as much as the ecosystem, it must be unfolded by and with the people and community structures who will enact it over the long term. As Wendell Berry (1992) notes in "Preserving Wildness," it is one thing to preserve wilderness apart from human activity, the more challenging endeavor is to integrate human culture and economic activity into the constraints and relationships of the natural world. Aware also of the broad elements set forth by the Brundtland Commission (WCED 1987) and considering our own history and recent experience, the CPSDI has come to view sustainable development as an evolving process of integrating the culture and economy of people into the broader ecosystem of which we are a part.

Specifically, sustainability appears to include the following elements:

1. *Sustainability is good for people, enhancing not only economic opportunities, but people's skills and resourcefulness.* CPSDI board member Paul Kuczko tells what happened when, in 1992, he spoke with a group of high school students. "How many of you plan to stay in the area when you graduate?" he asked. Of 100 or so students present, a handful raised their hands. He then asked, "How many would stay if you could find jobs?" Now all but five or six raised their hands. In this region's extraction-oriented economy, it is not only coal and timber that leave the area en masse, but the community's youth and collective capacity for creativity and resourcefulness. By encouraging ecologically sustainable practices and focusing on moderately scaled businesses such as portable band mills, CPSDI helps to make economic opportunities more accessi-

ble to people of modest means and help them to tap into their own ideas and experiences.

2. *Sustainability is first of all local, that is, rooted in the human and ecological realities of particular bioregions.* Sustainability is, of course, a global concern, and many ecological problems transcend the boundaries of bioregion or nation. But as Berry (1992, 30) said of the grand environmental problems, "The large abuses exist within and because of a pattern of small abuses." Gross negligence on the part of Exxon and the captain of the *Exxon Valdez* were certainly behind the worst oil spill in our nation's history. But so too were we, as rapacious consumers of petroleum, driving our automobiles 2 1/2 times more than our parents' generation, yet demanding that energy costs remain low (Durning 1992). A healthy economy and natural world begin with the rejuvenation or creation of human cultures that are deeply connected to their place, interwoven with its particular constraints and opportunities.
3. *Sustainability is ecologically sound, working within the three basic principles governing the biosphere: diversity, community, and regeneration.* Diversity is the basis of ecological health, adaptability, and productivity. It is what gives ecosystems resilience under stress. Our modern economic and social arrangements, by contrast, are based largely on monoculture and specialization. The community of interrelationships describes not a conflict- free environment, but an intricate web of life in which stresses on one element of the web are felt, to some degree, along all its strands. A fragmented economy and an individualistic culture often ignore these interrelationships, especially when they are distant as to either time or place.

   Recycling of decaying and dead tissues, along with leaching of mineral nutrients, describes the regeneration constantly at work throughout the biosphere. This third ecological reality is beginning to impact our thinking, spurring recycling and composting efforts. It also encourages what is sometimes called full-cycle processing (i.e., identifying our waste—energy, trash, capital flows—and figuring how to recirculate these within our households, enterprises, and bioregions).
4. *Sustainability promotes local self-reliance within the larger context of the globe*—that is, with respect for those downstream, downwind, and downtown. Most rural communities in Central Appalachia—and for that matter, around the globe—have made the extraordinary transition from being largely self-reliant for basic needs to being almost entirely dependent on outside help: factories, capital, and "expertise." We are net importers of almost everything other than the few primary products (coal, timber, gas) on which we based this transition. Simultaneously, most urban areas, including towns within rural regions, have become net exporters of garbage (trash, medical waste, sewage, etc.). It is a mutually unhealthy relationship.

Local or regional self-reliance does not mean setting up trade barriers, but becoming accountable for our own needs and for the by-products and side effects of our activities. Urban gardens, community composting, and tree plantings are all examples of steps toward such accountability. Microenterprise and economic literacy efforts are also essential parts of increased self-reliance, for they help bring economically marginalized people into the community's life and, eventually, we believe, into its decision-making structures as well.

5. *Sustainability lasts—indefinitely.* Virtually every definition of sustainable development includes *preserving the ecosystem for future generations* as a central element. Obviously, this is fundamental to any approach to sustainability. Yet our contemporary economic strategies devalue long-term resources and discount, in an economic sense, any future benefits in relation to present consumption. Overcoming this deep bias against the future will require entirely new tools for evaluating our social and economic health, such as the genuine progress indicator. This alternative to the gross domestic product provides a more accurate assessment of economic health by subtracting destructive activities (e.g., pollution) and by adding in the value of productive activities (e.g., gardening and volunteer work) that are done without remuneration (Cobb et al. 1995). But a healthy natural world 100 years from now will also require new skills and new relationships between town and country, producer and consumer, people and the ecosystem. As a local logger, observing a denuded hillside remarked, "We hate what's happening to our mountains, what we're doing to them. But we don't have no choice" (Ft. Blackmore, Virginia, October 1994). See Figure 19.1. Creating alternative choices will require these renewed relationships and rejuvenated skills, along with new forms of service and product exchange infrastructures and networks that enable us to act creatively and frugally.

As the Clinch Powell Initiative developed parameters of sustainability, participants struggled to reconcile the desire for significant, often fundamental, long-term changes with the need to implement and build support for short term strategies. Often, when far-reaching social goals are espoused without benefit of local experience, or even a clear plan of action, the broader community becomes understandably alienated. Thus, each of CPSDI's early ideas for sustainable development was subjected to three direct questions:

- Is it good for people, especially those who are marginalized?
- Is it good for our natural environment?
- Will it work?

Figure 19.1 Conventional logging site, cut fall/winter 1995–1996, near Duffield (Scott County), Virginia.

Although we had no definite answers for any of these questions, it did help to focus our thinking and prioritize our plans.

DEVELOPING AN INFRASTRUCTURE FOR SUSTAINABILITY: THE CLINCH POWELL EXPERIENCE

There are many elements of our present economic system and attitudes that hinder sustainable development. First, our assessment tools, principally the gross domestic product and the discount rate, promote indiscriminate, quantitative growth while devaluing long-term, slowly-regenerating assets such as topsoil, groundwater, and forests. These tools likewise ignore regional and local realities—a 19% unemployment rate in some coal counties and poverty rates in excess of 30% in others is a story untold by our prevailing measurements of economic well-being. This short-term, quantitative growth bias is deeply rooted in both liberal and conservative political thinking.

Second, our tax system and laws, including international trade agreements, reward short-term financial gain and the mobility of capital at the expense of sustaining natural resources and stable communities. The current public abhorrence of regulations further perpetuates the transfer, from the private realm to the public sphere, of certain costs of doing business, such as pollution. This "externalizing of costs" is exemplified when coal mining activity polutes groundwater sources causing families and even whole communities to lose their wells. Although some coal companies compensate affected communities, in many instances, communities must wait for expensive publicly financed projects to restore their source of water. Moreover, legitimate frustrations with cumbersome or counterproductive government regulations—and there are plenty—have been exploited to once again set business interests against health, wage, and environmental needs.

Finally, there is a well-developed infrastructure for growth-driven, export-oriented, and externally based economic development in rural regions such as Central Appalachia. Key to this system is the construction of physical facilities—roads, utilities, and shell buildings—which, along with local, regional, and state officials, are principally designed to attract companies from outside. This industrial recruitment strategy has been and continues to be the principal economic development tool used in the southern Appalachians. An analysis by the *Lexington Herald Leader* (28 March 1993) found that neighboring Kentucky spent as much as $300,000 per job created in the late 1980s to bring particular corporations to the state. Economic development officials throughout the region often end up in a lowest-bidder contest that pits state against state, even county against county, an approach that also helped to foster a dependency mentality. It is common to hear citizens exclaim, "We need to bring some jobs in here."

All of the impediments discussed here will have to be addressed soon if we are to make sustainable development viable at either a local or global level. Changes in our tools for economic measurement or in the thinking and actions of policymakers would make much of our work easier, or at least more economically and politically feasible.

Recognizing the critical role of policy and institutional support, CPSDI has chosen to focus its activities at the ground level, that is, on the creation of an infrastructure for sustainability, the expectation has been that building systems, networks, and, eventually, local institutions that spawn and magnify sustainable development efforts will lead to more thoughtful policy and more significant and accurate evaluative tools.

The following paragraphs highlight a number of elements of the sustainability infrastructure that CPSDI is developing in support of its three main strategies of sustainable forestry and wood products, diversified and sustainable agriculture, and nature tourism.

Sustainable Forestry and Wood Products

Much of the area included in the Clinch Powell region is forest land, in excess of 60% of total land area in many counties (Eubanks and Frame 1984). Employment in forest-related industries, however, has declined even as rates of timber harvesting have increased in many counties. Manufacturers of furniture, cabinetry, and other wood products throughout the region complain that local wood is very difficult to obtain in usable condition.

The relatively high rates of timber harvest, the steady decline in jobs, and the lack of locally available lumber for manufacturing are due primarily to two factors: the increasing scale of most wood-processing industries locating in the region (i.e., pulp mills, chip board plants) and the lack of value-adding businesses and facilities at the local level.

We have a lot of timber, "mountains of hardwoods," as one publication described it (Eubanks and Frame 1984). Although there are still red oak, white oak, and other more desirable woods, there are much more popular and other lower-value species, owing in large part to the vast, often indiscriminate, logging that occurred during the past 10 to 40 years. The pulp and chip mills utilize huge quantities of smaller, lower-grade logs, thereby encouraging clear cutting, the use of mechanical skidders, and expansive road building through the forests. Driven by this fast-growing demand for cheap, low-grade logs, red oak, white oak, and walnut have become, in many ways, a by-product of our wood products industry rather than its anchor. The best of these logs are shipped to Japan for veneer; others are sawed locally and exported for further processing.

The Clinch Powell Initiative's wood products strategy has thus far had four central components, all driven by the intent to increase employment while reducing stress on the forest ecosystem:

Figure 19.2 Conventional logging site, High Knob, Virginia.

Figure 19.3 Horse logging site—after logging completed, High Knob, Virginia.

1. *Training loggers in environmentally sensitive harvesting practices,* especially the use of horses as the primary power for skidding logs out of the forest. This old-fashioned technique has been updated to reduce damage to the forest and risk to the logger and the animals. The training also includes elements of forest ecology, along with directional felling, chain saw safety, and tree selection that balances profitability with long-term forest stewardship.

   Over the past 18 months, about a dozen individuals have completed this training—five from our immediate region and the rest from neighboring areas. Of the five, two have started horse logging businesses and the other three are working to start such a business. It costs about $25,000 to capitalize a horse logging business, as compared with $100,000 or more for a one- or two-person conventional logging operation. See Figure 19.4.

   A more established entrepreneur with whom we work has an environmentally sensitive logging business using machinery. He follows the best management practices recommended by professional foresters, exercising selective cutting, protection of waterways, careful, minimal road construction, and related practices.
2. *Maximizing the value of each log harvested by sawing, drying, and manufacturing it into finished products,* such as cabinets, furniture, specialty flooring, and so forth. Probably the single biggest gap in this value-adding

Figure 19.4 Horse logging in Jefferson National Forest, November 1994, performed by Chad Miano, a trainee at this time (trained by CPSDI's Jason Rutledge).

Figure 19.5 Solar dry kiln, built by CPSDI and local volunteers, Fort Blackmore, Virginia. This was the first load of wood going in on June 29, 1995.

dimension of our infrastructure has been dry kiln capacity. Dried oak boards can bring three times as much as green boards.[6] The solar dry kiln we constructed in 1995 has nearly paid for itself with the first two loads of wood it finished. See Figure 19.5. We are developing plans for additional dry kilns, identifying and working with local entrepreneurs interested in building and/or operating such facilities. Kilns are relatively inexpensive, if modestly scaled, and are easily built close to the source of material and in tandem with sawmills.

The other element of primary processing we are promoting is the use of portable band saws. Although considerably slower than circular mills, their portability means they can be taken directly to a logging site, saving both the expense and potential damage of a large logging truck. Because band mills cut an extremely high-quality board, they can be utilized for custom cutting. They also generate far less sawdust waste than a conventional mill and yield about 20% more boards per log. Several local entrepreneurs own and use portable band mills and have, to varying degrees, been part of CPSDI trainings and demonstrations. See Figure 19.6.

The final component of keeping and adding value locally is the manufracture of secondary wood products (i.e., wood manufacturing facilities). The Tennessee entrepreneur who does environmentally sensitive mechanical logging has also built a wood manufacturing facility

Figure 19.6 Woodmizer portable bandmill sawing oak log during CPSDI demonstration, May 1995.

in an old school house. Along with his crew of five, he makes furnishings, cabinets, trim, and wall and flooring boards. Much as Clinch Powell is attempting to do, he has focused particularly on making high-value products from lower-value species: tongue-and-groove wallboards from poplar and beech, cabinets incorporating sycamore, and bookshelves using scrap oak.

In addition, CPSDI has developed networks of other local entrepreneurs, ranging from crafts people to cabinet and furniture makers, who are beginning to purchase the locally produced, sustainably harvested, kiln-dried lumber. Working with public sector agencies and private funders, we are actively developing plans for a regional wood products manufacturing center dedicated to production from local, sustainably harvested timber.

3. *Developing and applying certification standards* to ensure that sustainable forestry is employed. Through the collective efforts of Jason Rutledge (a pioneer in modern animal-powered extraction), the Nature Conservancy, professional foresters, and landowners, certification standards have been created and approved for use by Clinch Powell members. These rigorous standards encompass general principles of sustainable forestry and address region-specific issues and constraints for the Central Appalachian area.[7]

Only a few sustainably harvested logs have been sold under our certification umbrella. Significant complexities in the application of certification standards, especially over the long haul, remain. However, a test marketing of lumber and wood products bearing the certification will begin soon, while the Nature Conservancy and other members complete surveys and research intended to further our understanding of the relationship between certification, markets, and pricing.

4. *Accessing markets within and beyond our region* is the final major component of our sustainable forestry infrastructure. Obviously, the size, dependability, and value of markets are closely related to the quality and diversity of products, the level of processing, and certification. To date, CPSDI has pursued or cultivated three kinds of markets: sales of finished, dried lumber (primarily oak) to small and medium-sized manufacturers within the region, development of regional crafts associations and quasi-flexible manufacturing networks using small quantities of diverse wood species (e.g., oak, poplar, basswood, cherry, recycled pallet and barn wood), and exports of sustainable lumber to initiatives that are developing projects that showcase sustainability. Although we promote full-cycle processing within the region, we realize that exports of some of our wood can help educate the broader public and generate much needed demand for certified sustainable wood products.

Diversified and Sustainable Agriculture

The mountainous-to-hilly terrain of Central Appalachia, along with its generally moderate climate and abundant rainfall, forged an agriculture based largely on beef and dairy cows, intensive plots of tobacco, hillside orchards, and small truck farms along the bottomlands. Household gardens, even in the more mountainous areas, were also an integral part of the overall agricultural economy. The absence of large tracts of flat land and rich, deep soils engendered more diversified, small-scale farming systems in which household consumption and commercial production were likely to overlap. At the same time, it should be noted that tobacco, the region's most important cash crop, has required chemically intensive cultivation for decades, in terms of both fertilizer and pesticide use.

As in most of the nation, the number of household gardens has steadily declined in the region, and, concomitantly, the tradition of canning and preserving foods. Although gardens are probably still more common here than in more urbanized areas of the country, it is increasingly uncommon to find young people who garden, let alone "put up" their food. The raising of family hogs, chickens, and laying hens has also declined dramatically in importance, as people have become increasingly dependent on wages and purchasing power. Bartering systems that were once nearly universal are now more an artifact of language, as when people speak of going to the grocery store "to trade."

It is well documented that American agriculture is among the most productive in the world as to yield per acre or per hour of labor. However, the driving force behind American productivity—the infusion of capital in the form of heavy machines, the widespread use of pesticides, and reliance on petroleum-based fertilizers—is largely responsible for the precipitous decline in farm health and farming communities. As Austin (1994) pointed out, most of the expansion of the agribusiness sector over the past 70 years has occurred off the farm. The industries of packaging, marketing, transport, and off-farm inputs now dwarf farming itself, accounting for 90% of total agriculture-related sales. Farmers, on average, receive 25% or less of the final value of their products. This trend holds true in the Appalachian region, where, since 1980, tobacco farmers have seen a 30% decline in the real price of burley leaf, while the wholesale price charged by manufacturers has increased more than threefold (Purcell 1994).

The other driving force in agriculture is consumer demand for cheap, readily available products in all seasons. Indeed, U.S. consumers have been enormously successful in this regard. Americans spend about 10% of their total household income for food, as compared with 16% to 22% in Western Europe and nearly 50% in much of the developing world (World Bank 1990).

Although evidence of significant pesticide residues on our food has been inconsistent, the widespread contamination of wells and groundwater is thoroughly documented (Soule and Piper 1992). Whatever the indirect or cumulative health impacts for U.S. consumers, these are eclipsed by the 20,000 worldwide farm worker deaths and the 750,000 poisonings attributed to pesticides each year (Wright 1990). The fact that fruits and vegetables travel an average of 1,200 miles between field and American table means that most of us have no idea where our food is coming from, let alone the conditions under which it is being grown and picked.

In this context, CPSDI's Sustainable Agriculture strategy promotes more diversified and regenerative farm practices while creating high-value markets for fresh, as well as processed, agricultural products. Key to these objectives is the rekindling of relationships between producers and consumers, which for us is both a means and an end. There are three central elements to the sustainable agriculture infrastructure we are working to build:

1. *Training in intensive, sustainable production practices* for crop and livestock farmers, market gardeners, and Extension agents is accomplished through on-farm demonstrations, workshops, and a soon-to-open demonstration farm. Training focuses on biological production practices, organic disease and pest management, rotational grazing for live-

stock, composting and soil fertility, and innovative marketing opportunities. Biological or organic certification— an outgrowth of some of the trainings programs—is an essential means of gaining access to higher-value markets within and beyond the region.

We also use field demonstrations to build relationships with our household and restaurant partners. Although it is difficult for busy people to take the time to come to our farms, those who do so respond very positively. Many people seem to enjoy seeing their food grow and learning firsthand what is involved in ecologically oriented agriculture.

2. *We are continually cultivating and expanding efficient marketing networks* that directly link agricultural producers with consumers, trying to evaluate and improve them as we gain experience. At this point, our marketing efforts have focused on community-supported agriculture, restaurants, and wholesale and retail outlets. (A fourth market, involving specialty preserved food items, is discussed separately in item 3.) The emphasis in the first three areas is largely on developing partnerships within our region.

   Community-supported agriculture (CSA) is an approach to farming that directly connects agricultural producers with consumers. CSA is based on the idea that healthy farms, thriving local communities, and a safe, healthful food supply are inextricably linked. Through CSA, town and suburban consumers gain access to fresh, safe, usually organic, food from farmers or gardeners in their region. In so doing, they help provide an expanded, more stable market for local farmers while reducing the enormous costs—economic and ecological—of packaging, advertising, refrigeration, and transport associated with our present international food system.

   Community-supported agriculture begins in the off-season. A farmers or a growers co-op recruits people to buy a share or subscription in their farm, the cost of which varies from $300 to $500. This share translates to a weekly basket of produce (and, in some cases, some preserved foods as well) for 22 to 26 weeks or more, depending on the growing season. On-farm events, ranging from hayrides to composting demonstrations, offer occasional outings and educational opportunities to participating households.

   In southwest Virginia and northeast Tennessee, the first CSA emerged in 1994, involving 13 households and one grower. In 1996, with help from the Clinch Powell Initiative, three separate CSAs involving more than 10 growers is providing 25 weeks of biological produce to 150 households in the region at an average cost of $13 per week.

   Regional restaurant marketing network. Now entering its third season, the Highlands Restaurant Marketing Network was initiated by three local growers who had been individually selling small quantities of produce to a handful of area eating establishments. In 1996, 18 to 20 restaurants—most of the finer establishments in the area—are expected to be regular buyers. Twenty different types of fruits and vegetables, plus herbs, are offered for a 26- to 30-week season, by 15 local growers. Organic beef and pork are being test-marketed to the restaurants this year as well.

Pricing is slightly higher than for conventional produce sold by major wholesalers, but considerably lower than the organic market price. A number of items previously offered to restaurants have been eliminated because of the producers' inability to be competitive with giant wholesalers. This is a difficult dance. Increasingly, our growers are focusing on those items in which they either have a compelling qualitative advantage or can offer diverse varieties otherwise unavailable to the restaurants. Tent cards and menu inserts that proclaim, "We buy local, biological produce . . ." are also being offered to the larger or more regular buyers. It is hoped that these will help solidify the restaurants' commitment to buy locally, as customers inquire and commend them for doing just that.

Retail and wholesale outlets. This is a new market for most Appalachian growers whose primary experience with retail has been farmers markets and small produce stands. Although sales in those arenas will continue, most of the newer outlets require certified organic produce and thus pay 10% to 20% more than the restaurants and 25% to 75% more than produce stands. The market in 1996 includes four or five health food stores, some specialty shops (ethnic and upscale stores), a major herb broker serving Knoxville, Tennessee, and a major grocery chain specializing in organic produce.

The grocery chain and about half of the health and specialty stores are located outside our region, representing our first significant "export" market. Expansion into these markets, particularly the grocery chain, may offer the simplest and greatest potential for increased sales at good prices to growers. However, this advantage must be weighed against the increased energy used in shipping.

Cooperative networks have opened these restaurant and household markets to smaller-scale farmers or to those entering the produce business for the first time. Most farmers do not want to undertake the more intensive, relational type of marketing these outlets require, often settling for more convenient but much lower-priced wholesale opportunities. Farmers participating in our cooperative networks get 80% of the final sale price of their goods, as compared with 25% to 50% in most other outlets. The 20% retained by the marketing network thus far has covered costs, although not without a good deal of volunteer time on the part of several growers.

3. *We promote value-added agricultural products.* In most respects, all of our marketing efforts that directly link producers with household or retail consumers are also value-adding or, perhaps more accurately, value-retaining endeavors. The way our food system works is that we typically pay a farmer the equivalent of $0.20 per pound for tomatoes, then pay another $0.40 to $0.60 per pound for shipping, packaging, refrigeration, and advertising, ultimately charging the farmer's neighbors $.80 to $1.00 per pound for tomatoes. Alternatively, through our regional marketing networks, the farmer receives 200% to 300% more for a tomato that ultimately costs his or her neighbors about the same as the grocery price, and is fresher and of higher quality. The economic value thus retained fits into the Rocky Mountain Institute's category of "plugging the leaks" (Kinsley 1994).

   Two other efforts under way in our region add economic value to local agricultural products. The first, conceived and launched by Dick Austin, one of the founders of our network, is an attractive, diverse assortment of peppers sold primarily through mail order, gourmet, and specialty retail shops. The "Peck of Peppers" includes more than two dozen types of biologically raised sweet, hot, and exotic peppers, each variety labeled and described. A custom-designed pepper cookbook and anthology is an essential component of this package. As demand for our Pecks has grown, Austin has begun to subcontract the pepper production to other growers in the region. The price paid to each grower is the same he or she receives for peppers bound for the CSA or retail outlets.

   The second value-adding effort involves processed foods, including such items as sun-dried tomatoes, dried herbs and peppers, garlic jelly, sorghum molasses, jams and honey, herbal teas, loofah sponges, and more. Most of these items have been produced and sold by various local farmers for some time. The Clinch Powell Initiative is augmenting their efforts in two ways. First, we are beginning to test-market various assortments of these products, in some cases packaged in a basket handmade from local materials. Second, we are working to develop the processing facilities that will allow for more efficient production of a broader array of high-value food products and personal items. One of CPSDI's members, the Jubilee Project, has nearly completed construction of a small business incubator that will include a commercial kitchen and offer production and marketing assistance to food products entrepreneurs. We are also assisting home-based entrepreneurs by providing information on health regulations, as well as our joint marketing efforts.

   These value-added efforts are quite young and need further testing and development. It seems clear, nonetheless, that they complement our marketing networks and our training and outreach efforts. These three elements—along with the cooperative seed, equipment, and fertilizer purchasing which has begun to emerge—represent the central components of a more localized, sustainable agriculture infrastructure.

Nature Tourism

The third strategy that has emerged from CPSDI's sustainable development efforts is nature tourism. Although it is just getting under way, it is important to consider, because tourism in our region and in other scenic rural areas has become a key element of many economic development plans.

Our nature tourism strategy arose from both positive and defensive motivations. Positive incentives were provided by our region itself, which includes an extraordinary wealth of natural beauty and unique resources. There are numerous caves; relatively pristine rivers, mountains, and hardwood forests; small farms; and historical landmarks. There is also an unusually high concentration of rare and endangered plant and animal species, especially associated with the Clinch River. The regional office of the Nature Conservancy, one of our founding members, is actively working to preserve these species while collaborating with local community organizations to create economic and civic benefits from conservation of their habitats.

Our nature tourism strategy is attempting to give people an opportunity for recreation—canoeing, hiking, caving, sightseeing—as well as for learning how they can contribute to the preservation of these natural wonders. Thus the upcoming cultural and natural heritage festival in Scott County will include not only hikes on mountain trails, but opportunities to visit or learn about horse logging and the solar kiln. Similarly, canoe races down the Clinch River will be followed by presentations about the endangered species of freshwater mussels and the importance of water quality.

Our defensive motivations were spawned by our close proximity to Pigeon Forge and Gatlinburg, Tennessee, two of the most overdeveloped, congested, and consumptive tourist destinations in the nation. Because tourism is clearly being touted by local and regional officials, CPSDI members have been working to preclude theme parks or outlet malls that caricature our culture and create congestion, pollution, and waste stream problems.

To date, CPSDI has helped with the planning and marketing efforts of two fledgling nature tourism efforts, one based along the Clinch River in Scott County, Virginia, the other focused more on mountain trails, biking, and sightseeing in a three-county area of the region.

Entrepreneurial Training and Support

Each of the three major strategies of the Clinch Powell Initiative—sustainable forestry, sustainable agriculture, and nature tourism—are undergirded by an emphasis on entrepreneurship and innovation. Three of the eight member organizations of CPSDI are directly involved in entrepreneurial literacy, training, and support. The Business Start microenterprise program, initiated in 1993, was one of the first microloan funds in the region. Since it began training prospective entrepreneurs in early 1994, 40 small business loans have been made and more than 200 individuals have completed a Business Basics course. The Lonesome Pine Office on Youth has begun a youth entrepreneurship program, using the structure of Rural Entrepreneurship Through Action Learning. Youth development and entrepreneurship is also a central element of the work of the Jubilee Project mentioned earlier. Working with Business Start, Jubilee staff have adapted the microenterprise training curriculum to further its accessibility to lower-income, lower-literacy individuals.

In an area in which 65% of all businesses employ 10 or fewer people, microenterprise is a critical component of economic revitalization (People, Incorporated 1992). It is well suited to individuals with limited means, both because of the extensive training and technical support provided and the generally modest scale of capital investment (the average loan size in the first two years of Business Start's program has been just over $12,000). Although problems sometimes arise in undercapitalizing a business, the loan limitations impose a modest scale which can also engender resourcefulness. This has certainly been the case with a reuse and consignment shop that has turned secondhand clothes, furnishings, and appliances into one of Dickenson County's most attractive, thriving businesses. Such resourcefulness is also represented by youth craft persons using the abundant local cedar and by a custom band sawing operation that utilizes small "left-over" oak logs as substrate for a successful shitake mushroom business.

The limitations of individualized small business loans include the need for extensive training and technical assistance, lack of access to higher-value markets, and a certain redundancy of ideas. This last point should not be overlooked. Most people who start their own businesses do so based on either a long-held dream or a set of skills and experience with which they feel comfortable. Although this approach can sometimes work, most conventional small businesses, be they restaurants, sewing shops, or video stores, cannot long stand up against the consumer-driven international market. And many are not particularly sustainable either.

Gaining access to higher-value markets and generating more innovative business ideas are both nurtured by three types of networks: those that link producers with consumers, those that link businesses with not-for-profit and public agencies, and those that link entrepreneurs with one another. Direct links between producers and consumers, as occur in community-supported agriculture and our restaurant marketing network, create loops for very rapid feedback. People tell producers what they like, what they do not like, and what they hope to see developed. This spurs innovation, diversification of products, and commitment to quality. Personalizing what are generally anonymous market relationships also seems to gradually instill a greater degree of commitment to local enterprises by some consumers.

Networks that link local businesses or potential entrepreneurs with not-for-profit organizations and public agencies help create economic opportunities out of basic community needs. In many parts of our region, for example, limited land availability and poor soil percolation make conventional sewage treatment approaches very difficult or expensive. Five years ago a local Community Action agency teamed up with biologists and private contractors to pioneer some of the first "constructed wetlands" to be used for sewage treatment in the region. Under most circumstances, these systems are less expensive to build and to operate than conventional treatment facilities and are more ecologically efficient at purifying wastewater. Similarly, the Nature Conservancy is subcontracting with a local fencing company—one that received microenterprise training and loan funds—to protect sensitive watershed areas by fencing out cattle. In both of these instances, a pressing community need has spurred innovative, ecologically sound, private businesses through linkages of public and private, not-for-profit networks.

The third type of network links entrepreneurs with one another in temporary, market-driven relationships called flexible manufacturing networks (FMNs). FMNs bring several small businesses together to design, manufacture, and market products that would be too expensive or sophisticated for any one of the businesses to produce on its own. Often FMNs develop in tandem with a product design center or incubator, which helps provide access to needed equipment and critical market information. The Appalachian Center for Economic Networks' Adaptable Designs-Adjustable Systems (ADAS) is an example of an FMN. Based in Athens, Ohio, ADAS manufactures handicap-adapted kitchen components for homes and apartments. First initiated by ACEnet, its not-for-profit parent, this FMN involved six local firms in the design and development of the original components and many more in their production.[8]

As yet, we have started no FMNs through the Clinch Powell Initiative. However, we are developing plans for a sustainable wood products manufacturing center that we hope to locate in an unused shell building originally constructed for industrial recruitment. This facility will link ecologically sustainable loggers, local sawyers, and dry kiln operators into a system that manufactures a line of high-quality, certified sustainable wood products. Closer to completion is Jubilee Project's small business incubator in Hancock County, Tennessee. In addition to support for individual business tenants, this facility and its management are being designed to serve as a hub for food products networks such as those described in the earlier disucssion of sustainable agriculture.

The entrepreneurial literacy and training program, the microloan program, the small business incubator, and the three forms of networks serve as the final components of our "infrastructure for sustainability." Do they work? Are they helping vis-á-vis industrial recruitment or export-driven growth? Our efforts clearly are too young to provide answers to these questions. However, what we have learned thus far and the insights we are gaining as to how healthy rural communities and environments can be cultivated are highlighted in the following section.

LESSONS OF EXPERIENCE

Although it is too soon to see a significant impact on the region's economy or natural environment, the CPSDI experience to date suggests that the key elements of sustainable development fit well within the culture and natural resources of rural regions such as Central Appalachia. There are strong historical and cultural roots for sustainability. Local self-reliance and individual resourcefulness were, until the early part of the twentieth century, the norm in Central Appalachia and much of rural America. This greater self-reliance arose primarily out of the need for frugality and the value attached to self-help, which usually spawned a diversity of household enterprises. There was also a more common use of skills to take raw materials and craft them into finished, useful products, activities that we now term adding value.

The world of commerce and human relations is, of course, considerably more complex today than in those more self-reliant times. CPSDI's strategies are based not on a yearning for simpler times, but on an intent to harness effectively the inherent strengths of our communities: close human relationships, celebrated connections to history and culture, and efficient, regenerative utilization of our forests, farms, and waterways. Our approach also recognizes that there are limits to both local resources and global ecological sinks (i.e., the capacity of the biosphere to absorb our waste and pollution).

Because of these limitations, in part, and because of the stresses of modern urban life, there appears to be a reawakening of interest in quality, durability, and health. The biophysical limits of our planet demand that we live more sustainably; the growing desire for health and meaningful relationships creates insights as to how we might achieve such sustainability. The tremendous growth observed in the demand for local, biologically raised products (from 13 households and 6 restaurants in 1994 to 150 households, 18 restaurants, and several retail outlets just two years later) is surely an indicator of growing public interest.

In terms of economic impact, our efforts are still on the periphery. It is instructive, nevertheless, to consider the effectiveness of the staff and organizational resources employed in stimulating economic development. The

Business Start microenterprise program has created 40 small enterprises, 80% of which are still operating. These businesses employ about 40 people full-time and an equivalent number of part-time workers. Not counting the business training provided to nearly 200 additional individuals, some of whom secured loans and capital from other sources, these 40 full-time and 40 part-time jobs were created with the expenditure of $550,000 in funds, much of which will eventually revolve within the loan pool. Discounting both the training benefits and future loans made from this pool, the microenterprise program has created jobs at an average cost of less than $10,000, as compared with the staggering costs of industrial recruitment mentioned earlier on page 389.

Within the agricultural arena, it is far too soon to determine whether our diversified, local marketing efforts can provide viable alternatives for the region's thousands of tobacco growers and other small farmers. It is clear, however, that these systems provide profitable alternatives for at least some local farmers and food processors. One local grower involved in our networks has averaged more than $5,000 in sales from one tenth of an acre these past two years. This compares favorably to an average of $4,000 to $4,500 per acre for tobacco farming in the region. Obviously, the total market for fresh produce and organic meats is limited, but it is also largely untapped. It could clearly accommodate many more producers.

The globalization of our economy has exacerbated the problem of price competitiveness. Consumer prices for food, timber and wood products, and many other basic commodities are low, in part because of technological advances and economies of scale, but also because of the externalizing of health and environmental costs. Although the average consumer supports the idea of sustainable farms and forests, people often do not understand how this externalizing of costs is largely responsible for the cheap and easy access to their food and fiber. Even the most efficient horse loggers and biological producers cannot win this price war, because they are largely internalizing the costs of conservation.

Full-cost pricing will be extremely difficult in the global marketplace. There seem to be three components to achieving something approaching price competitiveness in this environment. The first is certification. It is clear from both our local experience and national trends that a growing number of people are willing to pay more for foods certified as organic or otherwise more ecological in origin (e.g., meats free of steroids and antibiotics). Surveys also indicate a past willingness on the part of consumers to pay more for certified sustainable wood products, although there is considerably less experience with actual consumer behavior in the marketplace in the United States.

Judging from trends in both organic agriculture and wood product certification, these fledgling industries will increasingly be internationalized as well. That is why developing relationships between producers and consumers, the second element of price competitiveness, is probably the most critical. As mentioned earlier, developing and maintaining these relationships takes a good deal of work and creativity. In a society as fast paced and convenience oriented as ours, local producers must have a compelling advantage in the quality, variety, or healthfulness of their products if they are to reach more than a fringe of people accustomed to the ease and availability of Wal-Mart and Kroger. Certification can certainly reinforce this advantage, especially if it is somehow specific to or monitored within the local region, as is our Environmentally Harvested Forest Products certification. But most people are concerned more with how something looks, feels, or tastes, and, to a lesser extent, with who grew it or made it. Direct relationships between town and country, producer and consumer enable these associations.

The third element of price competitiveness is the availability of marketing systems or networks that allow producers to gain a larger share of the value of their product. We have been more successful in this regard with our agricultural efforts than with our wood products, because the latter is more complex and involves more processing steps between field and home, or even between logger and cabinetmaker. Yet providing primary producers—farmers, loggers, fisherman, and others—and related entrepreneurs with a greater value for their products is an essential component to more sustainable, less extractive practices.

The importance of product pricing goes beyond the viability of smaller scale producers, for it gets to the heart of our consumption habits. As our median income has risen, so too have our appetites for consumer goods, which now comprise nearly two-thirds of the entire U.S. GDP (Cobb et al. 1995). This consumerism is obviously fueled by advertising and enabled by malls and discount stores. But it is also becoming an ever greater part of our lives, as we relinquish our basic skills and capacity for productive activities in our own homes and communities. Confronting the question of product price—Can I afford an extra 15% for a table made from certified sustainable sources?—in fact confronts our life-style choices and our economic priorities. And this is not simply a question for the middle class. As David Korten (1994) has pointed out, achieving what he aptly calls "sustainable livelihoods" will require not only better employment opportunities, but more productive activities and more frugal behavior at home and in our local communities. We believe that this frugality and productivity will be enabled by the rekindling of community-based cooperative associations and the development of the types of regional networks described in this chapter.

The final lesson of our experience is the critical role that not-for-profit organizations must play in stimulating and coordinating sustainable development efforts. Not-for-profits are critical for several reasons. First of all, they help convene a diversity of participants ranging from low-income community leaders to public officials, because of the trust and credibility they have developed. This is particularly the case in regard to the more action-oriented not-for-profits that have developed a reputation for achieving results. Second, not-for-profits can act as catalysts, introducing new ideas or taking risks that public sector agencies, universities, and other larger institutions are more hesitant to approach. CPSDI for example, has found a helpful partner in the Cooperative Extension Service, but only after some of our pilot production and marketing efforts had proved successful.

Finally, not-for-profits have the potential to integrate diverse, ostensibly competing interests into common goals. They accomplish this by bringing together a broad base of participants, out of which are created networks of action and learning. These networks become the catalysts for innovation, the bridge across differing ideas, and the glue that helps solidify a sustainability infrastructure. They can be difficult to fund and require very strong, imaginative leadership, but they are almost certainly a sine qua non of community-based sustainable development.

NOTES

1. LENOWISCO Planning District Commission (Duffield, Virginia) supplied the poverty rate for Lee County, Virginia, using poverty data derived from the 1990 United States Census.
2. First Tennessee Development District (Kingsport, Tennessee) supplied the poverty rate for Hancock County, Tennessee, using poverty data derived from the 1990 United States Census.
3. Although industrial pollution is generally lower in this region, deep scars remain in the environment from the extraction of natural resources, particularly coal. Mining reclamation improved considerably with the passage of the Surface Mining Reclamation Act of 1975; however, many abandoned mine sites continue to contribute acid drainage to local streams and rivers.
4. Long wall mining, the newest and most "productive" form of deep mining, has created significant problems with subsidence, sinking wells, cracking foundations of homes, and fracturing water sources for livestock. Long wall mining has also contributed to the continued decline in coal mining employment, as it greatly reduces the number of miners needed to extract coal. The *Kingsport Times News* (14 April 1996) cited a report from the Virginia Department of Mines, Minerals and Energy documenting a nearly 50% decline in coal mining employment in Virginia between 1984 and 1995.
5. To order a copy of the Clinch Powell Sustainable Development Forum's publication, *Sustainable Development in Northeast Tennessee and Southwest Virginia,* or for more information on the activities described in this chapter, contact CPSDI, P. O. Box 791, Abingdon, Virginia 24212. Phone (540) 623-1121.
6. The current price for high-quality dried oak boards is about $1,500 per thousand board feet, as compared with a green board price of roughly $450 to $500 per thousand board feet.

7. For a copy of the Provisional Standards for Environmentally Harvested Forest Products in the Appalachian Region, contact CPSDI at (540) 623-1121 or Jason Rutledge of the Environmentally Sensitive Logging and Lumber Company at (540) 651-6155.
8. For more information on the Appalachian Center for Economic Networks and its flexible manufacturing networks, contact June Holly, president, at (614) 592-3854, ACEnet, 94 Columbus Road, Athens, OH 45701.

REFERENCES

Austin, D. 1994. *The Spiritual Crisis in Modern Agriculture.* Dungannon, Va.

Berry, W. 1992. *Sex, Economy, Freedom and Community.* New York: Pantheon Books.

Clinch Powell Sustainable Development Forum. 1994. *Sustainable Development in Northeast Tennessee and Southwest Virginia.* Abingdon, Va.: Clinch Powell Sustainable Development Initiative.

Cobb, C., T. Halstead, and J. Rowe. 1995. *The Genuine Progress Indicator.* San Francisco: Redefining Progress.

Durning, A. 1992. "The Conundrum of Consumption." *How Much Is Enough?* New York: W. W. Norton and Company, 19–25.

Eubanks, S., and E. Frame. 1984. *Mountains of Hardwoods: Virginia's Southern Mountain Region.* Charlottesville: Virginia Division of Forestry.

Kinsley, M. 1994. *Economic Renewal Guide: How to Develop a Sustainable Economy Through Community Collaboration.* Old Snowmass, Col.: Rocky Mountain Institute.

Korten, D. C. 1994. "Sustainable Livelihoods: Redefining the Global Social Crisis." *Earth Ethics (*Fall):8–13.

People, Incorporated. 1992. *Growing the Economy, Creating Economic Opportunity.* Abingdon, Va.: Commonwealth of Virginia Center on Rural Development.

Purcell, W. 1994. *The Virginia Tobacco Industry in a World of Change.* Virginia Tech Rural Economic Analysis Program. Blacksburg, Va.

Soule, J. D., and J. K. Piper. 1992. *Farming in Nature's Image: An Ecological Approach to Agriculture.* Washington D. C.: Island Press.

Virginia Employment Commission. 1996. *Estimated Labor Force Components for State, MSAS, LMAS, Cities, Single Counties.* Vec. Norton, Va: VEC.

WCED (World Commission on Environment and Development). 1987. *Our Common Future.* Oxford: Oxford University Press

World Bank. 1990. *World Development Report–1990: Poverty.* New York: Oxford University Press.

Wright, A. 1990. *The Death of Ramon Gonzales: The Modern Agricultural Dilemma.* Austin: University of Texas Press.

Twenty

Equity and Sustainable Development: Community Self-Empowerment in Three Indigenous Communities

Nola-Kate Seymoar

INTRODUCTION

The current working definition of sustainable development is that it is development that improves the economic, environmental, and social well-being of current and future generations. Since the Brundtland Commission's report (World Commission on Environment and Development 1987), the interplay between economic development and environmental protection has received much attention. International agreements negotiated at the Rio Summit and the Montreal Protocol have resulted in numerous government regulations and policies related to the environment. However, when the United Nation's (UN) Social Summit and the Population Conference focused on the social aspect of sustainable development, discussion usually centered on population, health, and education. Access to political power and redistribution of wealth or land—the major equity issues—were raised by women at the Beijing Conference and in Vienna at the meetings on human rights. However, in mainstream sustainable development literature, less attention has been paid to these social issues. This chapter examines three indigenous communities in Canada from the perspective of the "self-empowerment cycle." It offers a descriptive account of the communities' environmental, economic, and political fight for recognition and self-determination and discusses the lessons learned in terms of community change and empowerment.

The three Canadian communities—Walpole Island First Nation, the most southern "Indian reserve" situated between Ontario and Michigan; Sanikiluaq, an Inuit village in the eastern Hudson Bay; and Oujé-Bougoumou, a Cree community located in northern Quebec (see Figure 20.1)—were chosen in 1995 from close to 400 nominations from around

Figure 20.1 The communities: (1) Sanikiluaq; (2) Oujé-Bougoumou; (3) Walpole Island.

the world made to the We the Peoples: 50 Communities Awards program—a program in honor of the 50th anniversary of the United Nations.[1] These communities are part of a set of 50 award winners chosen as examples of success in development categories important to the UN. An independent international panel representing all sectors of society chose them according to nomination criteria that specifically expressed a preference for models that "provide a foundation for creating just, inclusive, and sustainable communities rooted in place and capable of cooperation with others" (We the Peoples 1993). The selection process aimed at choosing the communities to represent as wide a diversity as possible, to highlight solutions of economically poor communities, and to incorporate both UN-supported initiatives and independent citizens' initiatives.

What made these communities so special, and what lessons do they have for rural development?

WALPOLE ISLAND FIRST NATION

Walpole Island First Nation (WIFN) won in the category of Environment and Sustainable Development. The following data are drawn directly from the community's self-study and its presentation to the study conference in New York.[2]

Walpole Island Indian Reserve is nestled between Ontario and Michigan at the mouth of the St. Clair River. Occupied by aboriginal people for thousands of years, it is today home to 2,000 Ojibwa, Potawatomi, and Ottawa. Having a common heritage, in 1940 they formed the Council of Three Fires, modeled after the Three Fires Confederacy—a political and cultural compact that has survived the test of time.

The "Island" is blessed with a unique ecosystem including 6,900 hectares of the richest and most diverse wetlands in all of the Great Lakes region. Walpole Island is also known for its rare flora and fauna. Citizens of this First Nation, incredibly, can still support their families through hunting, fishing, trapping, and guiding activities. The number one industry in the community is recreation and tourism, The second largest is agriculture. Nearly 12,000 acres, mostly corn and soya beans, are under cultivation.

Over the past 20 years Walpole Island has been subjected to an onslaught of pollutants and environmental threats. First, upstream is Canada's major petrochemical and refining region, called Chemical Valley, in Sarnia, Ontario. Between 1974 and 1986 a total of 32 major spills, as well as hundreds of minor ones, involved 10 metric tons of pollutants. Since 1986, the Ontario Ministry of Environment has recorded an average of 100 spills per year. Although the number of these spills has been decreasing, Walpole residents remark that a reduction from three spills per week to one per week is not good enough. The reduction proves that industry can control spillage, and WIFN argues that zero discharge is imperative.

Second, passing ocean-going freighters have been a constant reminder that a *Valdez*-type disaster was possible. These ships were responsible for introducing the menacing and resilient zebra mussels to Lake St. Clair and its surrounding wetlands. Another prolific foreign invader, the purple loosestrife, is crowding out everything in its path.

Third, significant agricultural runoff of pesticides and fertilizers has been a major nonpoint pollution source. In addition, once-popular beaches are closed for weeks on end because of high levels of bacteria.

The dredging of contaminated sediments in the surrounding waters posed yet another serious environmental problem. Environmental degradation has had significant implications for wildlife and its habitat, human health and well-being, and the community's economic development, which depends to a large degree on the viability of the natural resource base.

Clearly, Walpole Island First Nation has under gone great environmental stress. To aboriginal peoples, environmental concerns cannot be separated from issues of self-government. The health of the ecosystem is the very basis of the indigenous community's cultural, spiritual, and economic life. To respect and protect the circle of life for future generations requires community control over its land over decisions affecting the ecosystem.

In 1958, Walpole Island First Nation was the second band in Canada to assume responsibility for its own finances and, in 1965, was the first to achieve a form of local self-government. This was the result of considerable lobbying and confrontation with the federal department responsible for Indian affairs. The band then launched its first "capacity-building" projects, constructing a modern school, library, and bridge; developing a police force; and improving agriculture. In 1969, when a federal government white paper proposed to abolish Indian reserves and integrate aboriginal peoples into the Canadian mainstream, Walpole Island was among the groups that organized and successfully fought this move. A few years later, when the Supreme Court established that aboriginal peoples had land rights, Walpole began to research its homeland to establish the basis of its claim. But the process was moving at a snail's pace, so WIFN began an active campaign to protect its homeland—a campaign that involved confronting a number of environmental issues.

In addition to research, the community has undertaken protests, abatement practices, and advocacy work, including:

- Years of lobbying governments and working with Chemical Valley industries to reduce discharges;
- Being instrumental in conducting air-monitoring studies;
- Investigating lead-shot poisoning;

- Experimenting with conventional tilling and low-till farming techniques; and
- Intervening against several development permit applications by large corporations, which would affect either the ecozone or the land claims of the First Nation. A rotary kiln (incinerator) application by Laidlaw, a pipeline project (Interprovincial Pipelines) crossing the bed of the St. Clair River, and a proposed electrical bulk transmission line by Ontario Hydro were all subsequently denied or dropped by the corporations.

The community is also the only one in the region that continues to formally oppose the proposed discharge of 750 million gallons of contaminated water from a former fertilizer plant near Courtright into the St. Clair River.

In 1983, WIFN established its own cooperative community-based research and scientific organization, Nin.Da.Waab.Jig, which means "those who seek to find." The objectives of the center are to support the efforts of the band council and community to:

- Preserve and restore Walpole Island First Nations' natural and cultural heritage within its homeland;
- Restore the rights and improve the capacity to manage and govern the Walpole Island First Nation fairly, effectively, and efficiently; and
- Promote the sustainable development of the Walpole Island First Nation into the next century.

The process of change requires significant institutional adjustments to make way for social, cultural, economic, technological, and political development. Walpole Island developed a statement of principles, drafted reports, and created a concrete action plan that is now under way.

> At Walpole Island, we believe sustainable development must be defined in practical terms. The people of Walpole Island view life in a spiritual, holistic and dynamic way. As our ecosystem knows no political boundaries, neither should sustainable development. We know that we cannot do it alone. Only through an integrated approach will society be able to reconcile the environment with economic development to complete the circle. We have committed ourselves to this end. The future depends on it (Jacobs 1995).

SANIKILUAQ

Sanikiluaq is a village in the Belcher Islands whose 600 inhabitants depend largely on traditional livelihoods of hunting, fishing, and trapping.

Sanikiluaq's efforts focused on protecting its ecozone and preserving the knowledge of its Elders.[3] This community was a winner in the category of Cultural Development.

The work of the Sanikiluaq Environmental Committee (SEC) began in 1990 when Hydro-Quebec—a large public utility of the province of Quebec—announced plans to construct the Great Whale Project in northern Quebec.

> We were aware that there was going to be a major Hydro-Quebec project involving the damming of rivers in northern Quebec and construction of power stations on the Great Whale River. We were told that we should try to learn more and do something about the project because it was only 90 miles east of the Belcher Islands, and the contaminants it created could damage our ecosystem since it is very sensitive to man-made contaminants. (Arragutainaq 1995)

As the community learned more of Hydro-Quebec's intentions and of other hydroelectric developments in the region, the people became increasingly concerned about how it would affect the marine environment on which Belcher Island Inuit depend for food and sustenance:

> If Hydro-Quebec goes ahead, it could cause sickness to those people living on sea life. Inuit can't survive on food imported from the South, so it is important to know how Hydro-Quebec will affect our land and its people. We are very worried because if the whales, birds, fish and seal contain methyl-mercury, there will come a day when we are told not to eat them. We want to keep having the food we like to eat. . . . What will happen to the large families who can't afford to buy food when the food from the sea can no longer be eaten? (Makivik Society and Laval University 1992).

Leading scientists in both the public and private sectors had few answers to many of the community's original concerns and questions and admitted to the lack of baseline data to answer questions such as, What will the changes in the annual cycle of water flow mean for whales, fish, and seals? How will it affect fish and animal foods? Will it change the sea ice cycle? Will these changes affect the travel routes of sea animals and fish?

It also became apparent that Sanikiluaq presented significant challenges to Hydro-Quebec and to the governments of Canada and Quebec. The primary reason was that the Belcher Islands—situated only 90 miles from the proposed development—are outside the province of Quebec and within the overlapping political jurisdictions of the government of Canada, the government of the Northwest Territories, and the Nunavut Land Claim settlement area. The rivers and marine environment are under federal jurisdiction. The federal government also has fiduciary responsibilities to the indigenous peoples and is currently involved in numerous negotiations on both the settlement and implementation of aboriginal land claims. Other governments, including those of the Northwest Territories and provinces of Quebec, Ontario, and Manitoba, exert additional influence by virtue of the industrial activities they control and the land usage they encourage. As a result, there are many stakeholders with different, often competing, political-economic interests and cultural perspectives.

From an Inuit perspective, provincial and federal governments have allowed and continue to allow large-scale industrial projects to proceed in ways that are harmful to the environment, wildlife, and indigenous ways of life, because decision makers value neither the environment nor the knowledge of the people who have lived in and been sustained by that environment for thousands of years. However, they recognize the mutual benefits of sharing indigenous and scientific knowledge.

> Policy-makers and government bodies tend to make decisions for us without asking what we know. If, however, we were to combine science and indigenous knowledge, so much time and money could be saved. The two knowledges may not always fit closely together, but both sides can benefit so much from each other (Novalinga 1994).

In the course of seeing that the community's concerns for the Great Whale Project were addressed, the SEC undertook a number of initiatives to address the need for better participation and representation in the joint federal as well as provincial environmental review process, and a solid information base that incorporated traditional ecological knowledge and scientific information.

In close collaboration with two Ottawa based nongovernment organizations—the Canadian Arctic Resources Committee and the Rawson Academy of Aquatic Science—and with funding from private foundations, provincial and federal agencies, public utilities, and regional aboriginal organizations, the SEC developed the Hudson Bay Program, a joint program designed to:

- Identify key cumulative impacts of human activities, including industrial development, on marine and freshwater ecosystems of the Hudson and James Bay bioregion using scientific data and traditional ecological knowledge;
- Examine and propose cooperative processes for decision making among governments, developers, aboriginal peoples, and other stakeholders that will foster sustainable development in the Hudson and James Bay bioregion.

Out of this effort, the Hudson Bay Traditional Ecological Knowledge and Management Systems (TEKMS) Study emerged in order to communicate indigenous knowledge and information to scientists and government policymakers on how the Hudson Bay functions ecologically, changes Inuit and Cree have observed in the Hudson Bay environment since the 1940s, and traditional values and systems for managing the use of living resources in Hudson and James Bays.

The Hudson Bay TEKMS Study was organized as participatory action research. Overall, 14 Cree and 15 Inuit communities participated in the study, and a total of 78 individuals shared and contributed valuable knowledge. The majority of participants were either Elders or active hunters, trappers, and gatherers who grew up in times of early contacts with whites.

The study was conducted with an appreciation of the community's persepctive, which was expressed as follows:

> We are attempting to tap into traditional knowledge from around Hudson and James Bay because we will need some reference that we can look back on. We will need records.. . . Because if there is to be mining, oil exploration or hydroelectric dams, we know that they will affect our wildlife, ice and marine areas. We have knowledge in those areas. Our knowledge is in our head: our fathers' and grandfathers' knowledge is in our heads. We are trying to put that knowledge to use . . . because we can get a much better and bigger picture of the area when we have people from all areas of Hudson and James Bay participating (Kattuk 1992).

The TEKMS study included a strong focus on values, as evident in the following statement:

> When I was a young man, I happened to visit my grandfather in his camp. In those days, the otter was a very respected animal, and I arrived when he had just finished cooking one. My grandfather usually didn't do such things but, because the otter was such a respected animal, it was a treat for them to eat. It was like a feast. My grandfather reached for this box sitting beside him and brought out a very clean white sheet which he put on the ground. Then, he took out his plate with a spoon, put the cooked food into it. That day, I witnessed how carefully and methodically my grandfather arranged his table and food because of his great respect for this wild animal as food. In those days, our Elders truly appreciated their wild foods, especially the otter (Solomon 1993).

The study contributors knew exactly what they were talking about and were willing to share what they knew because of their respect for the environment, interest in each other, and trust in the study coordinators. When the right question was asked, they spoke with a strong feeling and sense of purpose. As a result, untapped knowledge of sea ice, river systems, currents, weather, animals, environmental well-being, and traditional management was documented in writing and on maps.

The community of Sanikiluaq and others that fought the Great Whale Project achieved success in November 1994, when the Premier of Quebec announced that the Great Whale Project was officially "on ice" until completion of a provincial energy debate and inquiry. Subsequently, Hydro-Quebec terminated its research activities, the provincial and federal panels suspended the joint environmental review assessment, and Hydro-Quebec and the northern Quebec Inuit extended their agreement-in-principle for four years. It is uncertain, therefore, whether the Great Whale hydroelectric development will proceed at all.

Like Walpole Island, Sanikiluaq's conflict with the dominant nonaboriginal interests arose out of environmental and political concerns. Both communities used participatory action research to document their cases and, in the process, built pride in their own knowledge. Sanikiluaq today is one of the most active communities in the creation of the new northern territory of Nunavut.

OUJÉ-BOUGOUMOU

Oujé-Bougoumou, which means "place where people come together," won in the Human Settlements category of the We the Peoples: 50 Communities Awards.[4] This community's story is one of a displaced people whose perseverance eventually led to the establishment of a place worthy of their name.

It was in the 1920s that mineral exploration brought white prospectors to the territory of the Chibougamau Crees, as they were then known. Many hunters welcomed the newcomers and the wages that could be earned by guiding and provisioning them, but by 1927 the impact of the new mining companies was felt as trees and dwellings were cleared for drilling sites. In the early 1960s the situation worsened when the indigenous Chibougamau people were "eliminated" on paper because a federal Indian agent declared them to be "strays" of the Mistassini Crees, 90 kilometers to the north.

From the 1950s to the mid-1980s, this community was forcibly relocated no less than seven times by mining, forestry, and government interests. As Chibougamau grew into a mining and lumbering town, the original tribe (the Oujé-Bougoumou) moved to nearby Hamel Island. Road crews cleared the island and used the sand for roads, and the next spring the rest of the island washed away. The community then moved to Swampy Point—the only camping site not staked out by a mining company. When the swampy conditions and mine effluent endangered their health, the people moved again, this time to Doré Lake. There, with the help of a church organization, in 1962 they built a new village, erected a community hall, and elected a chief to begin an organized drive for band status and their own reserve. Chief Mianscum eventually obtained promises for a reserve, welfare payments, and access to regular federal programs that had previously been denied. But the promises were broken in 1970, when a mining company needed sand from Doré Lake. Indian Affairs officials revived the fiction that the group belonged at Mistassini. They ordered the people to move and had the village destroyed.

The Oujé-Bougoumou dispersed. Some moved to Mistassini, Chibougamau, and nearby Chapais. Almost all the rest left for their traditional trapping territories where, clustered in six family groups, they built shacks beside logging roads on the margins of Canadian social, economic, and political life. They became what Quebec authorities labeled "squatters on provincial Crown lands," or as one newspaper headline later put it, "the forgotten Crees."

In the mid-1980s, the Oujé-Bougoumou Crees cried "enough" and decided to live up to their name and establish a place where people come together. They staked out a village site at Lake Opémisca and erected a tent colony. They claimed jurisdiction over their traditional territory of 10,000 square kilometers and blockaded the main logging road north. They received support from several well-known native leaders, including Grand Chief Matthew Coon-Come of the Quebec Cree, Chief Bernard Ominayak of the Lubicon Cree of Alberta, and Chief Daniel Ashini of the Labrador Innu. They formed a native defense alliance, vowing to stand together in the event of a police or army attack. Quebec quickly reopened negotiations and committed $25 million to construction of a village and socioeconomic programs. Eventually, Ottawa added $50 million and agreed to a process for the community's acquiring status under the James Bay Agreement.

The Oujé-Bougoumou then hired renowned native architect Douglas Cardinal to design the village's school, clinic, and administration building. He worked with the Elders to chose the site and design the buildings.

> The results, when you put the village into its cultural context, are striking. The village design is circular, with the shaptuwan (traditional meeting place for feasts) central and at the top of the hill. The inner two rings are lined with community buildings, reflecting the culture of sharing. The homes are built in clusters, just like the old camps. . . . There are no steps to the front doors of the elders residences. . . . Every room and office has a big window. . . . The doors of the homes face east, where the sun rises, as the elders demanded (Picard 1993).

The people devised a home-ownership program as part of their stated drive toward self-sufficiency. Band members built their own homes, training first as carpenters so that they could build to their own specifications rather than having to use prefabricated houses.

Innovative aspects of the village include not only the design itself but also the district heating system, which reduces total energy consumption, utilizes the energy of industrial waste from the nearby sawmills, reduces the cost of heating, creates local employment, and contributes to the overall sense of community self-sufficiency. The community purchased a fishing camp for development into a year-round tourist destination to further increase employment opportunities and community pride. Instead of police, Oujé-Bougoumou has peacekeepers.[5]

> If the people of Oujé-Bougoumou feel empowered by their new home, it is because they played an intimate part in its conception, creation and construction, and because the village is a living reflection of their culture and lifestyle (Picard 1993).

OVERVIEW OF THE COMMUNITY SELF-EMPOWERMENT CYCLE

What are the lessons to be learned from these three communities? Are there common issues and processes? When one examines the details of the three cases, it is clear that each had unique problems, challenges, setbacks, and opportunities. Sanikiliuaq and Walpole Island encountered an environmental threat from outside forces. Oujé-Bougoumou had nothing but moral courage and the indomitable will to fight for its land base and economic and political rights. Yet their stories, and indeed those of the majority of the We the Peoples: 50 Communities Award winners, seem to fit a pattern that I have heuristically called the "Community Self-Empowerment Cycle" (Seymoar 1996).[6]

Any attempt to build a theory suffers from the problem of generalization. It may be correct in describing the overall tendency or process, but each situation is unique and involves a whole host of idiosyncratic events. Communities and the processes of community change are vastly complex. This is perhaps why there have been few attempts in the field of sustainable development to build a theory of community change and empowerment.[6]

The first overall observation is that these communities had or developed a holistic and long-term vision of their development process. Each tackled economic, environmental, and social/political issues simultaneously, perceiving them to be inextricably linked. The framework of sustainable development is simply a modern way of reflecting the traditional cultural values of these indigenous communities.

A second and most important insight gained from examining the award winners in general, and these three in particular, is the normality and necessity of conflict. Conflict occurred at some point in the process of empowerment for every community or group. Empowerment involved a significant struggle between these communities and those who held power over their choices or resources. One of the characteristics that distinguished them from other not-so-successful communities seems to have been that they were able to manage the conflict process in such a way as to achieve their goals. Although confronting the established powers, they had access to mediators from within and outside the community and commitment to a set of values that rejected violence. These were key factors that contributed to the successful resolution of conflicts.

The third important conclusion drawn from this study is that the process of community empowerment is a cyclical one, which proceeds in a series of steps or cycles toward increased sharing of power. None of these communities had arrived at an end state of equity and sustainable development, yet all had made great progress toward such a goal. As each major inequity was identified, acted upon, and resolved, they achieved increased "space to negotiate" and confidence to identify and tackle other problems more easily, handling conflicts better and negotiating settlements more effectively. It seems that rather than picturing empowerment as a straight linear progression, one should consider it more like a series of cycles, whereby the resolution of each issue increases the group's choices and power (see Figure 20.2).

The fourth observation is that there is a distinct stage of nonengagement, when the newly empowered group or community withdraws and looks inward. At some point Walpole Island, and each of the Sanikiluaq, and Oujé-Bougoumou communities focused on learning from their Elders and building their own capacity to achieve their goals. This focus on their

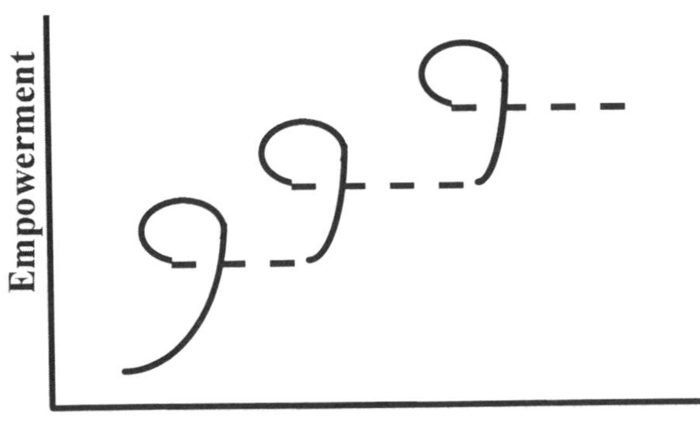

Figure 20.2 Increasing degrees of empowerment over time.

cultural identity was associated with a profound sense of spirituality and was followed by a reengagement with nonindigenous society. A stage of separation may be necessary before indigenous and nonindigenous groups can reengage in a more equitable partnership. Indeed, the selection of these three communities as successful had much to do with the fact that they have matured to a state of greater equality and mutual respect with their nonindigenous neighbors.

These observations—that holistic and long-term vision of community development is necessary, that conflict is an inevitable ingredient of empowerment, that achieving equity is a spiraling series of events not a linear progression, and that a period of separation occurs after the conflict—may seem almost self-evident. Yet the majority of workers and programs in the field of sustainable development proceed as though conflict will not occur and make no plans for conflict management. They are surprised and hurt when they become one of the targets of the community's expressed anger. Likewise, calls for separation are viewed as a failure of the group-building process, rather than as a demonstration of growth. Finally, development agencies of all kinds continue to believe the myth that they are working themselves out of a job. If empowerment is cyclical, then there will continue to be a need for engagement in an ongoing process of common development. The undertaking, which begins with providing external aid, if successful, will evolve into a partnership between equals. Thus a commitment to become involved in a process of change means a long-term commitment to stick around—to listen to grievances and make changes to resolve the conflicts, to respect separation, and eventually to enter into partnerships for mutual benefit.

The cycle involves at least four identifiable states or phases of relationship between two groups—one that begins with greater power (e.g., the dominant white society as represented in these three cases by governments and corporations) and one that begins from a state of lesser power (e.g., the indigenous communities). *Power,* in such instances, refers to access to resources (economic, social/political, and natural), the ability to influence decisions affecting one's life and that of future generations, and the degree of freedom of choice.

Inequity

The first phase or state of the relationship between the two groups (indigenous and nonindigenous) was one of *inequity*, or dependence. The process that triggered movement toward empowerment involved increased consciousness, on the part of each community, of its lack of equity or access to resources. Each of these communities perceived itself to be dependent and acted accordingly, as in the case of Oujé-Bougoumou, where some members were dependent on social assistance; or Sanikiluaq, at risk from a southern development over which it had no say; or Walpole Island First Nation, suffering from the results of upstream pollution.

These successful communities did not initiate changes until they experienced such severe difficulties (the threat to be integrated or to lose their livelihoods) and deprivation (including suffering from the avoidable diseases of poverty, alcoholism, and abuse) that they could no longer ignore their distress. The first stage of self-empowerment began with the community's feeling pain and arriving at a sufficiently clear analysis to envision a better way.

Conflict

Movement out of the phase of inequity inevitably involved blaming the power group. During this period the focus was outward, on the dominant white groups, and on analyzing the sources of inequity. Having analyzed the problems (in the cases of Walpole and Sanikiluaq, by using participatory action research) and identified a number of options, the indigenous groups made their proposals. Generally the dominant groups, whether governments or corporations, were unwilling to give up control over resources (natural, economic, or political), and the parties entered into a power struggle. This second state or phase of relationship was one of *conflict*.

Within these three Canadian communities, conflict took the form of protests, blockades, and soliciting media coverage, along with pressure on politicians and company executives. Legal briefs and actions were threatened and taken. The conflict stage lasted for several years before changes occurred. During this time the communities experienced a sense of exhilaration in that they were finally taking action for themselves. Certainly, they reported frequent disappointments and periods of discouragement, but the conflict helped forge their sense of identity and unity.

The significant elements in managing this part of the change process were conflict resolution and negotiation skills and structures. These skills and structures already existed within the communities' own culture, but needed to be accentuated and strengthened. Conflict was necessarily involved in the change process but had to be contained so as not to escalate into violence, which could have greatly harmed the parties to the dispute.

With the help of advisors (indigenous and nonindigenous), the communities became knowledgeable about their legal positions, both in regard to land claims and in regard to environmental laws and regulations. At the same time there was strong support for a moral code that valued traditional indigenous ways (based on cooperation and consensus) and the wisdom of the Elders. Two of the communities sucessfully appealed to the media and to activist groups in New York to mount a campaign, resulting in the cancellation of a major Hydro-Quebec power contract by New York state. Whether or not the dominant white companies and federal departments acknowledged it, during the conflict phase they lost their psychological advantage and authority (the moral high ground) and were pushed to negotiate greater economic and political participation with the indigenous peoples.

Independence

The move from the conflict phase to the *independence* phase involved the process of separation. Rather than defining themselves through simply rejecting and blaming the more powerful group, after their initial successes these communities withdrew from the active conflict, looked inward, and defined themselves from within. During this capacity-building phase of empowerment each one articulated and refined its "better" vision. Cultural awareness, values clarification, and self-help initiatives were undertaken during this phase. This separation and focus on independence and internal matters was the beginning of a vibrant cultural life, despite continued economic hardship and struggle.

Sanikiluaq, for example, having researched and tape-recorded the knowledge of its Elders, also assumed leadership in the move for an independent Inuit territory in the eastern Arctic to preserve its cultural heritage as well as to ensure its economic, political, and environmental well-being. The new focus among the Inuit themselves is to define what is unique about the Inuit way and to articulate it in a constitutional context.

Walpole Island First Nation also went through a period of focusing on the reserve itself, building a modern school, improving housing, and establishing training courses. The founding of Nin.Da.Waab.Jig served a purpose in building the community as well as arming it with research data to fight its battles.

Oujé-Bougoumou hired an Indian architect and concentrated on designing the community. The process was participatory and involved the entire community in identifying its members' own culture and values, which they wanted to be reflected in their public buildings and living spaces.

Partnership

Equality does not come from separation. If freedom from oppression came to these communities during their consciousness raising, conflict, and independence phases, freedom to be equals came from a form of reconciliation. Equality became real for most through a process whereby the community chose to interact with the dominant groups on matters of common interest.

All three communities have demonstrated a willingness and, often, a necessity to move beyond the past. In several situations this involved a recognition that there were other problems or superordinate goals that could be achieved only through joint efforts. The fourth phase was thus one of *partnership*.[7] This stage involved a real change in the distribution of political/social power and economic and environmental resources.

The attitude of the Oujé-Bougoumou toward governments, and white society in general, has matured over the years. From helpless dispossession, they have evolved into respected innovators and self-confident planners. In the words of Chief Abel Bosum, "Instead of winning people's sympathy, we are now gaining people's respect."

Walpole Island First Nation participates in various federal and provincial forums, along with other stakeholder groups. Many bureaucrats, politicians, and neighboring community leaders would no longer consider undertaking environmentally sensitive developments without consulting Walpole's leadership.

Finally, it must be noted that the self-empowerment cycle is drawn with a broken arrow leading to another phase of inequity (see Figure 20.3). Having worked through one set of issues successfully, each of the communities became aware of other inequities and tackled them in an ongoing process of development. It is hoped that as a new issue leads to a replay of the cycle, each time the process will be a little easier and the issue will be resolved more quickly. Progress toward increased empowerment

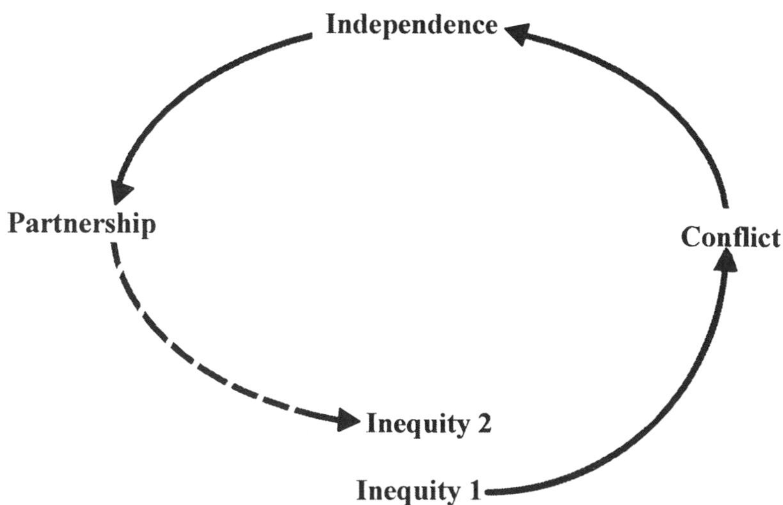

Figure 20.3 Phases of the self-empowerment cycle.

goes through a series of gains and periods of consolidation, followed by another round of conflict and negotiation.

IMPLICATIONS

What are the implications of this conceptual framework for those interested in sustainable human development or community empowerment—either professionals in the field or members of groups and communities who are lacking in power?

First, the concept of sustainable development is useful in examining community empowerment because it requires two things: a holistic approach that integrates social, economic, and environmental concerns and a long-term focus on current and future generations. Achievement of sustainable development involves dealing with issues of equity between and among generations, and between and among those considered developed and those described as developing or newly industrializing. Equity issues are about a community's power—power to make choices, power over resources, power to influence the decisions that affect it. Thus, commitment to sustainable development means learning to be a change agent, or if one really risks, becoming a changed agent.

The community self-empowerment cycle is a useful general framework for describing and predicting what will happen when the consciousness of inequity begins. With increased globalization of communications (TV and radio), commerce (transnational corporations and franchises), and the push for free trade and democratization, the poor are increasingly aware of their status as perceived by others. They experience being labeled *poor, underdeveloped,* or *powerless* and may react by withdrawing (the helpless and hopeless syndrome or escape into drugs, alcohol, violence). Or they may react by raising their own expectations to the degree that their leaders promise to eradicate poverty; achieve gender and racial equality; protect human rights, biodiversity, and cultural diversity; restore and protect eco-zones; and provide jobs for everyone. When their expectations are not met, it is no wonder that the "poor" become disenchanted and either sink into dependence or take action and enter the self-empowerment cycle.

The recognition of self-empowerment as a process that goes through predictable cycles is useful. It is not a failure when the community marches on the local legislature demanding change, blockades roads, seeks redress through the courts, or has a sit-in. Rather, such actions can be viewed as indicators of positive growth. Likewise, it is not a failure when demands for separation and partition occur. They are a predictable and necessary part of empowerment and establish the cultural identity of the community in a new and positive light. It is only after conflict and separation that a new and more equitable relationship can be developed. Partnerships and federations are the results of the recognition of both the independence and the separate identities of the parties and of a desire to achieve something together in association.

It is naïve to begin any rural development program without anticipating the inevitability of conflict and planning for it. Strengthening the experience of participants in negotiations, conflict resolution, and use of third party mediators is essential to success. This seems like common sense, yet the number of project proposals that include plans for training in conflict resolution are few and far between.

In keeping with the objectives of sustainable development, it is useful to focus the analysis of community needs on three areas—the economic, the social/political, and the environmental/natural—and the community's access to such resources. There are a number of techniques that help communities understand inequity. Participatory action research, which helps communities identify their local knowledge, was used in two of these communities. At the same time, the communities engaged in research into current science and technologies to identify important opportunities and options. This led the Walpole Island First Nation to become expert in chemicals and technologies for controlling discharges, and Oujé-Bougoumou to build a state-of-the-art biomass community heating system. The communities also tackled the influence of existing policies and regulations on their communities and articulated a compelling argument to change those policies.

These experiences reflect the model developed by the International Institute for Sustainable Development (IISD) in its program on Community Adaptation and Sustainable Livelihoods, based on work in 10 villages in five African countries (see Figure 20.4).

The working hypothesis is that successful adaptation to economic, social, or environmental shocks and stresses may be enhanced or inhibited by local knowledge, external policies, and various technologies. IISD's view is that sustainable development will be improved when communities draw on their own visions of their futures and define themselves in their own terms, but also make use of the policies and technologies of the rest of the world (Singh 1996).

But do these frameworks apply to nonindigenous rural communities seeking self-reliant development? The answer appears to be a qualified yes. Certainly other rural communities that also won the We the Peoples Awards reported a similar process of self-empowerment. These include ecovillages (like Findhorn in Scotland), the Finnish Village Action Movement, the Organization of Rural Associations for Progress in Zimbabwe, the Community Development Society of Alappuzha in India, and the Arkansas Land and Farm Development Corporation in the United States. Poor neighborhoods within larger cities also experienced a similar process of empowerment. Communities Organized for Public Service in San Antonio, Texas,

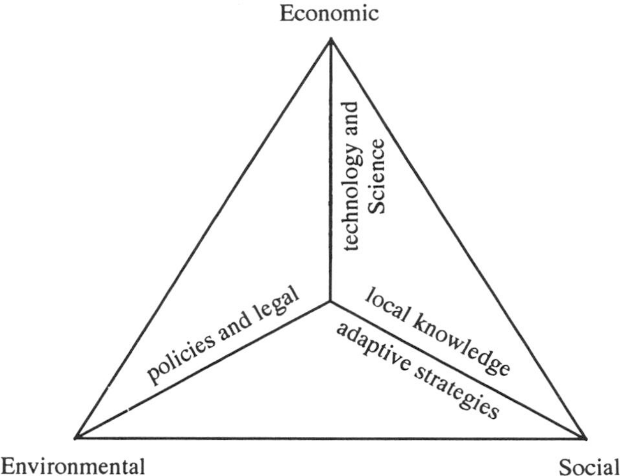

Figure 20.4 The community adaptation model.

and Whitfield in Dundee, Scotland, demonstrate the successful results of years of self-empowerment.[8]

There is a caveat, however. The indigenous communities described in this chapter were all in a state of real inequity. They were down and out. Generally, they were not middle class people relocated in a rural setting. Many North American rural communities, although facing economic stresses and difficulties, are not disempowered in the same way as these three indigenous communities. It is likely then that development agents working with comparatively well-off communities will find that the strategies suggested (in the following paragraphs) for the independence and partnership phases may be more applicabile than those given for the inequity and conflict stages.

Taking this self-empowerment cycle into account, in situations of real inequity, what then are the strategies that might be used to support change to sustainable development?

Analysis and Mobilization Strategies

Saul Alinsky (1972) used to say that people work on only "felt problems," and he advocated animating communities by "rubbing the sores raw." To raise consciousness about inequity and to initiate the process of self empowerment, awareness or consciousness-raising techniques are needed. Alinsky's books are full of examples, many of which are still relevant in the 1990s.

Participatory action research is also a valuable consciousness-raising tool, which highlights a community's strengths, as well as the problems it perceives itself to be facing. For the three indigenous communities discussed earlier, it was useful in clarifying and documenting the environmental threats facing them and in identifying how those threats affected the residents' livelihoods.

Free-floating anxiety about the future of the rural way of life must be channeled into much more specific analysis if a community is to be mobilized to take effective action.

Conflict Management and Resolution Strategies

In the three Canadian examples, the indigenous communities had a strong and deep-seated love of the natural world and respect for the process of consensus. Although they were torn by strife and social problems at various times, the appeal by community leaders to their traditional values and the wisdom of the Elders maintained nonviolence in situations that could easily have become violent. These native leaders were willing to discuss morality and spirituality and to reinforce the importance of dialogue.

All three communities required knowledge of their legal rights and of the rules and regulations and policies that had impact on their communities. The legal advisors for the First Nations had both a political and a constitutional perspective on the situation of the communities, and provided important insight into what was legal, what was outside the legal framework, and what was criminal. Strategic advice from activists on environmental, health, housing, and political issues was sought and used. Civil disobedience tactics were strategically employed to gain media and political attention, but were framed within a context of nonviolence. Those who were most trusted, and hence most useful, were people who had been involved with the community for some time on related issues.

The role of the media was crucial to the success of all three communities. CBC Radio and TV and *The Globe and Mail* exposed the plight of Oujé-Bougoumou and Walpole Island in particular and continued to follow the story over several months and years. CBC's northern news service followed the Sanikiluaq process. When the three communities were recognized with the We the Peoples: 50 Communities Awards, they received extensive coverage in the local press and significant coverage in the national media.

Strategies for Independence

During the independence phase, community capacity building benefited from visioning exercises, artistic events, cultural celebrations, and awareness programs. Building strategic plans and business plans was particularly important at this time because the communities were focusing not on what they opposed, but rather on how they wanted to be different. The importance of celebrating success was often noted.

For nonindigenous communities there is also a benefit to identifying and celebrating the community's own culture. The Finnish Village Action Movement, for example, grew out of a desire to strengthen the positive self-image and publicize the virtues of a rural culture.

Partnership Strategies

Third-party encouragement is often key to identifying areas of potential cooperation that will facilitate movement into the partnership phase. This involves encouraging frequent structured interaction between groups on neutral and potentially positive issues.

Farm-city vacation programs, cultural appreciation programs for nonindigenous visitors in indigenous communities, ecotourism, twinning of communities, and other projects that bring rural communities or rural-urban communities together may make a positive contribution to strengthening partnerships.

In Canada, as in many other countries, there is a consultation process that formalizes input to environmental impact assessments. There is also a formal process for land claims settlements and the recognition of treaty rights. All three communities became quite familiar with these procedures and used them to their advantage, while educating nonindigenous participants in the process.

After the Brundtland Commission completed its hearings across Canada, a multisectoral task force was formed. This was the predecessor of Canada's National and Provincial Roundtables on Environment and Economy. The roundtable process, as it has come to be known, has been extended to a variety of issues, including forestry, fishing, and community adaptation on the western plains. Multisectoral forums and events have come to have a recognized place in achieving partnerships. Like the President's Council on Sustainable Development in the United States, they provide a useful way of bringing diverse interests together to tackle large and overarching issues. Indigenous groups have a long history in the way of the circle and building consensus. There is much they can teach us if we are willing to learn.[9]

CONCLUSION

In discussing the lessons of these communities it is important to understand that they had different sets of circumstances and found different ways of overcoming their problems. Yet they identified with one another's struggles and applauded one another's successes. Although the specific content and actions were different, there were great similarities in the processes of change and empowerment they underwent.

This framework is one attempt to weave a conceptual net to capture the experiences of communities and groups involved in change. The value of any conceptual framework for social change is its ability to describe a process, predict behavior and outcomes, and provide guidance for interventions by local leaders or external facilitators. It is important to recognize that each community has unique strengths and gifts and faces different, increasingly complex factors as the forces of globalization impinge more and more on local reality. The concept of the Community Self-Empowerment Cycle will be valuable to communities, and to those of us involved in supporting citizens' initiatives toward sustainable development, to the extent that it provides an overall sense of a natural movement toward equity that involves a degree of conflict, separation, and reconciliation. If one accepts such a cyclical progression, then the timing of interventions such as consciousness raising, mediation, strategic planning, and roundtables becomes important to their success. Specific tactics and interventions, however, must always respond to the diversity of the individual circumstances and rest in the hands of the local people.

As pointed out at the beginning of this chapter, within the mainstream thinking on sustainable development there has not been as much focus on the issue of equity (the increased sharing of power and resources) as on the issues of the environment and the economy. The interrelated aspects of equity, the economy, and the environment are perhaps more apparent in the three indigenous communities discussed because they are so economically and culturally dependent on their environment. Yet we are all interdependent in this same way. Those of us who aspire to achieve sustainable development must understand and address social issues particularly from the perspective of increased equity. That is why we need to understand self-empowerment.

ACKNOWLEDGMENTS

The author wishes to acknowledge the contributions of many people to this work: representatives of the three communities and many others from among the We the Peoples Awards winners who shared their hearts and stories; Naresh Singh, Allen Tyrchniewicz, Stephan Barg, Susan Miskiman, and other colleagues at the International Institute for Sustainable Development; and, finally, the synthesizers, facilitators, and recorders for the We the Peoples program who gave freely of their time and energy to capture the stories and insights.

NOTES

1. For further information about the We the Peoples: 50 Communities Awards, go to the Communities section of IISD's World Wide Web site at http://iisd1.iisd.ca.
2. This description is drawn from Jacobs (1995).
3. This description is drawn from Sanikiluaq's presentation to the We the Peoples: 50 Communities Awards, New York, 1995.
4. This material is drawn from Oujé-Bougoumou's presentation to the We the Peoples: 50 Communities Awards, New York, 1995. See also Goddard (1994).
5. Oujé-Bougoumou's presentation to the We the Peoples: 50 Communities Awards, New York, 1995.
6. This conceptualization of community self-empowerment draws from a framework originally called the Dependency Cycle conceptualized by the author in the early 1970s while undertaking research on social action (see Seymor 1977).
7. In earlier versions of the cycle the fourth phase was described as *interdependence.* An Indian counselor, Joe Couture, pointed out that after the difficult battles for independence, *interdependence* implied mutual dependence and seemed to be a step backward. He suggested referring to this phase as inter-independence in order to better capture the sense of equality and strength won by native peoples. In keeping with the stress on partnerships in sustainable development, I have chosen to refer to the fourth phase as one of partnership.
8. See 50 Communities on Web site, http://iisd1.iisd.ca.
9. Readers interested in further information about the Community Self-Empowerment Cycle, the We the Peoples: 50 Communities Awards, or the work of the International Institute for Sustainable Development are invited to contact the author, Nola-Kate Seymoar, International Institute for Sustainable Development, 161 Portage Avenue East, 6th Floor, Winnipeg, Manitoba, Canada, tel: 1-204-958-7752, fax: 1-204-958-7710, e-mail: nkseymoar@iisdpost.iisd.ca. Readers interested in the stories and lessons of the three communities described in this chapter may obtain more information from Dean Jacobs, Walpole Island First Nation, Heritage Center RR #3, Wallaceburg, Ontario, Canada N8A 4K9, tel: 1-519-627-1475, fax: 1-519-627-1530; Abel Bosum, 207 Opemiska Street, Oujé-Bougoumou, Quebec, Canada G0W 3C0, tel: 1-418-745-3911, fax: 1-418-745-3168; Paul Wertman, c/o Grand Council of the Crees, 24 Baywater Avenue, Ottawa, Ontario, Canada K1Y 2E4, tel: 1-613-761-9635, fax: 1-613-761-1388; Lucassie Arragutainaq, HBP, Municipality of Sanikiluaq, Sanikiluaq, Northwest Territories, Canada X0A 0W0, tel: 1-819-266-8980, fax: 1-819-266-8837.

REFERENCES

Alinsky, S. D. 1972. *Rules for Radicals: A Practical Primer for Realistic Radicals.* New York: Vintage Book.

Arragutainaq, L.1995. Notes for Reference Issue. Sanikiluaq, Northwest Territories, Canada.

Goddard, J. 1994. "In from the Cold." *Canadian Geographic* 114(4).

Jacobs, D. M. 1995. "Sustaining the Circle of Life" Presentation to We the Peoples Awards, New York.

Kattuk, P. 1992. Proceedings of the Hudson Bay TEKMS Workshop 1, Transcript. Hudson Bay Program, Municipality of Sanikiluaq.

Makivik Society and Laval University. 1992. *Contaminants in the Marine Environment of Nunavik.* Collection Nordicana. No. 56. Laval, Quebec. Laval University, Center for Northern Studies.

Novalinga, Z. 1994 Personal Notes. Sanikiluaq, Northwest Territories, C anada.

Picard, A. 1993. "A Dispossessed People Comes Home." *The Globe and Mail* 4 December.

Symor, N. K. 1977. "The Dependency Cycle, Implications for Theory, Therapy and Social Action." *Transactional Analysis Journal* (Spring):37–43.

Seymoar, N. K. 1996. "The Community Empowerment Cycle" International Winter Cities' 96 Forum"Winter Cities Conference Proceedings." Winnipeg, Canada Feb 1996.

Singh, N. C. 1996. "Community Adaptation and Sustainable Livelihoods: Basic Issues and Principles," Working paper, International Institute for Sustainable Development, Winnipeg, Man.

Solomon, G. 1993. Proceedings of the Western James Bay Verification Meeting. Transcript. Hudson Bay Program. Sanikiluaq, Northwestern Territories, Canada.

We the Peoples: 50 Communities. 1993. Nominations materials. New York.

World Commission on Environment and Development. 1987. *Our Common Future.* Oxford: Oxford University Press.

About the Authors

Ivonne Audirac is assistant professor of urban and regional planning at the Florida State University, Tallahassee, where she teaches a seminar on sustainable development in the Americas. She has conducted research on the impacts of information technology in agriculture. Dr. Audirac cochaired the national conference, Rural Planning and Development: Visions of the 21st Century, held in Orlando, Florida, in February 1991. *e-mail: iaudirac@garnet.acns.fsu.edu.*

Charles H. Barnard joined USDA's Economic Research Service in 1980. A resource economist by training, he has worked on a variety of land economics issues, including real estate valuation and urban-fringe development. Most recently, his work has focused on farm financial analysis.

Lionel J. Beaulieu is professor of rural sociology with the Food and Agricultural Sciences at the University of Florida, Gainesville. As part of his activities in the Florida Cooperative Extension Service, Dr. Beaulieu has devoted his energies to community and rural development initiatives, particularly community leadership development, public policy education, and rural revitalization programs. He is the author of two books, *The Rural South in Crisis: Challenges for the Future* and *Investing in People: The Human Capital Needs of Rural America. email: LJB@nervm.nerdc.ufl.edu.*

Mark Benedict is adjunct research scientist at the University of Florida, Department of Landscape Architecture. He was the executive director of the Florida Greenways Commission, a joint program with 1000 Friends of Florida and the Conservation Fund. Dr. Benedict is a scientist with more than 17 year's experience in ecological research and administration.

Michael R. Boswell is a doctoral candidate in the Department of Urban and Regional Planning, Florida State University, Tallahassee. His doctoral work examines the evolvement of environmental planning in south Florida after the establishment of sustainable development and ecosystem management initiatives. Prior to entering doctoral studies, he worked as a planner for Brevard County, Florida. *email: mboswell@mailer.fsu.edu.*

Dan Chiras is adjunct professor at the University of Denver and at the University of Colorado, where he teaches courses on sustainable development. He is author and editor of 11 books, including several popular college textbooks of environmental science. His most recent trade books include *Voices for the Earth: Vital Ideas from America's Best Environmental Books* and *Lessons from Nature: Learning to Live Sustainably on the Earth.* Dr. Chiras is founder and president of the Sustainable Futures Society, Evergreen, Colorado. *e-mail: dchiras@aol.com.*

Richard M. Clugston is executive director of the Center for Respect of Life and Environment. The Center encourages the development of ecologically sensitive academic curricula and sustainable campuses through its programs in the arts, sciences, and applied professions, and through *Earth Ethics,* its quarterly publication. Dr. Clugston's most recent works include two books—*God, Nature, and Justice: Learning to Meet the Environmental Crisis,* with Dieter Hessel; and *Greening Higher Education: A Guide to Colleges and Universities*—and two video projects: *The Life and Thought of Thomas Berry,* and *Ethical Foundations for a Sustainable Society. e-mail: CRLE@aol.com.*

Erik Davies is Environmental Planner with ESSA Technologies Ltd, Ottawa. He has conducted work and research in environmental management and impact assessment both in Canada and abroad. His latest work focused on environmental cumulative effects assessment in Canada, as well as environmental policy and legislation in Mongolia, Malawi, and the Caribbean. *e-mail: eric@magi.com.*

Donn A. Derr is associate professor in the Department of Agricultural Economics and Marketing, Cook College, Rutgers, State University of New Jersey, New Brunswick. Dr. Derr's research interests include the economics of waste recycling of organic fractions in the municipal solid waste stream that can be utilized by agricultural operations to reduce traditional resource inputs. *e-mail: derr@aesop.rutgers.edu.*

Pritam S. Dhillon is associate professor in the Department of Agricultural Economics and Marketing, Cook College, Rutgers, State University of New Jersey, New Brunswick. Dr. Dhillon's specialty is farm management and production economics. He has done extensive work on the cost of producing farm products, as well as economic adjustments on farms, farmland preservation, and the recycling of urban waste through farming operations.

Anthony Flaccavento is executive director of the Clinch Powell Sustainable Development Initiative (Abingdon, VA 24212-0791). Before assuming his present position, he directed the Appalachian Office of Justice and Peace, where he designed and implemented a regional home ownership program for low-income families, a cooperative food buying network, and many environmental and community development initiatives. He is the author of *Habits of Creation: A Facilitators Handbook* and teaches a class on sustainable development at Emory and Henry College, Emory, Virginia.

Owen J. Furuseth is professor of geography and earth sciences at the University of North Carolina at Charlotte. His professional interests include rural land use and environmental planning. Dr. Furuseth is the U.S. editor of *Progress in Rural Policy and Planning*, an international serial addressing contemporary issues in rural planning and development. As a Fulbright Senior Research Scholar at the University of Auckland, he carried out research on the newly enacted Resource Management Act, which laid out a sustainable development guide for New Zealand's environmental and land use planning. *e-mail: fgg00ojf@unccvm.uncc.edu.*

Gary P. Green is professor of rural sociology at the University of Wisconsin-Madison. His teaching, research, and outreach activities focus on community and regional development. He has a joint appointment with University of Wisconsin-Extension and conducts applied research on labor market conditions, growth management activities, and economic development for communities throughout the state. He is currently working with several state, regional, and local organizations to develop programs addressing the needs of low-income workers. His current research focuses on economic restructuring of localities, urban inequality, and local economic development. *e-mail: GGREEN@ssc.WISC.EDU.*

Ralph E. Heimlich is a planner and regional scientist with USDA's Economic Research Service (ERS), Washington, D.C. Since joining ERS in 1976, Heimlich has worked on river basin planning, land use issues, and conservation provisions of 1985 and 1990 farm legislation. His work on land use included research on dynamics of land use change in fast-growth counties, conversion of environmentally sensitive lands to agricultural production, and the economics of agriculture in urbanizing areas. He has worked in the Environmental Protection Agency, Agricultural Policy Branch, during 1992. *e-mail: heimlich@ERS.bitnet.*

Julie Herman is a landscape architect and works as the transportation coordinator in the Environmental Affairs office of the city of Boulder, Colorado. Ms. Herman is a research associate at the Sustainable Futures Society and participates in the sustainable development workshops sponsored by the organization.

Glenn D. Israel is associate professor and extension specialist in program evaluation and organizational development at the University of Florida. Dr.

Israel's major responsibilities include providing leadership in the areas of program development and evaluation and rural development.

Mark B. Lapping is a professor in the Muskie Institute of Public Affairs and provost and vice president for Academic Affairs at the University of Southern Maine in Portland. Among other previous appointments, he served as founding director of the School of Rural Planning and Development at Canada's University of Guelph. He is author and coauthor of several books, including *A Long, Deep Furrow: Three Centuries of Farming in New England, Rural Planning and Development in the United States, The Small Town Planning Handbook,* and *Rural America; Legacy and Change. e-mail: lapping@usm.maine.edu.*

Joseph Luther is the Hyde Professor of Community and Regional Planning at the University of Nebraska-Lincoln. Dr. Luther has been involved in exploring, describing, and explaining the planning and community development of small towns over 20 years. In recognition of his teaching, research, public service, and professional practice, Dr. Luther has been honored as the Western Planner of the Year by the *Western Planner Journal* and with the President's Professional Award by the Nebraska Planning and Zoning Association in recognition for his contributions to small town planning. *e-mail: jluther@unlinfo.unl.edu.*

Thomas S. Lyons is associate professor of urban planning at California State University, San Bernadino. Dr. Lyons is coauthor of *Creating an Economic Development Action Plan,* and has written many articles on rural and small town incubation and economic development.

George Penfold is a partner with Westland Resource Group in Courtenay, British Columbia. Previously he was Professor in the University School of Rural Planning and Development at the University of Guelph, Ontario, Canada. He is a professional engineer and planner and has worked as an agricultural extension engineer and county planner prior to joining the University of Guelph in 1981. His primary research and teaching interests are in planning and development theory, agricultural land evaluation and policy, and rural community development. *e-mail: gpenfold@mars.ark.com.*

William E. Rees is professor and director of the School of Community and Regional Planning at the University of British Columbia. He is a founding member of the University of Toronto's Pollution Probe, one of Canada's foremost environmental nongovernmental organizations. Dr. Rees's teaching and research emphasize the policy implications of global environmental trends and the ecological requirements for sustainable socioeconomic development. His work in ecological economics focuses on the development and application of the "ecological footprint" and "appropriated carrying capacity" concepts, which provide land-based indicators of (un)sustainability. *e-mail: wress@unixg.ubc.ca.*

James A. Segedy is associate professor of urban planning at Ball State University, Muncie, Indiana. He was chair of the Small Town and Rural Planning Division of the American Planning Association and President of the Indiana Planning Association and Indiana Community Development society. Dr. Segedy is codirector, with Scott Truex, of the Indiana Total Quality of Life Initiative. *e-mail: 00jasegedy@bsuvc.bsu.edu.*

James Seroka is professor of political science and public administration at the University of North Florida, Jacksonville, where he is also director of the Center for Local Government Administration. He has published widely on issues relating to public policy and public administration in rural America, including *Rural Public Administration* (1985), and is the rural development editor for the *National Civic Review. e-mail: jseroka@gw.unf.edu.*

Matthew S. Sexton is the Conservation Fund's Florida Sustainable Projects Manager, and is developing the Fund's Florida Sustainable Communities Program. He has served as a planning and economic development consultant to regional heritage tourism and sustainable development projects throughout the south and southeast, including: the Roanoke River Greenways Project, the Tyrrell County Community Development Corporation, the Scuppernong River Greenways Project, the Partnership for the Sounds in North Carolina, the South Georgia Rivers Project, and the Roma Preservation Project in Texas.

Nola-Kate Seymoar serves as deputy to the president of the International Institute for Sustainable Development (IISD), Winnipeg, Manitoba. Prior to assuming her tenure at IISD, Dr. Seymoar served as executive director of the We the Peoples : 50 Communities Awards, a citizens' initiative in honor of the 50th anniversary of the United Nations. She is well known internationally for her role as executive director of ECO-ED (*Education and Communication on Environment and Development*), the World Congress followup to the Earth Summit. *e-mail: nkseymoar@iisdpost.iisd.ca.*

Raymond A. Shapek is professor of public administration at the University of Central Florida, Orlando. He has completed major research projects for the Florida Center for Solid and Hazardous Waste Management and the Florida Department of Commerce on recycling incentives, recycled commodities marketing, and public information and education programs for recycling. *e-mail: shapek@pegasus.cc.ucf.edu.*

Earl M. Starnes is professor emeritus, Department of Urban and Regional Planning, University of Florida. His public service has included county and state offices and memberships on various state advisory groups, including the recently concluded Florida Greenways Commission. His current professional work, along with individual practice in land use and growth management, includes working for the Florida Department of Environmental Protection in the development of the Florida Greenways System Plan. *e-mail: AFN51612@freenet.ufl.edu.*

Deborah S. K. Thomas is a Ph.D. candidate in the Department of Geography, University of South Carolina. While a student at the University of North Carolina, Charlotte, Ms. Thomas received the 1995 Outstanding Graduate Student Project Award given by the North Carolina Chapter of the American Planning Association in recognition of a coauthored report on her research effort to develop a sustainable planning framework for small towns.

Index